Toxicology of Essential and Xenobiotic Metals

Editors

João Batista Teixeira da Rocha

Departamento de Bioquímica, ICBS
Universidade Federal do Rio Grande do Sul
Porto Alegre, RS, Brazil
and
Departamento de Bioquímica e Biologia Molecular, CCNE
Universidade Federal de Santa Maria, RS, Brazil

Michael Aschner

Department of Molecular Pharmacology
Albert Einstein College of Medicine
Bronx, USA

Pablo Andrei Nogara

Federal Institute of Education
Science and Technology Sul-rio-grandense (IFSul)
Bagé, RS, Brazil

CRC Press
Taylor & Francis Group
Boca Raton London New York

CRC Press is an imprint of the
Taylor & Francis Group, an **informa** business
A SCIENCE PUBLISHERS BOOK

Preface

The toxicology of metals is a global concern. Acute
can induce overt or hidden toxic manifestations. Genei
intoxication depend upon the type (chemical species, ̦
composition, etc.), dose, pathway, and duration of
exposure to cadmium (Cd), mercury (Hg), lead (Pb), al
(Cr) can be associated with degenerative pathologies.
the immediate vital signs include nausea, vomiting, di
Humans' and animals' long-term exposure to toxic
lead to numerous physical, muscular, neurological, or
For instance, it has been suggested that toxic metals
development of multiple sclerosis, muscular dystropł
Parkinson's disease, and Alzheimer's disease.

The understanding of the metals' interactions wit
to identify their metabolism, distribution, and toxicity.
biology status by metals is a critical point in their toxic
mercury has been identified, mainly after the Minamat
the most toxic heavy metals, and its poisoning is known
The development of analytical sensitivity techniques
very small amounts of some toxic metals, at the molecu

This book discusses the toxicity, safety, and proper
and xenobiotic metals and metalloids (such as man
aluminum, mercury, cadmium, lead, zinc, chromium, c
The redox biology, neurodegeneration, molecular pa
associated with heavy metals and metalloids toxicity
11 chapters that describe the main biological features o
metabolism, and the use of animal models in toxicity s
this book is intended to be a reference for chemists, bi
professionals, and students of toxicology.

Joã

Contents

Chapter 1

Zebrafish as a Model Organism to Study Manganese Toxicity

Nilda Vargas Barbosa, Sabrina Antunes Ferreira* and *Matheus Mülling dos Santos*

1. Introduction

This chapter brings an overview of the physiological and toxicological effects of Manganese (Mn), emphasizing the neurotoxicity associated with excessive Mn levels. The review examined the current literature that investigated Mn toxicity using zebrafish as model organism, addressing the findings into developmental stages of fish (embryo/larvae and adults) as well as studies performed with SLC30A10/SLC39A14 zebrafish mutants and some Mn-derived compounds.

2. Manganese

Manganese (Mn) is the fifth transition metal and the twelfth most abundant element in the environment. It occurs naturally in the form of oxides, carbonates, or silicates in over 100 types of mineral rocks and is ubiquitously present in the soil, air, water, and foods. In living tissue, Mn is found as Mn^{2+} and Mn^{3+}. Mn^{4+}, Mn^{5+}, Mn^{6+}, Mn^{7+}, and other complexes of Mn. Manganese at lower oxidation states, is not observed in biological materials, but frequently found in natural and built environments (Takeda, 2003; Sachse et al., 2019).

Mn does not occur naturally in a pure state. It is found in both inorganic and organic compounds. Inorganic Mn has numerous applications, including the production of iron and steel, manufacture of dry cell batteries, production of potassium permanganate and other chemicals, as an oxidant in the production of hydroquinone, manufacture of glass and ceramics, textile bleaching, as an oxidizing agent for electrode coating in welding rods, adhesives, paint, matches and fireworks, and

Programa de Pós-graduação em Bioquímica Toxicológica, Universidade Federal de Santa Maria, Avenida Roraima, 1000, 97105-900 Santa Maria, RS, Brazil.
* Corresponding author: nvbarbosa@yahoo.com.br

tanning of leather (Aschner et al., 2007; Avila et al., 2013). Organic Mn compounds are part of the composition of fungicides (Maneb® and Mancozeb®), widely used in fruit, vegetable, and grain plantations and for conservation of ornamental trees (Santamaria et al., 2007). Mn has been used in the manufacture of other compounds, such as the antiknock MMT (Tricarbonyl methylcyclopentadienyl manganese), developed in 1974 in the USA to replace the tetraethyl lead added to fuels (Zayed et al., 1999).

In food, Mn is naturally present in legumes, rice, nuts, and whole grains. It is also found in seafood, seeds, chocolate, tea, leafy green vegetables, spices, soybean, and some fruits such as pineapple and acai. The greatest exposure to manganese is usually from food. Adults consume between 0.7 and 10.9 mg/day in the diet (Greger, 1999). The reference daily intake (DRI) established 9–11 mg/day Mn as the upper tolerable limit likely to pose no risk of adverse health effects for adults, and 2–6 mg/day Mn for children, depending on the age (Peres et al., 2016). However, manganese in drinking water is considered more bioavailable than manganese in food (Ljung and Vahter, 2007). Assuming that 20% of the tolerable daily intake comes from drinking water, the value of 400 μg/L has been considered a health-based guideline value for drinking water (WHO, 2011). Examples of drinking water or potential drinking-water supplies with Mn concentrations > 400 μg/L can be found worldwide. Manganese contamination in drinking water can originate from various natural and anthropogenic sources, including geological formations, industrial activities, and agricultural runoff.

In regions where manganese-rich geological formations predominate, such as areas with underlying manganese-rich bedrock or soils, natural weathering processes can leach manganese into groundwater and surface water bodies, leading to elevated concentrations in drinking-water sources. Additionally, industrial activities such as mining, metal processing, and chemical manufacturing may contribute to manganese pollution through the release of wastewater and effluents containing high levels of manganese compounds.

Agricultural practices involving the use of manganese-containing fertilizers and pesticides can also contribute to the contamination of water sources, as runoff from agricultural fields may transport dissolved manganese into nearby water bodies (Frisbie et al., 2015).

Naturally occurring concentrations of manganese in the air are low; however, the anthropogenic activities are significant sources of emission of various metals. The most common cases of Mn intoxication are found among workers in industries such as mining and manufacturing of metal alloys, steel and batteries, welding, and agricultural workers exposed to fungicides (Peres et al., 2016; Santamaria et al., 2007). In addition, parenteral nutrition also represents an important form of exposure to Mn, especially in newborns. Mn is commonly added to total parenteral nutrition, at significant concentrations, as an essential nutrient, and also in soy-based infant formulas (Aschner et al., 2005; Krachler and Rossipal, 2000). Drug abuse has also become a worrying source of Mn intoxication, mainly through illegal drugs such as ephedrone (metacathinone) (Peres et al., 2016) (Figure 1).

Figure 1. Main sources of manganese contamination. MMT = Tricarbonyl methylcyclopentadienyl manganese.

Mn is an indispensable trace element for the growth, development, and maintenance of the health of all prokaryotes and eukaryotes (Himeno et al., 2019). It is necessary for the activity of certain antioxidants and metalloenzymes, energy metabolism, and the functioning of immunological, nervous, and reproductive hormones (Sigel and Sigel, 2000; Avila et al., 2013). Moreover, Mn is an important cofactor for a variety of other enzymes, including those involved in neurotransmitter synthesis and metabolism.

Mn homeostasis depends directly on its uptake and efflux by transporters in the cell surface and by intracellular transporters located in the membrane of organelles (Lysosomes, Golgi apparatus, Nucleus) (Figure 2). The divalent metal transporter 1 (DMT1) is one of the best-known influx transporters of Mn. DMT1 is reported to have other substrates, including Fe^{2+}, Zn^{2+}, Mn^{2+}, Cu^{2+}, Co^{2+}, Cd^{2+}, Ni^{2+}, and Pb^{2+}. Besides Fe^{+2}, Mn has also been reported to compete with transporters of other divalent metals. Mn import can be carried by a family of Zn^{+2} transporters (ZIP14/ZIP8) and Ca^{+2} channels (TRPC3). Other putative Mn influx transporters include Dopamine transporter (DAT), choline transporter, transferrin/transferrin receptor (Tf/TfR), and citrate transporter (monocarboxylate transporter (MCT)) (Chen et al., 2015; Aschner et al., 2007). Less is known about the transporters involved in the Mn efflux. Currently, plasma membrane-localized exporters include ferroportin, SLC30A10 (carrier solute family 30 members 10), secretory pathway Ca^{2+}-ATPase 1 (SPCA1) and ATPase cation transporting 13A2 (ATP13A2 or PARK9), and sodium-calcium exchanger (NCX) (Tan et al., 2011; Leyva-Illades et al., 2014; Madejczyk and Ballatori, 2012; Mukhopadhyay et al., 2010; Chen et al., 2012).

Because of its widespread presence in human diets, Mn deficiency is generally not clinically recognized in humans. Just a few occurrences of Mn deficiencies have been reported in humans, with symptoms including dermatitis, hypocholesterolemia, alteration in hair and nail growth, decreased clotting proteins, increased serum

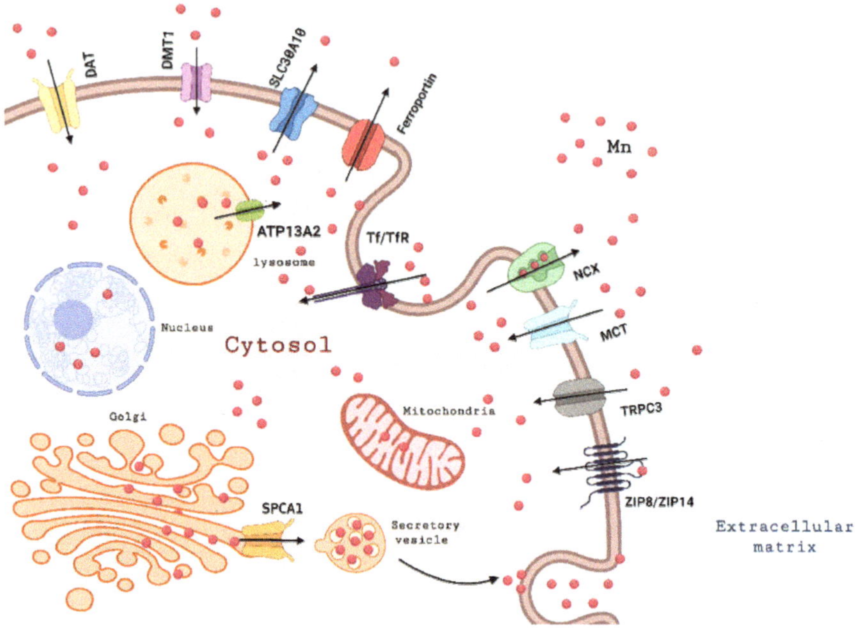

Figure 2. Diagram illustrating cellular structures and manganese (Mn) transport in a cell. Tf/TfR, DMT1, DAT, ZIP8/ZIP14, citrate transporters (MCT) and Ca channels (TRPC3) facilitate Mn influx into the cytosol. SLC30A10, NCX and ferroportin mediate efflux of Mn. SPCA1 on the Golgi membrane and ATP13A2 on the lysosome membrane facilitate Mn efflux into the extracellular matrix. Tf/TfR = transferrin/transferrin receptor; DMT1 = divalent metal transporter 1; DAT = dopamine transporter; ZIP8/ZIP14 = zinc transportes; MCT = monocarboxylate transporter; SLC30A10 = solute carrier family 30 member 10; NCX = sodium-calcium exchanger; SPCA1 = secretory pathway Ca^{2+}-ATPase 1; ATP13A2 = ATPase cation transporting 13A2.

calcium and phosphorus concentrations, and increased alkaline phosphatase activity (Keen et al., 1994; Finley et al., 1994). Reduced Mn status may also be observed in individuals with osteoporosis and epilepsy (Aschner et al., 2005). In animal studies, low manganese intake has been associated with impaired growth, skeletal defects, reduced reproductive function, birth defects, and abnormal glucose tolerance, as well as altered lipid and carbohydrate metabolism (Dobson et al., 2004).

The excess of Mn can be toxic to the organisms, and the ingestion of foods and water containing high levels of Mn, and occupational exposure are the main sources of contamination (Balachandran et al., 2020; Peres et al., 2016). Excessive intake and overexposure to Mn have already been associated mainly with neurological disturbances, namely parkinsonian-like movement disorder in adults and fine motor, attentional, cognitive, and intellectual deficits in children (Aschner et al., 2005; Bowler et al., 2007).

Mn neurotoxicity was first identified as an extra-pyramidal syndrome in miners exposed to high concentrations of Mn, today called manganism (Barbeau et al., 1984; Couper, 1837; Mena et al., 1967). Most cases were associated as a consequence of inhalation of manganese-containing dust in occupational settings. Manganism is characterized by symptoms similar, but not identical, to idiopathic

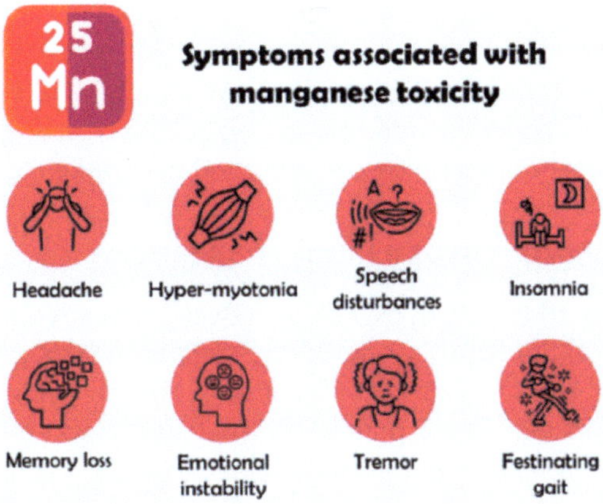

Figure 3. Disturbances associated with the Manganese toxicity.

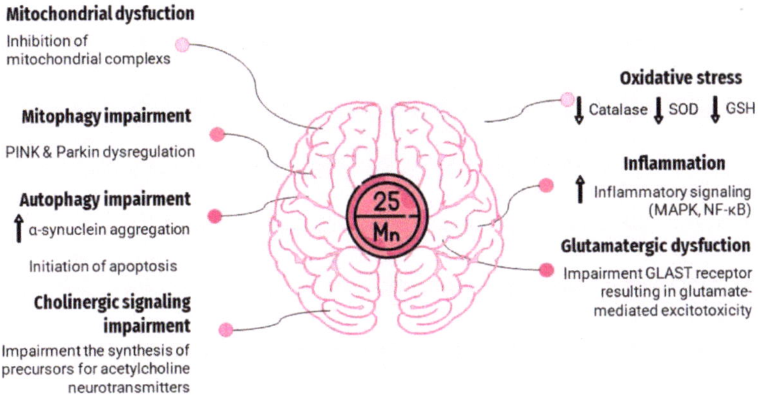

Figure 4. Scheme of mechanisms involved in Mn-induced neurotoxicity.

Parkinson's disease. The symptoms include bradykinesia, rigidity, tremor, gait disturbance, postural instability, and dystonia, and/or ataxia (Figure 3) (Josephs et al., 2005). Mounting evidence has indicated that Mn is also a risk factor for other neurodegenerative diseases, including Alzheimer's disease, Amyotrophic lateral sclerosis, and Huntington's disease (Avila et al., 2013; Bowman et al., 2011).

Psychiatric changes (i.e., depression, agitation, hallucinations), difficulty with visuomotor and visuospatial information processing, and cognitive deficits such as memory impairment, reduced learning capacity, and decreased mental flexibility have already been reported as disturbances linked to Mn intoxication (Mena et al., 1967; Josephs et al., 2005; Bowler et al., 2007). Overexposure to manganese also induces anxiety and depression-like behaviors. Studies have demonstrated that workers exposed to Mn show a dose-effect relationship between the level of exposure and the emergence of these neuropsychiatric disturbances (Bowler et al., 2007; Hong et al., 2014; Chen et al., 2019). Hong et al. (2014) reported that exposure to manganese may be partially responsible for the increased prevalence of mood disorders, as well as antisocial conduct and personality disorders.

Some mechanisms involved in Mn-induced neurotoxicity include impairment of mitochondrial dynamics and mitophagy, autophagy dysregulation, oxidative stress, excitotoxicity and inflammation (Figure 4) (Pajarillo et al., 2021). The evidence that Mn can accumulate in dopaminergic neurons suggests a critical role of dopaminergic signaling in the etiology of neurological abnormalities evoked by Mn.

3. Zebrafish (*Danio rerio*) as model organism

Zebrafish (*Danio rerio*) is a small tropical freshwater fish, measuring approximately 3–4 cm, belonging to the family Cyprinidae. Members of this group are mainly the genera Danio, Devario and Rasbora. There are currently 44 species of Danionin (Fang, 2001), distributed across South and Southeast Asia. In the wild, they are found in calm, shallow rivers and in flooded rice paddies, but are now available in pet stores around the world. The name "zebrafish" comes from the horizontal blue stripes on each side of its body (Spence et al., 2008) (Figure 5A).

The life cycle of zebrafish consists of several distinct stages, such as embryonic pre-hatching (0–72 hpf) and post-hatching (72–120 hpf); 'larval' (1–29 dpf; including a free-swimming larva from 5 dpf) to juvenile fish (30–89 dpf), adult zebrafish (90 dpf-2 years) and aging/aged zebrafish (from 2 years) (Figure 5B). Zebrafish can live for approximately 3 years on average and over 5 years maximally in laboratory conditions and remain sexually mature/active almost all this time.

Zebrafish's emergence as a model organism for modern biological investigation began with the pioneering work of Streisinger and colleagues (1981), who recognized many of the virtues of the experimental system. These include its short generation time, the large numbers of eggs produced by each mating, and the fact that, because fertilization is external, all stages of development are accessible (Briggs, 2002).

The main benefits of using zebrafish as a toxicological model over other vertebrate species are with regards to their size, husbandry, and early morphology. Zebrafish adults are only approximately 3–4 cm, as previously mentioned. This greatly reduces housing space and husbandry costs. Besides their size, this species is invaluable because of their high fecundity and transparent embryos. One pair of adult fish is capable of laying 200–300 eggs in one morning, and if appropriately maintained, they can provide this yield every 5–7 days (Hill et al., 2005). Unlike mammals, zebrafish develop from fertilized eggs to adults in transparent eggs, which enables monitoring of the developing embryos (and their organs), as well as manipulating these processes (e.g., by injecting drugs or genes) *in vivo*. This feature empowers neurodevelopmental studies using zebrafish models (Kaluef et al., 2014).

Zebrafish, as well as other fish species, may be considered a relatively simple organism when compared to mammals. However, this is not the case with respect to the zebrafish genome. The zebrafish shares a high degree of homology with the human genome (Howe et al., 2013). With a high-quality zebrafish genome now available, it appears that 71% of human proteins (and 82% of disease-causing human proteins) have an obvious orthologue in zebrafish. Furthermore, the basic structure of the CNS in fish has all the major domains found in mammalian brain, and the same neurotransmitters such as GABA, glutamate, dopamine, noradrenaline, serotonin, histamine, and acetylcholine are found in both interneuron systems and in

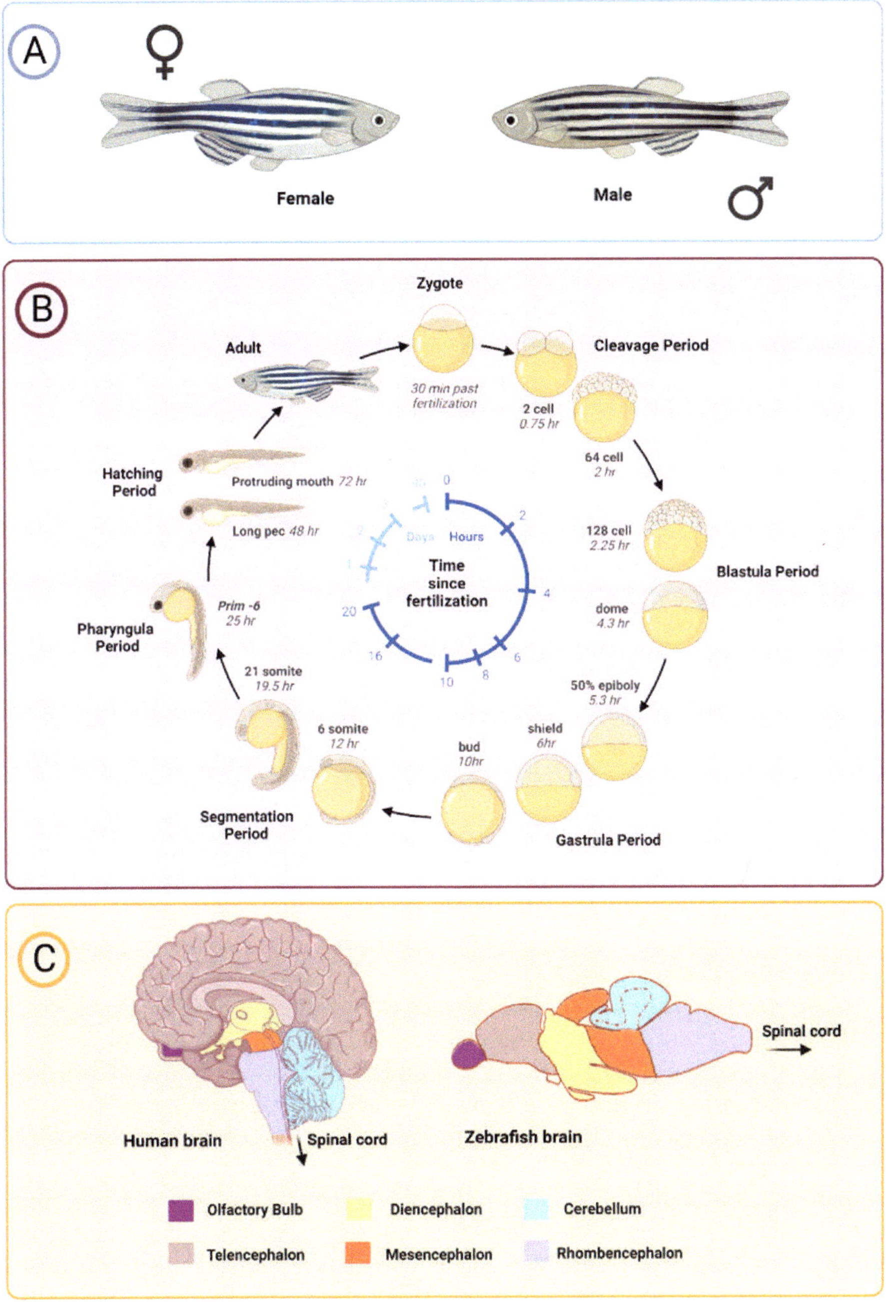

Figure 5. (A) Male and female zebrafish (*Danio rerio*) phenotype. (B) Zebrafish life cycle. (C) Conserved brain structures between zebrafish and humans.

long pathways (Panula et al., 2010). As vertebrates, human and zebrafish share basic central nervous system organization comprising forebrain, midbrain, hindbrain, and spinal cord (Figure 5C) (Kozol et al., 2016). Conservation of this structure, along with similarities in specialized neuronal populations, glial subpopulations, and circuitry organization, has allowed the zebrafish models to provide numerous insights into vertebrate nervous system development and physiology and are likely to allow similar elucidation of disease mechanisms (Senger et al., 2010).

4. Zebrafish as model organism to study Mn neurotoxicity

Although zebrafish is an organism already well established in the field of neuroscience, studies using zebrafish to investigate the neurotoxicity of manganese are still incipient when compared to rodents. However, the findings so far have emphasized the translational relevance of the model, since fish develop neurobehavioral phenotypes similar to those observed in humans and other animals overexposed to manganese. Similar to humans, wild zebrafish and Mn carrier mutants exhibit increased deposition of Mn in the brain and the neuromotor consequences of Mn metabolism disorders. Then, here we bring an overview of what exists in the current literature on Mn intoxication in zebrafish, highlighting the findings into the developmental stages of fish: embryo/larvae and adults.

4.1 Embryo/Larval stages

During zebrafish's development, the Mn content varies according to the life stage. Between the 5 and 30 dpf stages, Cu and Mn levels increased from 5 to 15 dpf, 5 to 30 dpf, and 7 to 30 dpf while Zn levels increased from 5 to 30 dpf and 7 to 30 dpf (Thomason et al., 2017). Results of the same study revealed that Mn was also more abundant in the body when compared to the yolk. In the work of Altenhofen and collaborators, zebrafish larvae exposed to $MnCl_2$ (0; 0.1; 0.25, and 0.5 mM) presented similar Mn content until 5 dpf. At 7 and 10 dpf, only groups exposed to higher concentrations (0.25 and 0.5 mM) maintained the Mn content higher than control (Altenhofen et al., 2017).

Behavioral, morphological and biochemical alterations evoked by Mn exposure highlight the usefulness of zebrafish embryos and larvae in this field of research, as well. A model of manganism using zebrafish larvae was firstly proposed by Bakthavatsalam et al. (2014). The authors initially exposed zebrafish larvae (3.5 hpf) to different Mn ($MnCl_2$) concentrations (0–2 mM) for 48 h to verify the LC50, which was obtained at 2 mM. Then, several experiments were done by incubating zebrafish larvae for 24 h or 48 h in Mn solutions at concentrations lower than 2 mM. The findings showed that Mn exposure caused postural and movement disorders similar to that observed in human manganism. The larvae exposed to Mn float on their sides, swim in circles and exhibited a spine curvature. The authors attributed the postural defects to the Mn interference on mechanotransduction at the neuromasts. The circling movements were related to the long-duration bursting in the motor neurons, which can culminate in long-duration tail bends larvae exposed to Mn. Another change found in this study was the decrease in the startle movements of the Mn exposed larvae. The authors showed that the intensity of tyrosine hydroxylase immunoreactivity was reduced in a reversible manner after Mn exposure. Based on this fact, the authors proposed that a reduction of dopamine neuromodulation drives the changes in the startle movements.

Changes in the locomotor activity and involvement of the dopaminergic system in the Mn toxicity in zebrafish larvae were also investigated by Altenhofen et al. (2017). For this proposal, zebrafish larvae (1 hpf) were exposed to $MnCl_2$ (0.1, 0.25 and 0.5 mM) for 5 days. These concentrations were chosen since the analysis during 10 days revealed survival percentages of 84%, 82% and 72%, respectively, which were not significantly different from the control group (88%). On the other hand, zebrafish larvae exposed to $MnCl_2$ at 1.0 and 1.5 mM had 0% of survival. Locomotor/exploratory analyses showed that Mn exposure was able to reduce the distance traveled and absolute turn angle. Moreover, the time immobile and number of immobile episodes significantly increased. Interestingly, no morphological changes were observed in Mn-exposed zebrafish larvae. Regarding the dopaminergic system, the authors showed an increase in tyrosine hydroxylase (TH) at 5 and 7 dpf, and a decrease at 10 dpf. However, no significant differences were observed in dopamine and DOPAC levels when compared to the control. Additionally, the same study investigated the effects of Mn exposure on proapoptotic biomarkers in zebrafish larvae. They showed, for the first time, that Mn was capable of increasing the protein levels of p53, capase-8, and Bax-α, evidencing that Mn exposure can lead to an increase in cellular death. In accordance, Xu et al. (2022) evidenced that zebrafish larvae exposed to $MnCl_2$ (0, 25, 50, and 100 μM) for 8 days (from 1 hpf to 8 dpf) presented an increase in the proportion of apoptotic dopaminergic neurons (mainly in the telencephalon and midbrain), accompanied by a decrease in the TH expression. Transcriptome analysis revealed that Mn exposure caused an up-regulation in the expression of genes associated with apoptosis and DNA damage repair. Moreover, western blot analysis demonstrated that $MnCl_2$ (100 μM) induced the activation of the MAPK pathway.

Several studies in the literature report an inverse association between childhood Mn exposure and intelligence quotient (IQ) and/or other indices related to mental development (Rodríguez-Barranco et al., 2013; Iyare, 2019; Liu et al., 2020; Irizar et al., 2021). In this context, Tu et al. (2017) investigated, in addition to locomotor activity, the effects of $MnCl_2$ exposure on the mRNA levels of neuroxin 2A (*nrxn2a*), which is involved in neurotransmission and differentiation of synapses, playing an important role in cognitive functions (Südhof, 2008; Dachtler et al., 2015; Saint-Martin et al., 2018). In that study, zebrafish larvae were exposed to $MnCl_2$ (0, 25, 50, and 100 μM) for 5 days. The findings showed a significant downregulation of *nrxn2a* mRNA levels evoked by Mn exposure. Regarding locomotor activity, $MnCl_2$ (25 and 50 μM) was able to reduce the swimming distance compared to the control. In addition, $MnCl_2$ exposure caused pericardial edema, increased mortality and decreased the hatching rate of zebrafish larvae.

In the aquatic system, Mn occurs mainly as a soluble divalent ion ($[Mn(H_2O)_6]_2$). Since Mn (II) solution can be obtained using $MnCl_2$, it is the predominant Mn source used in studies evaluating the effects of Mn exposure in zebrafish embryo/larvae.

However, environmentally and physiologically, Mn (II) can be complexed with different molecules including lactate, citrate, pyrophosphate and proteins. In this sense, Hernández et al. (2015) investigated the role of chemical speciation on the Mn toxicity in developmental stages (2–122 hpf) of zebrafish. The authors compared the effects of $MnCl_2$ to divalent (Mn II) and trivalent (Mn III) species prepared with lactate (Lac), citrate (Cit) and pyrophosphate (PPi). Firstly, they evaluated the possible toxic effects *per se* of the compounds to determine safe concentrations. The results showed that lactate at 25 mM, citrate at 12.5 mM and pyrophosphate at 3 mM induced ~28%, 10% and 50% of death, respectively. These previous observations limited the maximum Mn exposure concentrations to 10 mM (Mn (II) Lac), 6 mM (Mn (II) Cit), 2 mM (Mn (III) Cit), and 0.75 mM (Mn (III) PPi), respectively. Regarding divalent species prepared with citrate and lactate appeared to be slightly more toxic than $MnCl_2$. The authors observed that the toxicity was more prominent for post-hatched and dechorionated stages suggesting a possible potential barrier function present at the chorion. On the other hand, for longer periods of exposure, Mn (II) complexed with citrate, and lactate was more toxic. The authors suggested that these effects may partially be a result of mixture toxicity of Mn (II) and the correspondent ligands (citrate and lactate) present in the Mn (II) solutions. The toxicity evoked by Mn (II) was associated with calcium disruption, manganese speciation and metal fractionation, including bioaccumulation in tissue, organelles, granule fractions and proteins. Interestingly, no toxicity was observed in zebrafish larvae exposed to Mn (III) species.

4.2 Adult zebrafish

It has been demonstrated that exposures to environmentally relevant concentrations of Mn (from 4.0 to 16.0 mg L^{-1}) can cause motor, learning and psychiatric abnormalities in adult zebrafish. Fish acutely (96 h) and chronically exposed (21 or 30 days) to $MnCl_2$ display decreased exploratory behaviors, such as distance traveled, mean speed, absolute turn angle, number of transitions and time spent in the top area of the novel tank task (Marins et al., 2019; Rodrigues et al., 2020; Nadig et al., 2022). In terms of mechanisms, the locomotor deficits induced by Mn in adult zebrafish are linked with damage to the dopaminergic system since after acute and chronic exposure fish present decreased dopamine, and increased tyrosine hydroxylase and DOPAC levels in the brain (Altenhofen et al., 2017; Nadig et al., 2022). Moreover, oxidative stress-mediated apoptosis, decreased brain derived neurotropic factor (BDNF) and increased pro-inflammatory cytokines levels were found in the brain of zebrafish chronically exposure to the $MnCl_2$ (Nadig et al., 2022).

The brain purinergic system can also be affected by short-exposure to Mn. The study performed by Altenhofen and collaborators showed that the treatment, for 96 h, with $MnCl_2$ (0.5 to 1.5 mM) inhibited both NTPDase and ecto-ADA enzymes as well as affected the pattern of gene expression of the NTPDases, modulating the nucleotide and nucleoside levels in the brain (Altenhofen et al., 2018).

Cholinergic components have been gaining attention for their modulatory role in memory and motor processes, which are also affected by Manganism (Peres et al., 2016). Short-term Mn exposure in adult zebrafish has also been associated with poorer memory. Altenhofen and collaborators found that fish exposed to $MnCl_2$ (0.5, 1.0, and 1.5 mM), for 4 days, had impaired aversive long-term memory in the inhibitory avoidance task (Altenhofen et al., 2017). In addition, some findings have associated Mn neurotoxicity with disruption in brain AChE activity. The treatment of fish with Mn at 7.5, and 15, for 96 h, enhanced markedly the AChE activity in the brain, without modifying the activity of enzyme in muscles (Ferreira et al., 2022). Herein, it is important to highlight that the activity of brain AChE varies so much according to the exposure frequency. For example, Mairins and collaborators found that a chronic exposure to $MnCl_2$ (30 days) affected the Ache activity from zebrafish muscles without altering the brain enzyme. The lower concentration (0.2 mM) increased the activity of enzyme, whereas the higher (0.4 mM) decreased it (Marins et al., 2019).

Mounting evidence has associated excessive Mn with psychiatric disorders as well. Using adult zebrafish some studies have related Mn exposure with the onset of anxiety phenotypes. By the light/dark tank, a test that explores the natural preference of fish for the darkroom, it was possible to verify that fish acutely (96 h) and chronically treated with 4.0 mg L^{-1} of $MnCl_2$ (30 days) spent less time in the light and also presented a greater number of risk episodes, two anxiogenic behaviors (Rodrigues et al., 2020; Ferreira et al., 2022). The anxiogenic effect of Mn was also investigated on social preference test, a task that reflects the natural tendency of fish in establishing shoal. Fish exposed to 7.5 and 15 mg/L Mn developed behavioral patterns that mirror anxiety, namely a shorter time spent close to the conspecifics (Ferreira et al., 2022). In addition to anxiety and depression-like phenotypes, olfactory dysfunction has already been verified in fish exposed to 2 mM $MnCl_2$ for 21 days (Nadig et al., 2022).

The behavioral changes observed in adult zebrafish exposed to Mn have been accompanied by some endpoints of toxicity, among them: RS overproduction and loss of cell viability in the brain, liver damage, formation of micronuclei in the brain, liver, and muscle, thiol depletion in brain and liver, disruption in SOD and CAT activities from brain and liver, and increased whole-body cortisol levels (Altenhofen et al., 2017; Marins et al., 2019; Rodrigues et al., 2020; Ferreira et al., 2022; Nadig et al., 2022).

The magnitude of neurotoxic effects of Mn have been related with the levels of element accumulated in the brain. In the study performed by Ferreira and collaborators (2022), most of the deleterious effects evoked by Mn were observed in the groups exposed to the highest concentrations of Mn (7.5 and 15 mg/L for 96 h), which also induced higher bioaccumulation of the element in the brain. The bioaccumulation of Mn has already been demonstrated in the liver and muscle of zebrafish after long-term exposure (Marins et al., 2019).

5. Zebrafish mutants to study Mn metabolism disorders

It is known that naturally occurring mutations in genes from Carrier Family 39 members, which encode transporters that mediate the cellular uptake of divalent metals such as manganese, zinc, iron, and cadmium are associated with the onset of the Mn metabolism disorders in humans: Hypermanganesemia with Dystonia 1 and 2 (HMNDYT1 or 2) (Anagianni and Tuschl, 2019). Mutations in the human *SLC30A10* gene, which encodes a manganese efflux transporter, are associated with HMNDYT1, polycythemia, and cirrhosis, where the patients usually accumulate Mn in the liver, brain, and blood (Tuschl et al., 2016). HMNDYT2 occurs due to mutations in the *slc39a14* gene, that encodes transporters involved in the uptake of manganese, zinc, iron, and cadmium (Tuschl et al., 2022). The patients exhibit elevated blood and brain Mn concentrations, but unlike HMNDYT1, the liver is unaffected and polycythemia is absent.

Knockout animals have played an important role in elucidating the pathogenic mechanisms of HMNDYT 1 and 2 (Anagianni and Tuschl, 2019). In the zebrafish *slc30a10* mutant lines, Mn overexposure recapitulated aspects of human disease from the embryonic stages. After exposure, embryos accumulated Mn in the brain and liver, and developed neurological deficits in adulthood (Xia et al., 2017). Both embryos and adults presented impairments in dopaminergic and gabaergic signaling. Additionally, mutant animals developed steatosis, liver fibrosis, and polycythemia. These findings demonstrate that the zebrafish *slc30a10* mutant develops hepatic damage similar to the symptoms related to the HMNDYT1 with polycythemia, and cirrhosis (Xia et al., 2017).

In *slc3914* zebrafish mutants, excessive Mn causes locomotor alteration, and accumulation of Mn particularly in the brain, without affecting the levels of Fe, Zn, and Cd (Tuschl et al., 2016). In accordance, *slc3914* mutant larvae exposed to Mn (until 5 dpf at 50 μM) had impairments in neuronal activity within the telencephalon, locomotor, and visual functions, changes that were accompanied by calcium imbalance, activation of the unfolded protein response, oxidative stress, mitochondrial dysfunction, and apoptosis (Tuschl et al., 2022).

6. Exposures to Mn-derived compounds in Zebrafish

The exposure to Mn-based compounds, including fungicides such as Maneb® ($(C_4H_6MnN_2S_4)n$) and Mancozeb® (MZ) ($C_8H_{12}MnN_4S_8Zn$) and the fuel additive methylcyclopentadienyl manganese tricarbonyl (MMT) ($C_9H_7MnO_3$) have been associated with increased symptoms of manganism (Figure 6). Literature brings reports mainly about the impact of these agents on embryonic zebrafish development.

Embryos exposed to commercial formulation of mancozeb (0.5, 5, and 50 $\mu g\ L^{-1}$) presented a marked decrease in the hatching rate and several malformations (cardiac and yolk sac edema and spinal torsions), with higher prevalence in the highest concentration (Vieira et al., 2020). The exposures also caused impairments in the locomotor activity and response to an aversive stimulus of embryos. In this work, the toxicity of MZ was associated with oxidative stress and disruption in the activity of several enzymes from the antioxidant system (Vieira et al., 2020). In line with this, Costa-Silva and collaborators showed that zebrafish embryos exposed to MZ (from 5 hpf to 72 hpf) developed morphological abnormalities such as body axis distortion, and behavioral deficits during the development. DNA damage, cell death, and increased reactive species levels were highlighted as events involved in the MZ-induced embryotoxicity (Costa-Silva et al., 2018). The same research group also demonstrated that the behavioral changes in developing zebrafish exposed to MZ

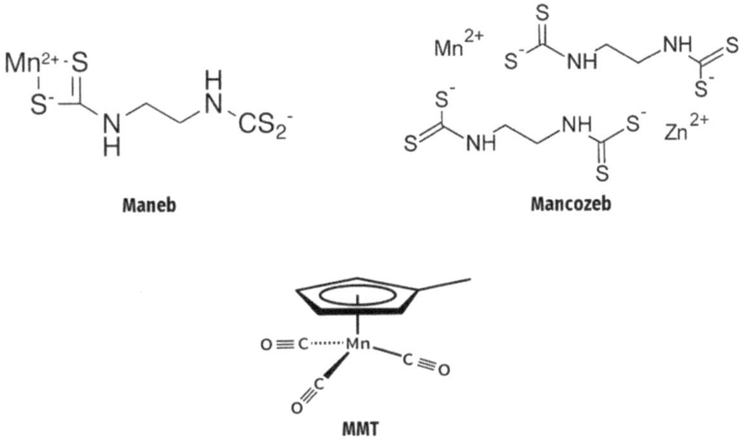

Figure 6. Chemical structure of Maneb, Mancozeb and MMT (methylcyclopentadienyl manganese tricarbonyl).

occurred earlier than redox alterations (Leandro et al., 2021). The embryos exposed to MZ (5, 10, and 20 μg/L) presented alterations in spontaneous movement, escape responses, swimming capacity, and exploratory behavior at 24, 28, 72 and 168 hpf, while the change in redox markers were observable only at 72 and 168 hpf (Leandro et al., 2021).

Interestingly, there is also evidence that environmentally relevant MZ concentrations (ranging from 1.88 μM to 7.52 μM) induced cardiac developmental toxicity in the larval stage (Wang et al., 2021). By transcriptome analyses, the abnormal development of zebrafish heart was indicated to be related to the activation of Notch and apoptosis-related signaling pathways, since related genes were significantly up-regulated after MZ treatment.

Regarding MMT exposure, there are findings showing that low-doses affect the development of diencephalic dopaminergic neurons in zebrafish. MMT (about 5 mg Mn/L) caused an up-regulation of dopamine-related genes in association with an increase in the number of diencephalic DA neurons, DA levels, and brain Mn accumulation. Also, there was locomotor hyperactivity. Although DA levels were restored at adulthood, a deficit in the acquisition and consolidation of memory (Fasano et al., 2021) was observed. Collectively, these findings suggest that developmental exposure to low-level MMT-derived Mn is responsible for the selective alteration of diencephalic DA neurons and with long-lasting effects on fish exploratory behavior in adulthood.

7. Conclusions

Mn deficiency is a rare condition in humans (Chen et al., 2018). In fact, the most common disturbances in humans are associated with the excess of Mn exposure and its neurotoxic effects. The mechanisms by which Mn induces neuronal damage are not well defined, but its neurotoxicity appears to be regulated by a number of factors, including oxidative injury, mitochondrial dysfunction, neuroinflammation and disruption in dopamine signaling. Alternative to mammalian models have been instrumental to study the gaps in our knowledge about the molecular processes involved the toxicity associated with Mn toxicity in humans (Manganism and Hypermanganesemia). Specifically, the use of zebrafish models has contributed to unravel biochemical changes associated with Mn neurotoxicity. Herein in this chapter, we have discussed findings that consolidate the importance of Zebrafish to study Mn-mediated neurotoxicity both in terms of neurodevelopmental analyses and neurotoxicity in adults. In general, the literature data have indicated that zebrafish exposure to Mn in the larval or adult stages re-capitulate key phenotypes and molecular or cellular findings observed in rodents and humans. Movement disturbances, developmental, morphological and neurochemical changes in the brain functioning are examples. Although the studies discussed in this chapter indicate the usefulness of zebrafish to study the pathology of Manganism and Hypermanganesemia, there are no detailed studies about on how the multiple Mn transporters described in mammals operate in the tissues of zebrafish. Importantly, since the proteins (transporters) involved in the uptake or efflux of Mn into or from cells can also transport other divalent metals (for instance, Fe^{2+}, Zn^{2+}, Cu^{2+}, Co^{2+}, Cd^{2+}, Ni^{2+}, and Pb^{2+}, etc.), the interaction of different cationic forms of Mn with these metals have also to be explored to clarify potential synergistic or antagonistic effects at level of brain and other tissues of zebrafish.

References

Altenhofen, S., Wiprich, M.T., Nery, L.R., Leite, C.E., Vianna, M.R.M.R. and Bonan, C.D. 2017. Manganese (II) chloride alters behavioral and neurochemical parameters in larvae and adult zebrafish. Aquat. Toxicol. 182: 172–183. doi: 10.1016/j.aquatox.2016.11.013.

Altenhofen, S., Nabinger, D.D., Pereira, T.C.B., Leite, C.E., Bogo, M.R. and Bonan, C.D. 2018. Manganese (II) chloride alters nucleotide and nucleoside catabolism in zebrafish (*Danio rerio*) adult brain. Mol. Neurobiol. 55: 3866–3874. doi: 10.1007/s12035-017-0601-8.

Anagianni, S. and Tuschl, K. 2019. Genetic disorders of manganese metabolism. Curr. Neurol. Neurosci. Rep. 19(6): 33. doi: 10.1007/s11910-019-0942-y.

Aschner, M., Erikson, K.M. and Dorman, D.C. 2005. Manganese dosimetry: species differences and implications for neurotoxicity. Crit. Rev. Toxicol. 35(1): 1–32. doi: 10.1080/10408440590905920.

Aschner, M., Guilarte, T.R., Schneider, J.S. and Zheng, W. 2007. Manganese: recent advances in understanding its transport and neurotoxicity. Toxicol. Appl. Pharmacol. 221(2): 131–47. doi: 10.1016/j.taap.2007.03.001.

Avila, D.S., Puntel, R.L. and Aschner, M. 2013. Manganese in health and disease. Met. Ions Life Sci. 13: 199–227. doi: 10.1007/978-94-007-7500-8_7.

Bakthavatsalam, S., Das Sharma, S., Sonawane, M., Thirumalai, V. and Datta, A. 2014. A zebrafish model of manganism reveals reversible and treatable symptoms that are independent of neurotoxicity. Dis. Model. Mech. 7(11): 1239–51. doi: 10.1242/dmm.016683.

Balachandran, R.C., Mukhopadhyay, S., McBride, D., Veevers, J., Harrison, F.E., Aschner, M. et al. 2020. Brain manganese and the balance between essential roles and neurotoxicity. J. Biol. Chem. 295(19): 6312–6329. doi: 10.1074/jbc.rev119.009453.

Barbeau, A. 1984. Manganese and extrapyramidal disorders (a critical review and tribute to Dr. George C. Cotzias). Neurotoxicology 5: 13–35.

Bowler, R.M., Roels, H.A., Nakagawa, S., Drezgic, M., Diamond, E., Park, R. et al. 2007. Dose-effect relationships between manganese exposure and neurological, neuropsychological and pulmonary function in confined space bridge welders. Occup. Environ. Med. 64: 167–177. doi: 10.1136/oem.2006.028761.

Briggs, J.P. 2002. The zebrafish: a new model organism for integrative physiology. Am. J. Physiol. Integr. Comp. Physiol. 282: R3–R9. doi: 10.1152/ajpregu.00589.2001.

Bowman, A.B., Kwakye, G.F., Hernández, E.H. and Aschner, M. 2011. Role of manganese in neurodegenerative diseases. Journal of Trace Elements in Medicine and Biology 25(4): 191–203. doi: 10.1016/j.jtemb.2011.08.144.

Chen, P., Chakraborty, S., Mukhopadhyay, S., Lee, E., Paoliello, M.M., Bowman, A.B. et al. 2015. Manganese homeostasis in the nervous system. Journal of Neurochemistry 134(4): 601–610. doi: doi:10.1111/jnc.13170.

Chen, P., Bornhorst, J. and Aschner, M. 2018. Manganese metabolism in humans. Front Biosci (Landmark Ed) 23: 1655–1679. doi: 10.2741/4665.

Chen, P., Totten, M., Zhang, Z., Bucinca, H., Erikson, K., Santamaría, A. et al. 2019. Iron and manganese-related CNS toxicity: mechanisms, diagnosis and treatment. Expert. Rev. Neurother. 19(3): 243–260. doi: 10.1080/14737175.2019.1581608.

Chen, Y., Payne, K., Perara, V.S., Huang, S., Baba, A., Matsuda, T. et al. 2012. Inhibition of the sodium–calcium exchanger via SEA0400 altered manganese-induced T_1 changes in isolated perfused rat hearts. NMR Biomed. 25: 1280–1285. https://doi.org/10.1002/nbm.2799.

Costa-Silva, D.G.D., Leandro, L.P., Vieira, P.B., de Carvalho, N.R., Lopes, A.R., Schimith, L.E. et al. 2018. N-acetylcysteine inhibits Mancozeb-induced impairments to the normal development of zebrafish embryos. Neurotoxicol. Teratol. 68: 1–12. doi: 10.1016/j.ntt.2018.04.003.

Couper, J. 1837. On the effects of black oxide of manganese when inhaled into the lungs. Br. Ann. Med. Pharmacol. 1: 41–42.

Dachtler, J., Ivorra, J.L., Rowland, T.E., Lever, C., Rodgers, R.J. and Clapcote, S.J. 2015. Heterozygous deletion of α-neurexin I or α-neurexin II results in behaviors relevant to autism and schizophrenia. Behav. Neurosci. 129(6): 765–76. doi: 10.1037/bne0000108.

Dobson, A.W., Erikson, K.M. and Aschner, M. 2004. Manganese neurotoxicity. Ann. N. Y. Acad. Sci. 1012: 115–28. doi: 10.1196/annals.1306.009.

Fang, F. 2001. Phylogeny and species diversity of the South and Southeast Asian cyprinid genus Danio Hamilton (Teleostei, Cyprinidae) (Doctoral dissertation, PhD Thesis, Stockholm University, Stockholm, Sweden).

Fasano, G., Godoy, R.S., Angiulli, E., Consalvo, A., Franco, C., Mancini, M. et al. 2021. Effects of low-dose methylcyclopentadienyl manganese tricarbonyl-derived manganese on the development of diencephalic dopaminergic neurons in zebrafish. Environ. Pollut. 287: 117151. doi: 10.1016/j.envpol.2021.117151.

Ferreira, S.A., Loreto, J.S., Dos Santos, M.M. and Barbosa, N.V. 2022. Environmentally relevant manganese concentrations evoke anxiety phenotypes in adult zebrafish. Environ. Toxicol. Pharmacol. 93: 103870. doi 10.1016/j.etap.2022.103870.

Finley, J.W., Johnson, P.E. and Johnson, L.K. 1994. Sex affects manganese absorption and retention by humans from a diet adequate in manganese. Am. J. Clin. Nutr. 60(6): 949–55. doi: 10.1093/ajcn/60.6.949.

Frisbie, S.H., Mitchell, E.J. and Sarkar, B. 2015. Urgent need to reevaluate the latest World Health Organization guidelines for toxic inorganic substances in drinking water. Environ. Health. 14: 63. doi: 10.1186/s12940-015-0050-7.

Greger, J.L. 1999. Nutrition versus toxicology of manganese in humans: evaluation of potential biomarkers. Neurotoxicology 20: 205–212.

Hernández, R.B., Nishita, M.I., Espósito, B.P., Scholz, S. and Michalke, B. 2015. The role of chemical speciation, chemical fractionation and calcium disruption in manganese-induced developmental toxicity in zebrafish (*Danio rerio*) embryos. J. Trace Elem. Med. Biol. 32: 209–217. doi: 10.1016/j.jtemb.2015.07.004.

Hill, A.J., Teraoka, H., Heideman, W. and Peterson, R.E. 2005. Zebrafish as a model vertebrate for investigating chemical toxicity. Toxicol. Sci. 86: 6–19. doi: 10.1093/toxsci/kfi110.

Himeno, S., Sumi, D. and Fujishiro, H. 2019. Toxicometallomics of Cadmium, manganese and arsenic with special reference to the roles of metal transporters. Toxicol. Res. 35: 311–317. doi: 10.5487/tr.2019.35.4.311.

Hong, S.B., Kim, J.W., Choi, B.S., Hong, Y.C., Park, E.J., Shin, M.S. et al. 2014. Blood manganese levels in relation to comorbid behavioral and emotional problems in children with attention-deficit/hyperactivity disorder. Psychiatry Res. 220: 418–425. doi: 10.1016/j.psychres.2014.05.049.

Howe, K., Clark, M.D., Torroja, C.F., Torrance, J., Berthelot, C., Muffato, M. et al. 2013. The zebrafish reference genome sequence and its relationship to the human genome. Nature 496(7446): 498–503. https://doi.org/10.1038/nature12111.

Irizar, A., Molinuevo, A., Andiarena, A., Jimeno-Romero, A., San Román, A., Broberg, K. et al. 2021. Prenatal manganese serum levels and neurodevelopment at 4 years of age. Environ Res. 197: 111172. doi: 10.1016/j.envres.2021.111172.

Iyare, P.U. 2019. The effects of manganese exposure from drinking water on school-age children: A systematic review. Neurotoxicology 73: 1–7. doi: 10.1016/j.neuro.2019.02.013.

Josephs, K.A., Ahlskog, J.E., Klos, K.J., Kumar, N., Fealey, R.D., Trenerry, M.R. et al. 2005. Neurologic manifestations in welders with pallidal MRI T1 hyperintensity. Neurology 64: 2033–2039. doi: 10.1212/01.wnl.0000167411.93483.a1.

Kalueff, A.V., Stewart, A.M. and Gerlai, R. 2014. Zebrafish as an emerging model for studying complex brain disorders. Trends Pharmacol. Sci. 35: 63–75. doi: 10.1016/j.tips.2013.12.002.

Keen, C.L., Zidenberg-Cherr, S. and Lonnerdal, B. 1994. Nutritional and toxicological aspects of manganese intake: an overview. pp. 221–236. *In*: Mertz, W., Abernathy, C.O. and Olin, S.S. (eds.). Risk Assessment of Essential Elements. Washington, DC: ILSI Press.

Kozol, R.A., Abrams, A.J., James, D.M., Buglo, E., Yan, Q. and Dallman, J.E. 2016. Function over form: Modeling groups of inherited neurological conditions in zebrafish. Front. Mol. Neurosci. 9: 1–15. doi: 10.3389/fnmol.2016.00055.

Krachler, M. and Rossipal, E. 2000. Concentrations of trace elements in extensively hydrolysed infant formulae and their estimated daily intakes. Annals of Nutrition and Metabolism 44(2): 68–74. doi: 10.1159/000012823.

Leandro, L.P., Siqueira de Mello, R., da Costa-Silva, D.G., Medina Nunes, M.E., Rubin Lopes, A., Kemmerich Martins, I. et al. 2021. Behavioral changes occur earlier than redox alterations in developing zebrafish exposed to Mancozeb. Environ. Pollut. 268(Pt B): 115783. doi: 10.1016/j.envpol.2020.115783.

Leyva-Illades, D., Chen, P., Zogzas, C.E., Hutchens, S., Mercado, J.M., Swaim, C.D. et al. 2014. SLC30A10 is a cell surface-localized manganese efflux transporter, and parkinsonism-causing mutations block its intracellular trafficking and efflux activity. J. Neurosci. 34(42): 14079–14095. doi:10.1523/jneurosci.2329-14.2014.

Liu, W., Xin, Y., Li, Q., Shang, Y., Ping, Z., Min, J. et al. 2020. Biomarkers of environmental manganese exposure and associations with childhood neurodevelopment: a systematic review and meta-analysis. Environ Health. 19(1): 104. doi: 10.1186/s12940-020-00659-x.

Ljung, K. and Vahter, M. 2007. Time to re-evaluate the guideline value for manganese in drinking water? Environ. Health Perspec. 115: 1533–1538. doi: 10.1289/ehp.10316.

Madejczyk, M.S. and Ballatori, N. 2012. The iron transporter ferroportin can also function as a manganese exporter. Biochim Biophys Acta. 1818(3): 651–657. doi:10.1016/j.bbamem.2011.12.002.

Marins, K., Lazzarotto, L.M.V., Boschetti, G., Bertoncello, K.T., Sachett, A., Schindler, M.S.Z. et al. 2019. Iron and manganese present in underground water promote biochemical, genotoxic, and behavioral alterations in zebrafish (*Danio rerio*). Environ. Sci. Pollut. Res. 26: 23555–23570. doi: 10.1007/s11356-019-05621-0.

Mena, I., Marin, O., Fuenzalida, S. and Cotzias, G.C. 1967. Chronic manganese poisoning. Clinical picture and manganese turnover. Neurology 17: 128–136. doi: 10.1212/wnl.17.2.128.

Mukhopadhyay, S., Bachert, C., Smith, D.R. and Linstedt, A.D. 2010. Manganese-induced trafficking and turnover of the cis-Golgi glycoprotein GPP130. Mol. Biol. Cell. 21(7): 1282–1292. doi:10.1091/mbc.e09-11-0985.

Nadig, A.P.R., Huwaimel, B., Alobaida, A., Khafagy, E., Alotaibi, H.F., Moin, A. et al. 2022. Manganese chloride ($MnCl_2$) induced novel model of Parkinson's disease in adult zebrafish; Involvement of oxidative stress, neuroinflammation and apoptosis pathway. Biomed Pharmacother. 20; 155: 1136972022. doi: 10.1016/j.biopha.2022.113697.

Pajarillo, E., Nyarko-Danquah, I., Adinew, G., Rizor, A., Aschner, M. and Lee, E. 2021. Neurotoxicity mechanisms of manganese in the central nervous system. Adv. Neurotoxicol. 5: 215–238. doi: 10.1016/bs.ant.2020.11.003.

Panula, P., Chen, Y.C., Priyadarshini, M., Kudo, H., Semenova, S., Sundvik, M. et al. 2010. The comparative neuroanatomy and neurochemistry of zebrafish CNS systems of relevance to human neuropsychiatric diseases. Neurobiol. Dis. 40: 46–57. doi: 10.1016/j.nbd.2010.05.010.

Peres, T.V., Schettinger, M.R., Chen, P., Carvalho, F., Avila, D.S., Bowman, A.B. et al. 2016. Manganese-induced neurotoxicity: a review of its behavioral consequences and neuroprotective strategies. BMC Pharmacol. Toxicol. 17(1): 57. doi: 10.1186/s40360-016-0099-0.

Rodrigues, G.Z., Staudt, L.B., Moreira, M.G., Dos Santos, T.G., de Souza, M.S., Lúcio, C.J. et al. 2020. Histopathological, genotoxic, and behavioral damages induced by manganese (II) in adult zebrafish. Chemosphere 244: 125550. doi: 10.1016/j.chemosphere.2019.125550.

Rodríguez-Barranco, M., Lacasaña, M., Aguilar-Garduño, C., Alguacil, J., Gil, F., González-Alzaga, B. et al. 2013. Association of arsenic, cadmium and manganese exposure with neurodevelopment and behavioural disorders in children: a systematic review and meta-analysis. Sci. Total Environ. 454-455: 562–77. doi: 10.1016/j.scitotenv.2013.03.047.

Sachse, B., Kolbaum, A.E., Ziegenhagen, R., Andres, S., Berg, K., Dusemund, B. et al. 2019. Dietary manganese exposure in the adult population in germany-what does it mean in relation to health risks? Mol. Nutr. Food Res. 63(16): e1900065. doi: 10.1002/mnfr.201900065.

Saint-Martin, M., Joubert, B., Pellier-Monnin, V., Pascual, O., Noraz, N. and Honnorat, J. 2018. Contactin-associated protein-like 2, a protein of the neurexin family involved in several human diseases. Eur. J. Neurosci. 48(3): 1906–1923. doi: 10.1111/ejn.14081.

Santamaria, A.B., Cushing, C.A., Antonini, J.M., Finley, B.L. and Mowat, F.S. 2007. State-of-the-science review: Does manganese exposure during welding pose a neurological risk? J. Toxicol. Environ. Health B. Crit. Rev. 10(6): 417–65. doi: 10.1080/15287390600975004.

Senger, M.R., Rosemberg, D.B., Seibt, K.J., Dias, R.D., Bogo, M.R. and Bonan, C.D. 2010. Influence of mercury chloride on adenosine deaminase activity and gene expression in zebrafish (*Danio rerio*) brain. Neurotoxicology 31(3): 291–6. doi: 10.1016/j.neuro.2010.03.003.

Sigel, A. and Sigel, H. 2000. Manganese and its role in biological processes. Met. Based Drugs. 7(3): 167. doi:10.1155/mbd.2000.167.

Spence, R., Gerlach, G., Lawrence, C. and Smith, C. 2008. The behaviour and ecology of the zebrafish, *Danio rerio*. Biol. Rev. 83: 13–34. doi: 10.1111/j.1469-185X.2007.00030.x.

Streisinger, G., Walker, C., Dower, N., Knauber, D. and Singer, F. 1981. Production of clones of homozygous diploid zebrafish (*Brachydanio rerio*). Nature 291: 293–296. doi: 10.1038/291293a0.

Südhof, T.C. 2008. Neuroligins and neurexins link synaptic function to cognitive disease. Nature 455(7215): 903–11. doi: 10.1038/nature07456.

Takeda, A. 2003. Manganese action in brain function. Brain Res. Rev. 41(1): 79–87. doi: 10.1016/s0165-0173(02)00234-5.

Tan, J., Zhang, T., Jiang, L., Chi, J., Hu, D., Pan, Q. et al. 2011. Regulation of intracellular manganese homeostasis by Kufor-Rakeb syndrome-associated ATP13A2 protein. J. Biol. Chem. 286: 29654–29662. doi: 10.1074/jbc.m111.233874.

Thomason, R.T., Pettiglio, M.A., Herrera, C., Kao, C., Gitlin, J.D. and Bartnikas, T.B. 2017. Characterization of trace metal content in the developing zebrafish embryo. PLoS One 12(6): e0179318. doi: 10.1371/journal.pone.0179318.

Tu, H., Fan, C., Chen, X., Liu, J., Wang, B., Huang, Z. et al. 2017. Effects of cadmium, manganese, and lead on locomotor activity and neurexin 2a expression in zebrafish. Environ. Toxicol. Chem. 36(8): 2147–2154. doi: 10.1002/etc.3748.

Tuschl, K., Meyer, E., Valdivia, L.E., Zhao, N., Dadswell, C., Abdul-Sada, A. et al. 2016. Mutations in SLC39A14 disrupt manganese homeostasis and cause childhood-onset parkinsonism-dystonia. Nat Commun. 7: 11601. doi:10.1038/ncomms11601.

Tuschl, K., White, R.J., Trivedi, C., Valdivia, L.E., Niklaus, S., Bianco, I.H. et al. 2022. Loss of slc39a14 causes simultaneous manganese hypersensitivity and deficiency in zebrafish. Dis. Model Mech. 15(6): dmm044594. doi:10.1242/dmm.044594.

Vieira, R., Venâncio, C.A.S. and Félix, L.M. 2020. Toxic effects of a mancozeb-containing commercial formulation at environmental relevant concentrations on zebrafish embryonic development. Environ. Sci. Pollut. Res. Int. 27(17): 21174–21187. doi: 10.1007/s11356-020-08412-0.

Wang, Y., Yu, Z., Fan, Z., Fang, Y., He, L., Peng, M. et al. 2021. Cardiac developmental toxicity and transcriptome analyses of zebrafish (*Danio rerio*) embryos exposed to Mancozeb. Ecotoxicol. Environ. Saf. 226: 112798. doi: 10.1016/j.ecoenv.2021.112798.

World Health Organization (WHO). 2011. Guidelines for Drinking-Water Quality. 4th ed. Geneva: p. xx, 3, 161–162, 164, 177, 183, 223–224, 311, 323–324, 343, 387, 389–390, 394, 398–399, 413–415, 430, 471.

Xia, Z., Wei, J., Li, Y., Wang, J., Li, W., Wang, K. et al. 2017. Zebrafish slc30a10 deficiency revealed a novel compensatory mechanism of Atp2c1 in maintaining manganese homeostasis. PLoS Genet. 13(7): e1006892. doi: 10.1371/journal.pgen.1006892.

Xu, Y., Peng, T., Xiang, Y., Liao, G., Zou, F. and Meng, X. 2022. Neurotoxicity and gene expression alterations in zebrafish larvae in response to manganese exposure. Sci. Total Environ. 825: 153778. doi: 10.1016/j.scitotenv.2022.153778.

Zayed, J., Hong, B. and L'Espérance, G. 1999. Characterization of manganese-containing particles collected from the exhaust emissions of automobiles running with MMT additive. Environ. Sci. Technol. 33: 3341–3346. doi: 10.1021/es990709+.

Chapter 2

The Influence of Tellurium and Selenium Compounds in Biological Systems and Selenium (Se) and Tellurium (Te) Toxicity

Silvia Vávrová,[1] *Eva Struhárňanská,*[1] *Ján Turňa*[1,2] and *Stanislav Stuchlík*[1,2,*]

1. Introduction: Tellurium and selenium occurrence

With its unique properties and versatile applications, tellurium has become an intriguing subject of scientific exploration and technological innovation. However, its compounds, especially the oxyanions, exert numerous negative effects on both prokaryotic and eukaryotic organisms. This chapter aims to unravel the multifaceted nature of tellurium, delving into the current knowledge on the mechanisms of tellurium compounds' toxicity in bacteria and humans, and to summarize the various ways organisms cope and detoxify these compounds. Besides the existence of specific resistance mechanisms, tellurium and its toxic compounds interact with molecular systems, mediating general detoxification and mitigation of oxidative stress. This chapter offers also basic information on selenium as it belongs to the same chalcogen group as tellurium and yet it is an important element to organisms as it is a cofactor of various enzymes, therefore the similarity of tellurium and selenium biochemistry and the impact of their compounds on humans is discussed.

Tellurium (Te) and selenium (Se) are together with oxygen, sulphur and polonium members of the chalcogen group (Cooper, 1971). Se has crucial functions in biochemistry, biology, and medicine, whereas Te is a strange element with no apparent role in biological systems. The element tellurium Te^0, similar to selenium, has the chemical properties of a nonmetal or metalloid and the physical properties of

[1] Department of Molecular Biology, Faculty of Natural Sciences, Comenius University, Ilkovičova 6, 842 15 Bratislava, Slovakia.
[2] Science Park, Comenius University, Ilkovičova 8, 841 04 Bratislava, Slovakia.
* Corresponding author: stanislav.stuchlik@uniba.sk

a metal. Its chemistry is similar to that of Se in many ways (Ba et al., 2010). There are many not yet fully understood mechanisms of Te compounds in organisms but Se, with many similarities in properties, can provide a considerable example for its study.

These elements were classified as rare elements of the Earth's crust. They are only about as abundant as platinum or gold with a low global and heterogeneous distribution and natural occurrence (Figure 1). This is only about 10^{-2} to 10^{-8} ppm for tellurium (Cooper, 1971; Belzile and Chen, 2015) and 10^{-6} ppm for selenium (Figure 1) (Haynes, 2016). Because of human industrial activities, Te-containing compounds can become environmental pollutants (Cunha et al., 2009; Presentato et al., 2019). In areas adjacent to gold mines, the concentration of Te reaches extreme values of up to 14.8 ppm (Wray, 1998). Tellurium is also a by-product of the electrolytic refining of copper. The world tellurium producers are the USA, Peru, Canada, and Japan. The annual worldwide production is around 215 tons, while 1000 tons of copper ore is needed to obtain 1 kg of tellurium (Goldfarb, 2014). Its occurrence in places with high anthropogenic industrial activities (Sandoval et al., 2010) causes environmental problems that not only destroy the microbial ecology of soil but also pose a serious risk to human health since Te compounds are highly toxic with a teratogenic effect in rats (Perez-D'Gregorio and Miller, 1988). The elementary Te^0 has been classified as non-toxic for living forms (Chasteen et al., 2009) in comparison with soluble Te oxyanions (Taylor, 1999).

Se, a rare element on Earth, is found in the lithosphere, atmosphere, hydrosphere, and biosphere, in that order (Reich and Hondal, 2016). Volcanic gases are responsible for its atmospheric discharge. The atmospheric release of this element from industrial sources (burning coal and crude oil) is another significant impact (Tan et al., 2018). Se is released into the atmosphere as a result of human activities, accounting for 37.5–40.6% of all atmospheric emissions. The combustion of coal and crude oil is the main source of this element's atmospheric concentration in metropolitan areas, which ranges from 1 to 10 ng/m^3 (Wen and Carignan, 2007). Se is concentrated in the Earth's crust and is spread in an uneven way. This element is mainly present in parent rocks, sedimentary rocks with volcanic origin, and it also enters the soil when rain falls as a result of water evaporating from seas and oceans.

The tellurium compounds can be divided into three groups: (a) inorganic tellurides; (b) Te-containing complex-like structures; and (c) organotellurides (Ba et al., 2010). The Te-containing complex-like structures have been studied

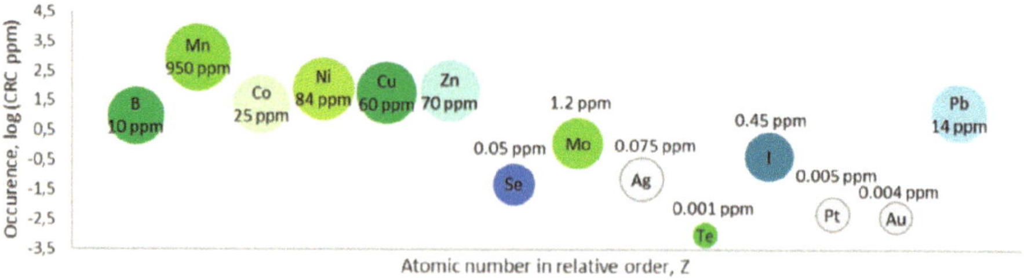

Figure 1. The occurrence of rare-Earth elements. The size of an element's bubble represents occurrence in Earth's crust and sea, graphically designed according to data from (Haynes et al., 2016).

extensively in a biological context. The small tellurium (IV) compound, ammonium trichloro (dioxoethylene-o,o')tellurate (AS101) has central Te atom with three active Cl^- species which have biomedical and pharmacologic potential (Chiaverini et al., 2022). The central Te atom can bond with a range of ligands. Cl^- species can be used for synthesis of bromido or iodide-replaced analogues which have higher tendency to react with different families of cysteine proteases. Ligand changes are responsible for the majority of the biological activities of AS101 analogues that can result in the differences in inactivation abilities of cysteine protease (Turner et al., 1999; Carmely et al., 2009; Chiaverini et al., 2022). For instance, an active site change at the Te atom results in the inactivation of cysteine protease. Organic Te agents have GPx(glutathione peroxidase)-like activity, as shown by tests *in vitro* in which organotellurides have been used as mimics of the antioxidant GPx, because of their resemblance to the human Se-containing enzyme GPx (Giles et al., 2001; Giles et al., 2003). Some organotellurides have a biochemical activity that mimics vitamin E (Kumar et al., 2010).

Mineral trace elements such as Se, Zn and Cu are members of the group of micronutrients with antioxidative ability. Organic Se in food is found in selenomethionine and selenocysteine, whereas inorganic Se occurs as SeO_3^{2-} (a more toxic form) and selenate (SeO_4^{2-}, a less toxic form). These inorganic forms are mobile, toxic and water-soluble. The bioavailability of these compounds allows them to be easily absorbed by plants and animals from Se-rich soil or water. Hence, Se is an essential trace element for both prokaryotic and eukaryotic biological systems at low concentrations but is toxic at higher levels (Kessi et al., 1999). Thus, Se can enter the food chain and can pose a potential threat to animals and humans (Whanger, 2002; Sors et al., 2005). Since the chemistry of Te slightly resembles that of sulphur, Te can be incorporated into amino acids such as cysteine and methionine and subsequently into proteins and enzymes. This incorporation does not require special machinery. It can occur naturally via bioincorporation (Liu et al., 2009). Te-containing fatty acids, Te-amino acids and many fluorescent Te particles (e.g., CdTe, CdSeTe, CdHgTe and CdTe-ZnTe) and nanoparticles are used as biomarkers for imaging methods and could have potential applications as nanosensors. Telluromethionine and tellurocysteine which are suitable for highly sensitive fluorescent imaging methods as biomarkers (Wang et al., 2010).

2. Effects of tellurium, selenium, and their compounds on prokaryotic systems

2.1 Mechanism of toxicity

Tellurites are highly toxic, even at micromolar levels (1 μg/mL) (Summers and Jacoby, 1977; Turner et al., 1999; Bajaj and Winter, 2014); however, there are no recorded thresholds according to the OSHA and NIOSH databases. The toxicity of potassium tellurite (K_2TeO_3) was recorded for the first time by Sir Alexander Fleming in 1932 (Fleming, 1932). These compounds were used for the treatment of some medical conditions such as syphilis, tuberculosis, cystitis, dermatitis, eye infections, and leprosy. Therefore, Te-containing soluble salts were historically used as antimicrobial and therapeutical agents before the era of antibiotics (Cooper, 1971). Several microorganisms, especially pathogenic, evolved mechanisms enabling the reduction of the noxious effects of Te toxicants accompanied by black colony formation. This chromogenic-based selective cultivation has been used for the diagnosis of the presence of antibiotic-resistant pathogenic bacteria for decades and, despite the emergence of genomic approaches, remains widely used (Taylor et al., 2002; Blaskovic and Turna, 2012; Perry, 2017).

Upon the entry of TeO_3^{2-} into a cell, the transmembrane proton gradient is disrupted in *E. coli*, irrespective of the level of resistance. This effect is accompanied by the inhibition of ATP synthesis, resulting in the depletion of intracellular ATP stores during aerobic growth (Lohmeir-Vogel et al., 2004). The same detrimental effect of Te compounds has been found in protein synthesis with regard to proteins containing amino acids with reduced thiol groups (Burkholz and Jacob, 2013) of both low-level and high-level resistant microbes. Two high-resistant bacteria, namely *Erythromonas ursincola* strain KR99 and *Erythromicrobium ramosum* E5, are distinct from other microbes, both showing an increase in protein and ATP synthesis in the presence of Te compounds (Maltman and Yurkov, 2015). The mechanism of this increase is uncertain, but it was implied (Moore and Kaplan, 1992; Moore, 1994; Maltman and Yurkov, 2015; Csotonyi et al., 2015) that reduction of such oxyanions could help with retaining optimal redox balance.

2.2 Toxicants containing tellurium and microbial resistance

Through the glutathionylation and methylation of Se molecules, several bacteria reduce Se toxicity. Glutathione as one of the most abundant thiols in cells was described in connection with Se reduction since late 60ies (Ganther, 1968; Ganther, 1971). Methyl-selenides and methyl-selenoxides are typical breakdown and detoxifying products for harmful Se oxyanions in addition to glutathionylation. Dimethyl selenide (DMSe) and dimethyl diselenide (DMDSe), which are volatile forms, are 500–700 times less hazardous than other Se compounds (Ganther, 1966).

Te probably exploits the metabolic machinery of Se, while tellurodiglutathione (GSTeSG), which is an analog of selenodiglutathione (GSSeSG), is produced during H_2Te formation. The toxicity mechanism of thiol-binding metal(loid)s is based on the interaction and subsequent inhibition of essential thiol groups of enzymes and proteins (Kinraide and Yermiyahu, 2007). The similar physical and electrochemical features of Te and Se with that of sulphur lead to their substitution in proteins (Moroder, 2005). The erroneous incorporation of the resulting tellurocysteine and/or telluromethionine into the protein structure leads to changes in protein activity or protein inactivity. It is important to note, that about 30% of proteins of all organisms require a bound metal for their function. These metals can be altered by tellurium with subsequent changes in protein/enzyme activity.

Many bacterial strains are able to resist the adverse effects of tellurium compounds. The mechanisms of bacterial resistance to toxicants are extraordinarily diverse (Vávrová et al., 2021).

Intrinsic resistance comprises the inherent properties of a microorganism that limit the action of antimicrobials. Bacteria can actively expel the toxicant from the cell by efflux or they can reduce its entry into the cell by cell surface reduction in the number or modification of entry channels, such as porins. In the case of no permeability changes, the generally reduced uptake of toxicant acts in synergy with other resistance mechanisms such as enzymatic degradation and efflux (Fernández and Hancock, 2012).

Acquired resistance is the next type of resistance that occurs when an originally sensitive microbe becomes resistant by acquisition and incorporation of new mobile genetic elements (e.g., plasmids, transposons, integrons, and naked DNA), by lateral gene transfer. Here microbes are organized into genomic islands consisting of gene clusters or operons with various gene compositions (Coombs and Barkay, 2003). The genomic islands are often named pathogenic islands since they create genomic features of pathogens and are characteristic tools of microbial evolution. Acquired resistance can also occur as a result of mutation. These mutational events may lead to a large increase in the minimal inhibitory concentration (Baquero, 2001; Fernández et al., 2011; Fernández and Hancock, 2012).

Adaptive resistance is the third type of resistance which involves a temporary increase in the ability of a bacterium to survive in the presence of a toxicant by alterations in gene and/or protein expression as a result of exposure to an environmental trigger, e.g., stress, nutrient conditions, growth state and sub-inhibitory levels of the toxicant or antibiotics themselves. Intrinsic and acquired resistance mechanisms are stable and can be transmitted vertically to subsequent generations. Adaptive resistance has a transient character and usually reverts upon the removal of the inducing condition (Fernández and Hancock, 2012).

3. Impact of tellurium, selenium, and their compounds on humans

The metabolic pathways involving Te, the mechanisms of its toxicity and its impact on human health, have been studied poorly to date. Te compounds can be taken in by inhalation or ingestion but these pathways are not often described. The clinical manifestation of the ingestion of metal-oxidizing solutions containing substantial concentrations of Te includes vomiting, nausea, metallic taste, black discoloration of the oral mucosa and skin, corrosive gastrointestinal tract injury and a characteristic garlic-like odour of the breath (Yarema and Curry, 2005). As mentioned above, organotellurium compounds are, in general, less toxic than inorganic Te compounds. They have in part different pharmacological and pharmacokinetic profiles from each other, and they overcome different metabolic conversions in the human body. A few publications mention Te toxicity connected with microorganisms (Chasteen et al., 2009) and, thus, a possible bridge between microbial and human cells can reasonably be considered. One experimental observation is the analogy of the discoloration of skin versus that of bacterial cells. Phenotypic changes attributable to the reduction of TeO_3^{2-} to elemental Te (Te^0) are accompanied by the blackening of colonies and/or media as a consequence of the presence of nanocrystals or nanoparticle formation in the periplasmic space of cells (Turner et al., 2012). Yarema and Curry (2005) suggest that the phenomenon of discoloration of the skin is a result of the deposition of elemental Te in the dermis and subcutaneous tissue. Exposure to gaseous hydrogen telluride is different from exposure to other forms of Te. When exposure is limited, the mucous membrane and pulmonary system might become irritated. In animals, extensive exposure to hydrogen telluride have resulted in hemolysis, hemoglobinuria, anuria, jaundice and pulmonary oedema, symptoms similar to toxicity from the inhalation of arsine or the poisonous gas stibine (Cerwenka and Cooper, 1961). The toxicity of Te, Se and As is associated with analogical secondary metabolite production, e.g., dimethyl telluride ($(CH_3)_2Te$), dimethyl selenide ($(CH_3)_2Se$) and monomethyl ($CH_3AsO(OH)_2$) and dimethyl arsenic acid ($(CH_3)_2AsO_2H$), all of which have a characteristic garlic odor. Their toxicity is exerted via strong interaction with cysteine-containing proteins and enzymes (Ba et al., 2010).

The activity of many more enzymes is affected by Te compounds in animals. Experiments on rats and mice have revealed that, after TeO_3^{2-} ingestion, the peripheral nerves become transiently demyelinated because of the inhibition of squalene epoxidase (squalene monooxygenase) (Abe et al., 2007). This enzyme is involved in the biosynthesis of cholesterol, a crucial component of myelin. The mechanism of squalene epoxidase inhibition resides in the binding of methyltellurium compounds or TeO_3^{2-} itself to the thiol groups of the cysteine residues at the active site (Laden and Porter, 2001). Te atoms attack sulphur-containing cysteine-proteases, such as cathepsin B and caspases, which can lead to changes in cellular metabolism resulting in cell death via apoptosis. The same consequences can be observed after selenium-tellurium interaction in less abundant selenocysteine-containing biomolecules, such as selenoprotein P, thioredoxin reductase (TrxR) and various GPx enzymes (Garberg et al., 1999). Inhibition of these enzymes by interaction with Te compounds triggers a loss of antioxidant defense and widespread oxidative stress. Another group of enzymes affected by organotellurium compounds' toxicity belongs to the glutamatergic system. Neurotoxic diphenyl ditelluride can alter enzymes of the glutamatergic system because of its interaction with the thiol groups of cysteine-containing enzymes and proteins (Stangherlin et al., 2009). The abovementioned group of enzymes use tellurium-sulphur chemistry to exert the toxicity of Te compounds. Tellurium-selenium interaction also has a significant effect on the activity of enzymes involved in Te toxicity. One member of this group is the human selenoenzyme thioredoxin reductase (TrxR), whose activity is modified by tellurium-selenium interaction (McNaughton et al., 2004). These tellurium-sulphur and/or tellurium-selenium bonds can equally change the protein function and abolish normal enzymatic activity and can cause extensive damage to the cell via oxidative stress leading to cell death. Both Te compounds and Se compounds (I) weaken the cell's antioxidant defense (II) actively generating ROS.

The main selenoenzymes, which are involved in antioxidant defense, detoxification, and thyroid activities, are glutathione peroxidase, iodothyronine deiodinase, and thioredoxin reductase (Palomo-Siguero and Madrid, 2017). Se can enter the body through food consumption; the primary food sources of Se are fish and vegetables. Se intake varies across the world depending on factors including the amount of Se in each country's soil, how much Se vegetables can store, the type of crop grown and consumed, Se speciation, soil pH, organic matter content, etc. (Rayman, 2012). Although Se was formerly thought to be harmful, there is a small concentration gap between Se essentiality and toxicity that depends on its speciation (Zhang et al., 2009). Human Se deficiency is linked to immune system dysfunction, cardiovascular illness, and hypothyroidism (Pedrero et al., 2006).

The human body is known to metabolize and excrete Te, even if the exact metabolic pathways are not understood. However, a parallelism with the pathways of Se is apparent. After ingestion of TeO_3^{2-} and TeO_4^{2-}, both are reduced in the liver. TeO_3^{2-} is methylated resulting in dimethyltellurium ($(CH_3)_2Te$) and finally trimethyltellurium ($(CH_3)_3Te^+$). Dimethyltellurium can bind to hemoglobin, which then accumulates in the blood cells of the rat. This interaction resembles the action of arsenic (Kobayashi and Ogra, 2009). Ba et al. (2010) suggest that these methylated species are the most abundant forms of Te in the human body; they are found in the kidney, the spleen and in the lungs. Finally, Te leaves the human body via urine and via breath as volatile $(CH_3)_2Te$, which is responsible for the garlic-like odour.

4. Tellurium and selenium toxicity vs potential benefits for prokaryotes and eukaryotes

When Te compounds enter cells, they can induce (I) changes in the integrity of cellular membrane structures (Kim et al., 2012), (II) changes in glutathione metabolism, (III) substitution of metal in enzymes and (IV) oxidative stress (Valdivia-González et al., 2012). The common feature of these metal(loid)s is their chemical affinity to proteins and to non-protein thiols and their ability to generate cellular oxidative stress by the Fenton reaction. Oxidative stress is induced by their interaction with the cell thiolome (Kell, 2010; Rubino, 2015), which represents the entity of the cellular thiol pool. Molecules carrying thiol groups are mostly derivatives of the amino acid cysteine and are called low molecular weight (LMW)-thiols: coenzyme A (CoA), glutathione (GSH), or bacillithiol (BSH). They serve to balance a reduced cell environment, act as cofactors in enzymatic reactions or help in the detoxification of reactive oxygen or nitrogen species, electrophilic compounds, or thiophilic metalloids (AsO_3^{2-}, TeO_3^{2-}). Te oxyanions (TeO_3^{2-}) are involved in the thiol redox system of the cell and interfere with thiol redox enzymes (glutathione reductase and thioredoxin reductase) and with their metabolites (glutathione, glutaredoxin and thioredoxin) (Turner et al., 1995). The key target for SeO_3^{2-} and TeO_3^{2-} cellular processing is glutathione, which participates in TeO_3^{2-} to Te (Te^0) reduction (Turner et al., 2001) accompanied by reactive oxygen species (ROS) formation (Kessi et al., 2004).

Glutathione is an endogenous tripeptide consisting of cysteine, glutamate and glycine and has antioxidative and other metabolic functions (Hopkins, 1929). Glutathione and sodium glutathione are used to prevent neurotoxicity associated with cisplatin or oxaliplatin during cancer treatment and can also prevent adverse effects of antineoplastic and radiation therapy. They can additionally be used in the treatment of a wide range of other disorders including poisoning with heavy metals and other compounds and even in COVID-19 disease therapy (Horowitz et al., 2020). The biosynthesis and metabolism of glutathione are directly related to stress tolerance.

5. Tellurium and selenium nanoparticles formation

Several microorganisms evolved mechanisms enabling cellular enzymes to be used for the reduction of the noxious effects of toxicants. The reduction of Te/Se compounds in such modified cell media is accompanied by black/red colony formation, respectively (Taylor et al., 2002; Perry, 2017). The first reports describing bacterial colonies with a black phenotype were published by Klett (1900) and Scheurlen (1900). These authors described the production of black or grey insoluble Te^0 in microorganisms treated with TeO_3^{2-}. This phenomenon is caused by the reduction of TeO_3^{2-} or Te^{4+} to the less toxic Te^0. Regardless of the concentration of TeO_3^{2-}, this results in the generation of black or grey deposits of metallic Te^0 inside the cell (Chasteen et al., 2009). The same observation was made by Yurkov et al. (1996) who also pointed out that TeO_3^{2-} reduction accompanied by black/grey colony formation and metallic Te crystal formation in obligate aerobic photosynthetic bacteria was not essential for the maintenance of high TeO_3^{2-} resistance.

Many non-tellurite resistant microorganisms are able to reduce TeO_3^{2-} at low concentrations, one such example being *E. coli* K12 (Summers and Jacoby, 1977; Avazeri et al., 1997). A major mechanism of resistance involves the inactivation of the toxicant by enzymes. The reduction of TeO_3^{2-} to Te^0, in this case, can be also carried out by the activity of the plasma membrane flavin-dependent reductase (Moore and Kaplan, 1992; Harrison, 2004). Several types of non-specific/housekeeping enzymes are involved in detoxification process of reduction, e.g., the thiol:disulphide oxidoreductase of *Rhodobacter capsulatus* (Borsetti et al., 2007) and GutS of *E. coli* (Guzzo and Dubow, 2000), periplasmic and membrane-associated nitrate reductases of *E. coli* and *Rhodobacter sphaeroides* (Avazeri et al., 1997; Sabaty et al., 2001). Details of only three specific TeO_3^{2-} reductases have been published to date: the cytoplasm-localized reductase of *Bacillus* spp. (Soudi et al., 2009; Etezad et al., 2009), the periplasmic TeO_3^{2-} reductase of the Gram-negative facultative anaerobe *Shewanella fridigimarina* (Maltman et al., 2017a) and the membrane-associated reductase of an aerobic anoxygenic phototroph *Erythromonas ursincola* (Maltman et al., 2017b). All these strains and the most recent strain *Shinella* sp. (Wu et al., 2019) started to be attractive due to their reduction potency which shows great promise in the bioremediation of toxic Te(IV) contamination. High toxicities of Se and Te oxyanions cause environmental problems by contaminated soils and waters (Harada and Takahashi, 2009). The ability of microorganisms to reduce Te and Se compounds in polluted and industrial areas is greatly appreciated in biometallurgy and bioremediation. These biotic methods involving the microbial retrieval of elements have become increasingly popular. Many bacterial strains have been described to be able to reduce these toxic oxyanions by producing elemental Se^0/Te^0 and by their ability to form nanoparticles containing Se/Te as a result of detoxification. These nanoparticles have various shapes (nanospheres, nanorods, nanoneedles, nanowires and nanotubes) (Wang et al., 2010), sizes and localizations (extracellular, intracellular). *Duganella violacienigra* can reduce both Se and Te oxyanions. It can be exploited in bioremediation and eco-friendly approaches to produce rare element nanoparticles, rather than synthesizing them by chemical means (Bajaj and Winter, 2014). Se, Te and other elements such as cadmium (Cd) and sulphur (S) are also used in the production of semiconductor nanoparticles (NPs) or quantum dots (QDs) with unique fluorescent properties and great technological potential (Deng et al., 2007; Monrás et al., 2014). *Shewanella oneidensis* MR-1 is another metal-reducing bacterium that reduces TeO_3^{2-} giving intracellularly accumulated needle-shaped crystalline Te^0 nanorods. These metal-reducing bacteria, in general, play an important role in the recycling of toxic Te elements and can be applied as a novel selective biological filter for the cellular accumulation of industry-applicable rare elements such as Te^0 nanorods (Kim et al., 2012). Te-containing nanoparticles also exhibit antibacterial properties against *E. coli*, with no apparent cytotoxicity against eukaryotic cells (Pugin et al., 2014). These nanoparticles provide a huge field for research with potential applications in medicine, pharmacy, optic, metallurgy, chemistry, and electronics. Not only bacteria can reduce Te, but other organisms such as plants or fungi are also able to accumulate toxic Te elements that can be applied as novel selective biological filters for use in bioremediation processes (Kim et al., 2012).

Microorganisms, in general, also play a crucial role in the biological transformation of SeO_3^{2-} and SeO_4^{2-} via metabolic reactions. The reduction of both forms, SeO_3^{2-} and SeO_4^{2-}, to Se^0 has been identified as an ideal strategy for Se detoxification and Se recovery in contaminated water, soil, and industrial effluent (Nancharaiah and Lens, 2015). A variety of microorganisms are used as microbial factories for the bioproduction of Se nanoparticles, e.g., *Enterobacter cloacae* (Yee et al., 2007), *Bacillus cereus* (Dhanjal and Cameotra, 2010), *Duganella* sp., *Agrobacterium* sp. (Bajaj et al., 2012), *Bacillus mycoides* (Lampis et al., 2014), *Shewanella oneidensis* (Li et al., 2014), *Pseudomonas putida* (Avendaño et al., 2016) and *Vibrio natriegens* (Fernández-Llamosas et al., 2017).

The concentration of Se in food depends on environmental conditions and thus its concentration in soil (Geering et al., 1968). Bacteria can convert an inert form of elemental Se to soluble SeO_3^{2-} and/or SeO_4^{2-}. The food intake of nutrient antioxidants is therefore considerably low in comparison with recommendations in Europe. The best way to supplement food lacking antioxidants is their addition in an inorganic form or rather as selenoproteins. The yeasts *Saccharomyces cerevisiae* and bacterial strains *Escherichia coli* and *Lactobacillus* spp. can accumulate, metabolize and convert SeO_3^{2-} and SeO_4^{2-} into Se and incorporate it into cysteine and/or methionine (Kieliszek et al., 2015). The presence of Se^0 in yeast cell structures has also been reported by Jiménez-Lamana et al. (2018). The various bacterial strains can also reduce and deposit metal(loid)s, e.g., Te and/or Se in a crystalline form (Burian et al., 1998). Some acetic acid bacteria can produce an amorphous metabolizable form of Se or incorporate it into metalloproteins (Grones et al., 1999; Macor et al., 2001; Macor et al., 2003).

6. Conclusions and perspectives

Selenium is a controversial but essential trace element of all organisms, which exerts beneficial as well as detrimental biological effects with a very narrow margin between the lowest acceptable levels of intake and toxicity. It is essential for the normal physiological functions of cells as a part of metal-binding proteins which have a central role in maintaining life processes such as catalysis, DNA/RNA binding, protein structure stability, etc. (Goldhaber, 2003). Tellurium is an element with no apparent role in biological systems. Despite considerable advancement in the clarification of the biochemistry of tellurite metabolism, tellurium function in biological systems as well as the molecular basis of tellurite toxicity, universal biomolecular mechanism of biochemical detoxification, mechanism of resistance and its interference with cellular processes, have not been elucidated to date. Thus, selenium can be chosen as an experimental model so that this element is more profoundly examined. Despite the fact that tellurium is much less explored, it is clear that inorganic salts of Te and a wide range of diverse organotellurium compounds show potential in diagnostics, pharmacology, and therapy. They may provide the basis for innovative drug development in the future (Ba et al., 2010). They are powerful agents in protein and enzyme inhibition; they can kill a wide range of microorganisms including bacteria and plasmodia and they are able to induce apoptosis of specific cancer cells.

Toxic chemicals, such as heavy metal(loid)s or toxic compounds of some elements can contribute to changes in human gut microbiota composition and metabolic profile due to their uptake. It is important to understand from future works, how these chemicals affect gut microbiota in relationship with metabolic and environmental diseases.

References

Abe, I., Abe, T., Lou, W., Masuoka, T. and Noguchi, H. 2007. Site-directed mutagenesis of conserved aromatic residues in rat squalene epoxidase. Biochem. Bioph. Res. Co. 352: 259–263.

Avazeri, C., Turner, R., Pommier, J., Weiner, J., Giordana, G. and Vermeglio, A. 1997. Tellurite reductase activity of nitrate reductase is responsible for the basal resistance of *Escherichia coli* to tellurite. Microbiology 143: 1181–1189.

Avendaño, R., Chaves, N., Fuentes, P., Sánchez, E., Jiménez, J.I. and Chavarría, M. 2016. Production of selenium nanoparticles in *Pseudomonas putida* KT2440. Sci. Rep. 6: 37155.

Ba, L.A., Döring, M., Jamier, V. and Jacob, C. 2010. Tellurium: An element with great biological potency and potential. Org. Biomol. Chem. 8: 4203–4216.

Bajaj, M., Schmidt, S. and Winter, J. 2012. Formation of Se(0) nanoparticles by *Duganella* sp. and *Agrobacterium* sp. isolated from Se-laden soil of north-east Punjab, India. Microb. Cell Fact. 13: 168.

Bajaj, M. and Winter, J. 2014. Se (IV) triggers faster Te (IV) reduction by soil isolates of heterotrophic aerobic bacteria: Formation of extracellular SeTe nanospheres. Microb. Cell Fact. 13: 1–10.

Baquero, F. 2001. Low-level antibacterial resistance: A gateway to clinical resistance. Drug Resist. Update. 4: 93–105.

Belzile, N. and Chen, Y.-W. 2015. Tellurium in the environment: A critical review focused on natural waters, soils, sediments and airborne particles. Appl. Geochem. 63: 83–92.

Blaskovic, D. and Turna, J. 2012. Tellurite-reduction-based assay for screening potential antibiotics. J. Medic. Microbiol. 61: 160–161.

Borsetti, F., Francia, F., Turner, R. and Zannoni, D. 2007. The thiol:disulfide oxidoreductase DsbB mediates the oxidizing effects of the toxic metalloid tellurite (TeO_3^{2-}) on the plasma membrane redox system of the facultative phototroph *Rhodobacter capsulatus*. J. Bacteriol. 189: 851–859.

Burian, J., Tu, N., Klucar, L., Guller, L., Lloyd-Jones, G., Stuchlik, S. et al. 1998. *In vivo* and *in vitro* cloning and phenotype characterization of tellurite resistance determinant conferred by plasmid pTE53 of a clinical isolate of *Escherichia coli*. Folia Microbiol. 43: 589–599.

Burkholz, T. and Jacob, C. 2013. Tellurium in nature. pp. 2163–2174. *In*: Kretsinger, R.H., Uversky, V.N. and Permyakov, E.A. (eds.). Encyclopedia of Metalloproteins, 1st ed. Springer: New York, USA.

Carmely, A., Meirow, D., Peretz, A., Albeck, M., Bartoov, B. and Sredni, B. 2009. Protective effect of the immunomodulator AS101 against cyclophosphamide-induced testicular damage in mice. Hum. Reprod. 24: 1322–1329.

Cerwenka Jr, E.A. and Cooper, W.C. 1961. Toxicology of selenium and tellurium and their compounds. Arch. Environ. Health. 3: 189–200.

Chasteen, T.G., Fuentes, D.E., Tantaleán, J. and Vásquez, C.C. 2009. Tellurite: History, oxidative stress, and molecular mechanisms of resistance. FEMS Microbiol. Rev. 33: 820–832.

Chiaverini, L., Cirri, D., Tolbatov, I., Corsi, F., Piano, I., Marrone, A. et al. 2022. Medicinal hypervalent tellurium prodrugs bearing different ligands: A comparative study of the chemical profiles of AS101 and its halido replaced analogues. Int. J. Mol. Sci. 23(14): 7505.

Coombs, J. and Barkay, T. 2003. Molecular evidence for the evolution of metal homeostasis genes by lateral gene transfer in bacteria from the deep terrestrial subsurface. Appl. Environ. Microb. 70: 1698–1707.

Cooper, W.C. 1971. Tellurium. Van Nostrand Reinhold Company, New York.

Csotonyi, J., Maltman, C., Swiderski, J., Stackenbrandt, E. and Yurkov, V. 2015. Extremely "vanadiphilic" multiply metal-resistant and halophilic aerobic anoxygenic phototrophs, strains EG13 and EG8, from hypersaline springs in Canada. Extremophiles 19: 127–134.

Cunha, R.L., Gouvea, I.E. and Juliano, L. 2009. A glimpse on biological activities of tellurium compounds. An. Acad. Bras. Cienc. 81: 393–407.

Deng, Z., Zhang, Y., Yue, J., Tang, F. and Wei, Q. 2007. Green and orange CdTe quantum dots as effective pH-sensitive fluorescent probes for dual simultaneous and independent detection of viruses. J. Phys. Chem. B. 111: 12024–12031.

Dhanjal, S. and Cameotra, S.S. 2010. Aerobic biogenesis of selenium nanospheres by *Bacillus cereus* isolated from coalmine soil. Microb. Cell Fact. 9: 52.

Etezad, S.M., Khajeh, K., Soudi, M., Ghazvini, P.T.M. and Dabirmanesh, B. 2009. Evidence on the presence of two distinct enzymes responsible for the reduction of selenate and tellurite in *Bacillus* sp. STG-83. Enzyme Microb. Tech. 45: 1–6.

Fernández, L., Breidenstein, E.B.M. and Hancock, R.E.W. 2011. Creeping baselines and adaptive resistance to antibiotics. Drug Resist. Update. 14: 1–21.

Fernández, L. and Hancock, R.E.W. 2012. Adaptive and mutational resistance: Role of porins and efflux pumps in drug resistance. Clin. Microbiol. Rev. 25: 661–681.

Fernández-Llamosas, H., Castro, L., Blázquez, M.L., Díaz, E. and Carmona, M. 2017. Speeding up bioproduction of selenium nanoparticles by using *Vibrio natriegens* as microbial factory. Sci. Rep. 7: 16046.

Fleming, A. 1932. On the specific antibacterial properties of penicillin and potassium tellurite. Incorporating a method of demonstrating some bacterial antagonisms. J. Pathol. Bacteriol. 35: 831–842.

Ganther, H.E., Levander, O.A. and Bauman, C.A. 1966. Dietary control of selenium volatilization in the rat. J. Nutr. 88(1): 55–60.

Ganther, H.E. 1968. Selenotrisulfides. Formation by the reaction of thiols with selenious acid. Biochemistry 7(8): 2898–2905.

Ganther, H.E. 1971. Reduction of the selenotrisulfide derivative of glutathione to a persulfide analog by gluthathione reductase. Biochemistry 10(22): 4089–4098.

Garberg, P., Engman, L., Tolmachev, V., Lundqvist, H., Gerdes, R.G. and Cotgreave, I.A. 1999. Binding of tellurium to hepatocellular selenoproteins during incubation with inorganic tellurite: Consequences for the activity of selenium-dependent glutathione peroxidase. Int. J. Biochem. Cell B. 31: 291–301.

Geering, H.R., Cary, E.E., Jones, L.H.P. and Allaway, W.H. 1968. Solubility and redox criteria for the possible forms of selenium in soils. Soil Sci. Soc. Am. Pro. 32: 35–47.

Giles, G.I., Tasker, K.M., Johnson, R.J.K., Jacob, C., Peers, C. and Green, K.N. 2001. Electrochemistry of chalcogen compounds: Prediction of antioxidant activity. Protective effect of the immunomodulator AS101 against cyclophosphamide-induced testicular damage in mice. Chem. Commun. 23: 2490–2491.

Giles, G.I., Giles, N.M., Collins, C.A., Holt, K., Fry, F.H., Lowden, P.A.S. et al. 2003. Electrochemical, *in vitro* and cell culture analysis of integrated redox catalysts: Implications for cancer therapy. Chem. Commun. 16: 2030–2031.

Goldfarb, R.J. 2014. Tellurium: The Bright Future of Solar Energy. US Department of the Interior, US Geological Survey.

Goldhaber, S.B. 2003. Trace element risk assessment: essentiality vs. toxicity. Regul. Toxicol. Pharm. 38(2): 232–242.

Grones, J., Macor, M., Siekel, P. and Bilska, V. 1999. Capability of *Escherichia coli* and *Lactobacillus* spp. to accumulate selenium in a biologically utilisable form. Bull. Food Res. 38: 45–53.

Guzzo, J. and Dubow, M. 2000. A novel selenite- and tellurite inducible gene in *Escherichia coli.* Appl. Environ. Microb. 66: 4972–4978.

Harada, T. and Takahashi, Y. 2009. Origin of the difference in the distribution behavior of tellurium and selenium in a soil-water system. Geochim. Cosmochim. Ac. 72: 1281–1294.

Harrison, S. 2004. Whither structural biology? Nat. Struct. Mol. Biol. 11: 12–15.

Haynes, W.M., Lide, D.R. and Bruno, T.J. 2016. Abundance of Elements in the Earth's Crust and in the Sea. CRC Handbook of Chemistry and Physics 97th, Boca Raton.

Hopkins, F.G. 1929. On glutathione, a reinvestigation. J. Biol. Chem. 84: 269–320.

Horowitz, R.I., Freeman, P.R. and Bruzzese, J. 2020. Efficacy of glutathione therapy in relieving dyspnea associated with COVID-19 pneumonia: A report of 2 cases. Respir. Med. Case Rep. 21: 101063.

Jiménez-Lamana, J., Abadálvaro, I., Bierla, K., Laborda, F., Szpunar, J. and Lobinski, R. 2018. Detection and characterization of biogenic selenium nanoparticles in selenium-rich yeast by single particle ICPMS. J. Anal. Atom. Spectrom. 33: 452–460.

Kell, D.B. 2010. Towards a unifying, systems biology understanding of large-scale cellular death and destruction caused by poorly liganded iron: Parkinson's, Huntington's, Alzheimer's, prions, bactericides, chemical toxicology and others as examples. Arch. Toxicol. 84: 825–889.

Kessi, J., Ramuz, M., Wehrli, E., Spycher, M. and Bachofen, R. 1999. Reduction of selenite and detoxification of elemental selenium by the phototrophic bacterium *Rhodospirillum rubrum*. Appl. Environ. Microb. 65: 4734–4740.

Kessi, J. and Hanselmann, K.W. 2004. Similarities between the abiotic reduction of selenite with glutathione and the dissimilatory reaction mediated by *Rhodospirillum rubrum* and *Escherichia coli.* J. Biol. Chem. 279(49): 50662–50669.

Kieliszek, M., Błazejak, S., Gientka, I. and Bzducha-Wróbel, A. 2015. Accumulation and metabolism of selenium by yeast cells. Appl. Microbiol. Biot. 99: 5373–5382.

Kim, D.H., Kanaly, R.A. and Hur, H.G. 2012. Biological accumulation of tellurium nanorod structures via reduction of tellurite by *Shewanella oneidensis* MR-1. Bioresource Technol. 125: 127–131.

Kinraide, T.B. and Yermiyahu, U. 2007. A scale of metal ion binding strengths correlating with ionic charge, Pauling electronegativity, toxicity, and other physiological effects. J. Inorg. Biochem. 101: 1201–1213.

Klett, A. 1900. Zur Kenntniss der reducirenden Eigenschaften der Bakterien. Z. Hyg. Infectionskrank. 33: 137–160.

Kobayashi, A. and Ogra, Y. 2009. Metabolism of tellurium, antimony and germanium simultaneously administered to rats. J. Toxicol. Sci. 34: 295–303.

Kumar, S., Johansson, H., Kanda, T., Engman, L., Muller, T., Bergenudd, H. et al. 2010. Catalytic chain-breaking pyridinol antioxidants. J. Org. Chem. 75: 716–725.

Laden, B.P. and Porter, T.D. 2001. Inhibition of human squalene monooxygenase by tellurium compounds: Evidence of interaction with vicinal sulfhydryls. J. Lipid. Res. 42: 235–240.

Lampis, S., Zonaro, E., Bertolini, C., Bernardi, P., Butler, C.S. and Vallini, G. 2014. Delayed formation of zero-valent selenium nanoparticles by *Bacillus mycoides* SelTE01 as a consequence of selenite reduction under aerobic conditions. Microb. Cell Fact. 13: 35.

Li, D.B., Cheng, Y.Y., Wu, C., Li, W.W., Li, N., Yang, Z.C. et al. 2014. Selenite reduction by *Shewanella oneidensis* MR-1 is mediated by fumarate reductase in periplasm. Sci. Rep. 4: 3735.

Liu, X., Silks, L.A., Liu, C., Ollivault-Shiflett, M., Huang, X., Li, J. et al. 2009. Incorporation of tellurocysteine into glutathione transferase generates high glutathione peroxidase efficiency. Angew. Chem. Int. Edit. 48: 2020–2023.

Lohmeir-Vogel, E., Ung, S. and Turner, R. 2004. *In vivo* 31P nuclear magnetic resonance investigation of tellurite toxicity in *Escherichia coli*. Appl. Environ. Microb. 70: 7324–7347.

Macor, M. and Grones, J. 2001. Genetic basis of selenium incorporation into proteins in bacterial cells. Bull. Food Res. 40: 101–118.

Macor, M., Kretova, M., Korenovska, M., Siekel, P. and Grones, J. 2003. Utilisation of selenium and distribution into bacterial cell structures. Bull. Food Res. 42: 205–212.

Maltman, C. and Yurkov, V. 2015. The effect of tellurite on highly resistant freshwater aerobic anoxygenic phototrophs and their strategies for reduction. Microorganisms 3: 826–838.

Maltman, C., Donald, L. and Yurkov, V. 2017a. Two distinct periplasmic enzymes are responsible for tellurite/tellurate and selenite reduction by strain ER-Te-48 isolated from a deep sea hydrothermal vent tube worms at the Juan de Fuca Ridge black smokers. Arch. Microbiol. 199: 1113–1120.

Maltman, C., Donald, L. and Yurkov, V. 2017b. Tellurite and tellurate reduction by the aerobic anoxygenic phototroph *Erythromonas ursincola*, strain KR99 is carried out by a novel membrane associated enzyme. Microorganisms 5: 20.

McNaughton, M., Engman, L., Birmingham, A., Powis, G. and Cotgreave, I.A. 2004. Cyclodextrin-derived diorganyl tellurides as glutathione peroxidase mimics and inhibitors of thioredoxin reductase and cancer cell growth. J. Med. Chem. 47: 233–239.

Monrás, J.P., Collao, B., Molina-Quiroz, R.C., Pradenas, G.A., Saona, L.A., Durán-Toro, V. et al. 2014. Microarray analysis of the *Escherichia coli* response to CdTe-GSH Quantum Dots: Understanding the bacterial toxicity of semiconductor nanoparticles. BMC Genomics 15: 1099.

Moore, M. and Kaplan, S. 1992. Identification of intrinsic high-level resistance to rare-earth oxides and oxyanions in members of the class Proteobacteria: Characterization of tellurite, selenite, and rhodium sesquioxide reduction in *Rhodobacter sphaeroides*. J. Bacteriol. 174: 1505–1514.

Moore, M.D. 1994. Members of the family *Rhodospirillaceae* reduce heavy-metal oxyanions to maintain redox poise during photosynthetic growth. ASM News. 60: 17–23.

Moroder, L. 2005. Isoteric replacement of sulfur with other chalcogens in peptides and proteins. J. Pept. Sci. 11: 187–214.

Nancharaiah, Y.V. and Lens, P.N.L. 2015. Selenium biomineralization for biotechnological applications. Trends Biotechnol. 33: 323–330.

Palomo-Siguero, M. and Madrid, Y. 2017. Exploring the behavior and metabolic transformations of SeNPs in exposed lactic acid bacteria. Effect of nanoparticles coating agent. Int. J. Mol. Sci. 18(8): 1712.

Pedrero, Z., Madrid, Y. and Cámara, C. 2006. Selenium species bioaccessibility in enriched radish (*Raphanus sativus*): a potential dietary source of selenium. J. Agr. Food Chem. 54: 2412–2417.

Perez-D'Gregorio, R.E. and Miller, R.K. 1988. Teratogenicity of tellurium dioxide: Prenatal assessment. Teratology 37: 307–316.

Perry, J.D. 2017. A decade of development of chromogenic culture media for clinical microbiology in an era of molecular diagnostics. Clin. Micro. Rev. 30: 449–479.

Presentato, A., Turner, R.J., Vásquez, C.C., Yurkov, V. and Zannoni, D. 2019. Tellurite-dependent blackening of bacteria emerges from the dark ages. Environ. Chem. 16: 266–288.

Pugin, B., Cornejo, F.A., Muñoz-Díaz, P., Muñoz-Villagrán, C.M., Vargas-Pérez, J.I., Arenas, F.A. et al. 2014. Glutathione reductase-mediated synthesis of tellurium-containing nanostructures exhibiting antibacterial properties. Appl. Environ. Microb. 80: 7061–7070.

Rayman, M.P. 2012. Selenium and human health. Lancet. 379: 1256–1268.

Reich, H.J. and Hondal, R.J. 2016. Why nature chose selenium. ACS Chem. Biol. 11(4): 821–841.

Rubino, F.M. 2015. Toxicity of glutathione-binding metals: A review of targets and mechanisms. Toxics. 3: 20–62.

Sabaty, M., Avazeri, C., Pignol, D. and Vermeglio, A. 2001. Characterization of the reduction of selenate and tellurite by nitrate reductases. Appl. Environ. Microb. 67: 5122–5126.

Sandoval, J.M., Levêque, P., Gallez, B., Vásquez, C.C. and Buc Calderon, P. 2010. Tellurite-induced oxidative stress leads to cell death of murine hepatocarcinoma cells. Biometals 23: 623–632.

Scheurlen. 1900. Die Verwendung der selenigen und tellurigen Säure in der Bakteriologie. Z. Hyg. Infektionskr. 33: 135–136.

Sors, T.G., Ellis, D.R. and Salt, D.E. 2005. Selenium uptake, translocation, assimilation and metabolic fate in plants. Photosynth. Res. 86: 373–389.

Soudi, M., Ghazvini, P., Khajeh, K. and Gharavi, S. 2009. Bioprocessing of seleno-oxyanions and tellurite in a novel *Bacillus* sp. strain STG-83: A solution to removal of toxic oxyanions in the presence of nitrate. J. Hazard. Mater. 165: 71–77.

Stangherlin, E.C., Ardais, A.P., Rocha, J.B.T. and Nogueira, C.W. 2009. Exposure to diphenyl ditelluride, via maternal milk, causes oxidative stress in cerebral cortex, hippocampus and striatum of young rats. Arch. Toxicol. 83: 485–491.

Summers, A.O. and Jacoby, G.A. 1977. Plasmid-determined resistance of tellurium compounds. J. Bacteriol. 129: 276–281.

Tan, L.C., Nancharaiah, Y.V., van Hullebusch, E.D. and Lens, P.N. 2018. Selenium: Environmental Significance, Pollution, and Biological Treatment Technologies. Anaerobic Treatment of Mine Wastewater for the Removal of Selenate and its Co-Contaminants, Leiden.

Taylor, D.E. 1999. Bacterial tellurite resistance. Trends Microbiol. 7: 111–115.

Taylor, D.E., Rooker, M., Keelan, M., Ng, L.K., Martin, I., Perna, N.T. et al. 2002. Genomic variability of O islands encoding tellurite resistance in enterohemorrhagic *Escherichia coli* O157:H7 isolates. J. Bacteriol. 184: 4690–4698.

Turner, R.J., Weiner, J.H. and Taylor, D.E. 1995. Neither reduced uptake nor increased efflux is encoded by tellurite resistance determinants expressed in *Escherichia coli*. Can. J. Microbiol. 41: 92–98.

Turner, R.J., Weiner, J.H. and Taylor, D.E. 1999. Tellurite mediated thiol oxidation in *Escherichia coli*. Microbiology 145: 2549–2557.

Turner, R.J., Aharonowitz, Y., Weiner, J.H. and Taylor, D.E. 2001. Glutathione is a target in tellurite toxicity and is protected by tellurite resistance determinants in *Escherichia coli*. Can. J. Microbiol. 47: 33–40.

Turner, R.J., Borghese, R. and Zannoni, D. 2012. Microbial processing of tellurium as a tool in biotechnology. Biotechnol. Adv. 30: 954–963.

Valdivia-González, M., Perez-Donoso, J.M. and Vásquez, C.C. 2012. Effect of tellurite-mediated oxidative stress on the *Escherichia coli* glycolytic pathway. Biometals 25: 451–458.

Vávrová, S., Struhárňanská, E., Turňa, J. and Stuchlík, S. 2021. Tellurium: A rare element with influence on prokaryotic and eukaryotic biological systems. Int. J. Mol. Sci. 22(11): 5924.

Wang, Z., Wang, L., Huang, J., Wang, H., Pan, L. and Wei, X. 2010. Formation of single-crystal tellurium nanowires and nanotubes via hydrothermal recrystallization and their gas sensing properties at room temperature. J. Mater. Chem. 20: 2457–2463.

Wen, H. and Carignan, J. 2007. Reviews on atmospheric selenium: Emissions, speciation and fate. Atmos. Environ. 41(34): 7151–7165.

Whanger, P.D. 2002. Selenocompounds in plants and animals and their biological significance. J. Am. Coll. Nutr. 21: 223–232.

Wray, D.S. 1998. The impact of unconfined mine tailings and anthropogenic pollution on a semi-arid environment—an initial study of the Rodalquilar mine district, south east Spain. Environ. Geochem. Hlth. 20: 29–38.

Wu, S., Li, T., Xia, X., Zhou, Z., Zheng, S. and Wang, G. 2019. Reduction of tellurite in *Shinella* sp. WSJ-2 and adsorption removal of multiple dyes and metals by biogenic tellurium nanorods. Int. Biodeter. Biodegr. 144: 104751.

Yarema, M.C. and Curry, S.C. 2005. Acute tellurium toxicity from ingestion of metal-oxidizing solutions. Pediatrics 116: 319–321.

Yee, N., Ma, J., Dalia, A., Boonfueng, T. and Kobayashi, D. 2007. Se(VI) reduction and the precipitation of Se(0) by the facultative bacterium *Enterobacter cloacae* SLD1a-1 are regulated by FNR. Appl. Environ. Microb. 73: 1914–1920.

Yurkov, V., Jappe, J. and Vermeglio, A. 1996. Tellurite resistance and reduction by obligately aerobic photosynthetic bacteria. Appl. Environ. Microb. 62: 4195–4198.

Zhang, B., Zhou, K., Zhang, J., Chen, Q., Liu, G., Shang, N. et al. 2009. Accumulation and species distribution of selenium in Se-enriched bacterial cells of the *Bifidobacterium animalis*. Food Chem. 115: 727–734.

Chapter 3

Aluminum Toxicity

Marilina de Sautu,[1] *Nicolás A. Saffioti*[1,2] and *Irene C. Mangialavori*[1]

1. Introduction

Aluminum (Al) is one of the most abundant elements in the Earth's crust (Exley and Mold, 2015), Exposure to Al occurs through various sources in everyday life. The relationship between Al exposure and neurodegenerative disorders has been widely reported (Igbokwe et al., 2020). This toxic metal is ubiquitous distributed, and in addition to the brain and the central nervous system, it affects the liver, kidney, and, hematopoietic systems (Rahimzadeh et al., 2022).

The cellular toxicity of aluminum has been explained through several mechanisms. The primary interaction of Al with ligands is electrostatic having a slow dissociation rate. Among other ligands, Al binds strongly to ATP, phosphate groups, and cation-binding sites of proteins (Kawahara and Muramoto, 2011). Thus, Al alters protein conformation, induces protein aggregation, and displaces cations such as Fe^{3+}, Mg^{2+}, and Ca^{2+} from its binding sites in biological molecules. Hence, alterations of Ca^{2+} homeostasis, protein phosphorylation, amyloid formation, and lipid peroxidation have been proposed as mechanisms that would explain Al toxicity. Although these mechanisms are supported by experimental evidence, the causality between human exposure to Al and the development of pathologies remains a controversial point in the scientific community.

In this chapter, we will summarize the mechanisms by which aluminum affects the function of key proteins in cell pathophysiology. Given the complex Al chemistry, we also address this topic in the context of the design and analysis of experiments.

[1] Instituto de Química y Fisicoquímica Biológicas, Facultad de Farmacia y Bioquímica, Universidad de Buenos Aires, CONICET. Buenos Aires. Argentina.

[2] Instituto de Nanosistemas, Universidad Nacional de San Martin, CONICET. Buenos Aires. Argentina.

* Corresponding author: irenem@qb.ffyb.uba.ar

2. Aluminum overview

After oxygen and silicon, aluminum is the most abundant elements in the Earth's crust (Exley and Mold, 2015). The oxidation state of aluminum is +3, and its effective ionic radius in sixfold coordination is 0.54 Å. The primary interaction of Al with ligands is electrostatic and has a slow dissociation rate due to its high mass to charge ratio (Martin, 1986). The radius of Al is similar to that of Fe^{3+} (0.65 Å) thus Al tends to replace it at its binding sites. Al also displaces magnesium (Mg) in biological systems, especially in the interaction with phosphate groups (Rezabal et al., 2006). Al binds 107 times more strongly to ATP than Mg, and Al at a nanomolar concentration can compete with Mg for phosphate groups (Martin, 1986). Calcium (Ca) is more frequently found in eightfold coordination and it is nine-fold greater than that Al, thus competition between Al and Ca for protein binding sites is less frequent (Kawahara and Kato-Negishi, 2011).

In biological systems, Al associates with oxygen donor ligands, such as carboxylate and phosphate groups, inorganic phosphate, nucleotides, and polynucleotides forming only weak complexes with amines and sulfhydryl ligands (Macdonald and Martin, 1988). Al forms strong complexes with multi-dentate amino-carboxylate ligands such as EDTA (Fulgenzi et al., 2014).

2.1 Aluminum in aqueous solutions

In aqueous solutions, Al forms different coordination complexes with water molecules reaching an equilibrium of species. The relative concentration of these species in equilibrium is highly dependent on pH (Figure 1A). In acidic solutions (pH < 5), Al exists as the octahedral hexahydrate complex, $Al(H_2O)_6^{3+}$ -or simply Al^{3+}- and, as the pH increases, this species undergoes successive deprotonations to yield $Al(OH)_2^+$ and $Al(OH)^{2+}$. Al precipitates as $Al(OH)_3$ and is redissolved as pH rises to form $Al(OH)_4^-$. Al dominates at pH < 5, $Al(OH)_4^-$ dominates at pH > 6.2, while there are a mixture of species between pH 5 and 6.2 (Martin, 1986).

In addition to complexes with only one Al atom (mononuclear species), Al forms complexes with different oligomerization grades (Al_n) and charges (Maki et al., 2017). These polynuclear species form slower than mononuclear ones and depending on the Al concentration, take a long time to reach equilibrium with the mononuclear species. When the equilibrium is perturbed, the polynuclear species can release Al slowly (from minutes to hours). In biological systems, the concentration of Al ligands is high and the pH is neutral keeping a relatively low concentration of free Al and hampering the formation of polynuclear complexes (Macdonald and Martin, 1988).

On the other hand, Al decreases the medium pH, thus a solution of 100 μM $AlCl_3$ when prepared in water (pH ~ 6) contains more Al than a prepared buffer solution at pH 6 (Martin, 1986), and this is because $AlCl_3$ drops the pH of water and then favors the Al species. Thus, in an aqueous solution, the Al concentration is not linear with respect to the $AlCl_3$ concentration (Figure 1B and C). This is shown in Figure 1B, where different $AlCl_3$ concentrations were dissolved in water, then, the pH of each solution was measured and the Al concentration was predicted as in Figure 1A. Note that the

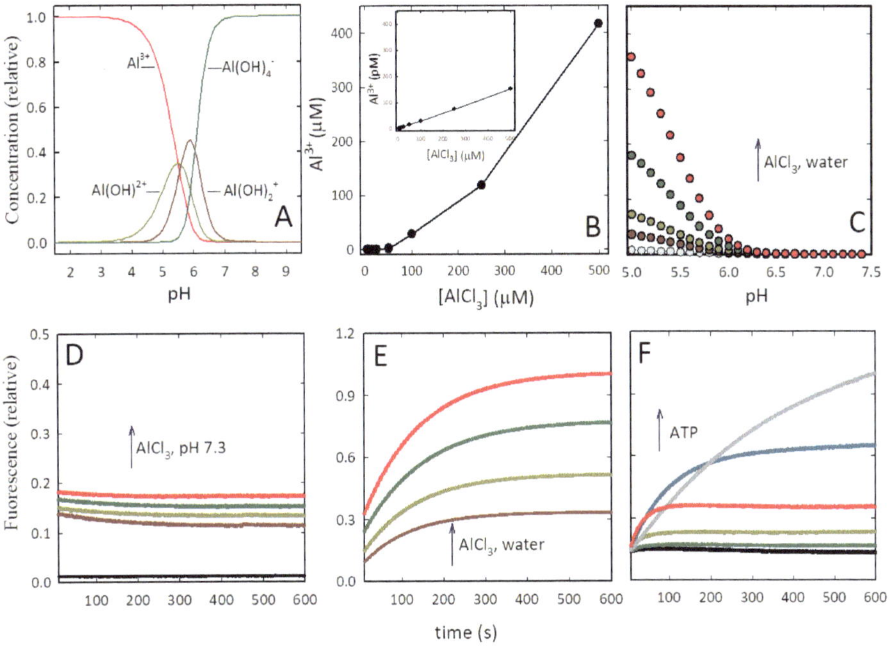

Figure 1. Aluminum in aqueous solutions. (A) Distribution of soluble mononuclear aluminum ion species in aqueous solution. The associated equilibrium constants were taken from (Martin, 1986): $K_1 = (H^+)[AlOH^{2+}]/[Al^{3+}] = 10^{-5.5}$; $K_2 = (H^+)[Al(OH)_2^+]/[AlOH^{2+}] = 10^{-5.6}$; $K_3 = (H^+)^2[Al(OH)_4^-]/[AlOH^{2+}] = 10^{-12.1}$. (B) Predicted Al concentration as a function of AlC13 concentration in a water or buffer (pH 7.3; 25°C) (inset) solution. (C) Predicted Al concentration in AlC13-water solutions as a function of pH (from top to bottom: 500; 250; 100; 50 and 10 μM $AlCl_3$). (D and E) Fluorescence of lumogallion-Al complex (565 nm) as a function of time in $AlCl_3$-buffered (pH 7.3 at 25°C) (D) or water solutions (E) (from bottom to top: 5; 10; 25 and 35 $AlCl_3$ μM). The black line in D corresponds to the fluorescence in the absence of $AlCl_3$. (F) Effect of ATP (from top to bottom: 1000; 500; 250; 100; 25; 0 μM ATP) on the lumogallion-Al fluorescence in an $AlCl_3$-buffered solution (pH 7.3 at 25°C).

concentrations indicated in the ordinate axis assume that all Al is in mononuclear form, i.e., the figure must be analyzed qualitatively since polynuclear species can form under the experimental conditions used. When the same $AlCl_3$ concentrations were dissolved in a buffered solution at pH 7.3, the Al predicted concentration was proportional to the $AlCl_3$ concentration (Figure 1B, inset). Figure 1C shows the predicted Al concentration in an AlC13-water solutions as a function of pH. If we qualitatively approximate a linear behavior for the predicted Al concentration between pH 5 and 6, we can see that the slope increases with increasing $AlCl_3$ concentration. The equilibrium between Al species is slow, therefore, it is not correct to assume a rapid equilibrium will be reached between Al complexes as with other metals in the solution (de Sautu et al. 2018). Thus, when studying the effect of Al on biological systems, experimental results may depend on whether Al was previously equilibrated or not with the experimental medium. In addition, the presence of ligands such as phosphate groups can increase the availability of Al (Luque et al., 2014). This is shown in Figure 1D–F where the Al is determined with lumogallion as a function of time in different conditions (de Sautu, 2021). Lumogallion is a fluorescent probe with a high affinity for Al forming a 1:1 stoichiometric complex (Mirza et al., 2016). Upon excitation at 500 nm, the lumogallion·Al complex has a maximal emission wavelength at 565 nm, and the fluorescence increases linearly with the Al concentration. Different concentrations of $AlCl_3$ were equilibrated for 30 min in a buffered solution at pH 7.3 (Figure 1D) or water (Figure 1E), then, 10 μM lumogallion was added and the fluorescence of lumogallion-Al complex was recovered as a function of time. In the $AlCl_3$-water solutions, the fluorescence of the lumogallion-Al complex increases as a function of time and then stabilizes, whereas in $AlCl_3$-buffered solution remains approximately constant. For each $AlCl_3$ concentration, the fluorescence in panel E is higher than in panel D, because the available Al in water is higher than at pH 7.3. When different ATP concentrations were added to an $AlCl_3$-buffered solution (20 μM $AlCl_3$, 20 μM lumogallion, pH 7.3 at 25°C), the fluorescence of lumogallion-Al complex increases as a function of time, indicating that ATP increases the available Al. A solution of $AlCl_3$ in the absence of a buffer, has a high relative concentration of Al. The addition of this solution to a biological system, such as a protein, may produce a greater effect in comparison with the addition of $AlCl_3$ dissolved in a buffer at pH 7, particularly if the Al effect is irreversible. In the first case, there is a high amount of transient Al ready to interact with the protein before the system equilibrates, sequestering Al into inactive complexes. An example of this has been described for the inhibition of plasma membrane calcium pump (PMCA) by Al (see below). When 100 μM $AlCl_3$ in water was added to a sample containing the protein at pH 7.3, the inhibition of enzyme activity as a function of time showed two phases. The amplitude of the fast phase was proportional to the concentration of $AlCl_3$ in the stock solution (between 5 and 40 μM) and decreased when $AlCl_3$ was dissolved in buffer at pH 7.3. Therefore, the complexity of Al coordination chemistry must be taken into account when designing *in-vitro* experiments.

2.2 Uptake of aluminum by cells

The presence of Al inside mammalians cells has been clearly documented (Cheng et al., 2018). However, the mechanism by which Al crosses the plasma membrane remains controversial. It is unlikely that Al crosses the membrane passively and there is no evidence of a transmembrane transporter of trivalent cations in humans (Exley and Mold, 2015).

In plants, several transporters play a key role in Al tolerance, either by mediating its extrusion or its sequestration (Kar et al., 2021). Nrat1 belongs to a divalent metal transporter family and is located in the plasma membrane of all plant root cells. Nrat1 can transport trivalent cations, including Al, and is required for sequestering Al in vacuoles. ABC transporters, aquaporins, and H^+-ATPases carry out vacuolar sequestration of Al contributing to Al tolerance in plants. The Al-activated malate transporter (ALMT) and toxic compounds and multidrug extrusion (MATE) family of transporters export malate and citrate, and their activity is increased in the presence of Al (Rahman et al., 2018). The ALMT and MATE transporters are essential to the Al tolerance in plants through the transport of citrate-Al and malate-Al.

In mammalians cells, several studies suggest that Al would interact with other molecules and cross membranes through diffusion, ion symports/antiports, active transport, and adsorptive or receptor-mediated endocytosis (Exley and Mold, 2015). Similar to plants, Al could cross the membrane forming complexes with amino acids, citrate, phosphates, and other molecules (Nagasawa et al., 2005).

2.3 Aluminum exposure

Al is a ubiquitous element and people are daily exposed to it. The main source of Al is food (~ 95%) and to a lesser extent drinking water (1–2%). Al intake reaches an amount of 4–9 mg daily (Nie, 2018). Incorporation through breathing is much lower but can increase in industrial environments. The incorporation of Al can increase with the consumption of antacids or the exposure to adjuvants and deodorants containing Al (Rahimzadeh et al., 2022).

Al is absorbed in the small intestine where it enters the bloodstream (Nie, 2018). In the plasma, most of Al is bound to transferrin (Tf) and to citrate in a lesser extent (Zhang et al., 2016). In cells, Al interacts with the phosphate group of phospholipids affecting the biophysical properties of the plasma membrane (Jones and Kochian, 1997). Inside the cells, it is found inside lysosomes and in the perinuclear space, and a lesser extent in the cytoplasm (Tenan et al., 2021). In the body, Al is mainly stored in the bones and lungs, but also has been identified in muscle, liver, and brain (Rahimzadeh et al., 2022).

Toxicokinetic studies showed that only a minor fraction of the ingested Al is absorbed (around 1%) (Schreeder et al., 1983), and the absorption depends on the Al formulation (Bilkei-Gorzó, 1993). Al is excreted mainly in urine. People with normal renal function are able to excrete all absorbed Al during the first week after exposure, but a minority of the absorbed dose (around 15%) is excreted slowly (Priest et al., 1995). Studies have reported that 4% of the absorbed Al remains in the body after 3 years (Connor, 1988). When people are exposed to high levels of Al, some of the absorbed dose can accumulate in the body due to binding to proteins which prevents its ultrafiltration (Rahimzadeh et al., 2022).

It has been reported that intake of silicone-containing compounds decreases absorption and facilitates excretion of Al (Krewski et al., 2007). Fluoride increases the excretion of Al through urine and feces (Glynn et al., 2001). In addition, absorption of Al is generally higher in people with iron (Fe) and Ca deficiency (Zhang et al., 2017).

The most widely used treatment for Al poisoning is chelation therapy. One of the most used chelators is ethylenediaminetetraacetic acid (EDTA) which is administered intravenously. Combinations of chelators including ascorbate (vitamin C), desferrioxamine, and Feralex-G—a effective compound for removing Al from different human brain cells—are also used (Kruck et al., 2004; Shin et al., 2003). The combined treatment with N-(2-Hydroxyethyl) ethylenediaminetriacetic acid (HEDTA) and propolis—a bee natural product—can help the recovery of damaged cell membranes and organ disfunction induced by Al. In Al administered mice, the treatment with HEDTA and propolis mitigates oxidative stress and improves liver, kidney, and brain functions (Bhadauria, 2012).

Al has been linked to various pathologies including neurological disorders (Exley and Clarkson, 2020; Nie, 2018), hypertension (Cheng et al., 2018; Ferguson et al., 2019; Zhang et al., 2016), anemia (Kaiser et al., 1984), renal dysfunction (Krewski et al., 2007), and cancer (Jeong et al., 2020; Temel et al., 2017). The relationship between Al and Alzheimer's disease (AD), amyotrophic lateral sclerosis, and Parkinsonism dementia has been extensively documented (Igbokwe et al., 2020). However, the hypothesis that Al contributes to neuropathological development has been discussed for decades and is still controversial in the scientific community (Inan-Eroglu, 2018). It has also been suggested that Al exposition can be implicated in the etiology of breast cancer (Mousavi, 2021). Al produces a malignant cell transformation and has the capacity to form tumors in mammary epithelial cells (Mandriota et al., 2016; Tenan et al., 2021). While the mechanism is still poorly understood, those cancerous cells suffer an increase in the number of structural chromosome rearrangements, which is a well-known hallmark of cancer. These findings are in line with other studies that related Al to carcinogenesis due to its genotoxicity (Paz et al., 2017).

2.4 Aluminum toxicity in the nervous system

Al does not have any known physiological function in biological processes (Lai and Blass, 1984). In humans, the first case of Al intoxication was reported in 1921, where neurological symptoms like loss of memory and impaired coordination were observed (Miu and Benga, 2006). Later evidence supported the Al neurotoxicity but indicated that other systems are affected as well.

Al intoxication was related to several diseases in dialysis patients. For example, Al can induce osteomalacia, a pathology in which softening of the bones is caused by inadequate mineralization. Osteomalacia was observed in patients with renal failure (Yiannopoulou and Papageorgiou, 2020) and/or high Al consumption (Palop and Mucke, 2010). Moreover, osteomalacia can be induced in rats by the administration of high doses of Al. Microcytic anemia is another pathology induced by Al in dialysis patients which can be elicited in rats by chronic administration of Al (Nehru and Anand, 2005). The relationship of Al with Fe is not surprising as the biochemical interactions of these two elements are related (see below). Noteworthy, Al is the main etiological factor of dialysis encephalopathy, a neurological disorder characterized by alterations in speech, disordered muscle action and seizures (Parkinson et al., 1981). Oral administration of Al phosphate to control serum phosphate in dialysis patients was reported as the principal etiological factor (Palop and Mucke, 2010). In these cases, insufficient Al clearance leads to its accumulation, despite its low gastrointestinal absorption. Other studies showed that high Al concentration in dialysis water was a risk factor for dialysis encephalopathy (Alfrey et al., 1976; Yiannopoulou and Papageorgiou, 2020). Thus, a link between Al intoxication and the nervous system was ascertained.

3. The effect of aluminum on protein in the nervous system

The study of the neurotoxicological effects of Al was started by Klatzo et al. in 1965, who found that intracerebral injection of Al phosphate in rabbits induced motor disorders, ataxia and seizures several days after the administration (Klatzo et al., 1965). Since then, many studies helped to characterize the effect of Al on the nervous system. For example, high oral doses of Al (100 mg/kg) in rats induced learning impairment (Bilkei-Gorzó, 1993; Connor et al., 988). The administration of Al in drinking water of mice for several months increased the production of inflammatory cytokines in neurons and glia, showing that Al induces inflammation in the brain (Becaria et al., 2006). Cultured glia also develops an inflammatory response upon the addition of Al in culture media (Campbell et al., 2002). At high concentrations (1 mM), Al is capable of inducing apoptosis in astrocytes (Suarez-Fernandez et al., 1999). However, the mechanism for damage in neuronal tissues is complex and can be explained by the different effects of Al in the cell. Figure 2 shows a simplified scheme of the effects of Al in neurons which are explained in the following paragraphs.

Al is known to induce a pro-oxidative response in neuronal tissues. Lipid peroxidation is augmented upon administration of Al (Becaria et al., 2006;

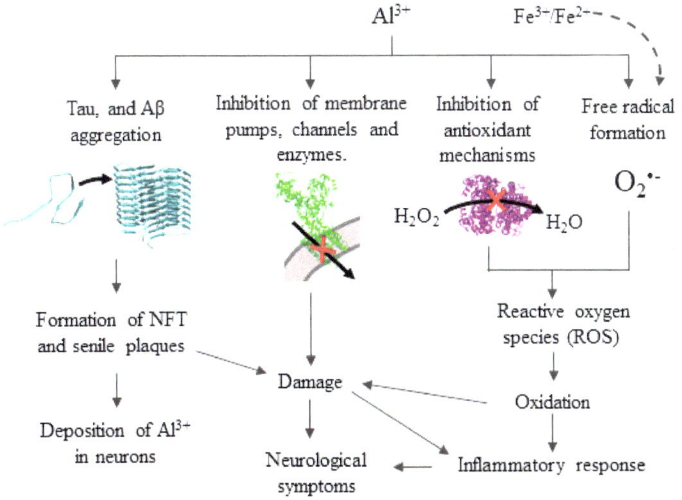

Figure 2. A simplified scheme of the effects of Al in neuronal tissues. Tau and β amyloid aggregation may result in the formation of neurofibrillary tangles (NFT) and senile plaques toxic to neurons. Al is found mainly in NFT within neurons. Al can inhibit a diversity of proteins like membrane pumps, enzymes involved in energetic metabolism and ionic channels, affecting cellular homeostasis mechanisms. This damage occurs simultaneously with the pro-oxidant effect of Al by inhibiting antioxidant mechanisms and by affecting Fe metabolism. Neuronal and neuroglia damages trigger inflammation responses. These effects can explain the neurotoxic effects of Al.

Nedzvetsky et al., 2006) and myelin is one of the main targets (Verstraeten et al., 1997). Three mechanisms may explain the pro-oxidant effect of Al:

1. Al is capable of inhibiting oxidation defense mechanisms like superoxide dismutase (Kumar et al., 2009; Yuan et al., 2012), catalase (Nehru and Anand, 2005) and glutathione peroxidase (Julka and Gill, 1996).
2. Al induces the activation of the Nitric Oxide Synthase (NOS) resulting in an increased production of free radical species like NO (Bondy et al., 1998).
3. It is suggested that Al by itself can catalyze the production of reactive oxygen species (ROS) by the formation of an Al superoxide radical ion ($AlO_2^{2+}\bullet$) (Exley, 2012).

Al inhibits different enzymes probably due to its interaction with metal binding sites in proteins. Neurotransmission can be affected, as voltage-activated calcium channel currents in neurons are inhibited by Al (Platt and Büsselberg, 1994). The plasma membrane Ca^{2+}-ATPase (PMCA) and sarco(endo)plasmic Ca^{2+}-ATPase (SERCA) are inhibited by Al (see below) which could explain the alterations in calcium homeostasis in neurons (Kaur and Gill, 2005; Marilina de Sautu et al., 2018). The Na^+/K^+-ATPase is another P-type ATPase inhibited by Al contributing to defects in neurotransmission (Silva and Gonçalves, 2003). Al blocks the glycolysis in neurons mainly by inhibition of the hexokinase (Lai and Blass, 1984). Also, Al alters the Krebs cycle activating the α-ketoglutarate dehydrogenase and the succinate dehydrogenase, and inhibiting the aconitase (Zatta et al., 2000).

Notably, the concentrations at which Al exerts these effects are between 1 and 100 μM, which are comparable to its concentration in the brain of patients with dialysis encephalopathy (Lai and Blass, 1984). However, Al availability could be hard to estimate because it depends on the equilibrium between different Al complexes (Macdonald and Martin, 1988; Maki et al., 2017).

The contradictory results started a debate about whether Al can be a risk factor for Alzheimer's disease or other neurodegenerative diseases. Epidemiological studies carried out showed the correlation of the Al level in drinking water in different areas with the incidence of Alzheimer's disease (Inan-Eroglu, 2018). Despite a high number of studies, the evidence is not conclusive (Flaten, 2001; Inan-Eroglu, 2018; Wang et al., 2016). The World Health Organization (WHO) has concluded that the evidence provided by the studies was not enough of a proof to support a causal association between Al in drinking water and Alzheimer's disease. Therefore, WHO recommends a maximum concentration of Al of 0,2 mg/ml in drinking water and a provisional tolerable weekly intake of 1 mg/kg body weight in all Al compounds in food, including additives.

The neurotoxicity of Al has been established and solid evidence demonstrated its effects in humans. The toxic mechanism is not fully understood but seems to involve different biological targets including enzymes and proteins associated with neurodegenerative diseases. Al inhibits homeostasis mechanisms and elicits a pro-oxidant effect resulting in damage to neuronal tissues, triggering an inflammatory response. The neurotoxicity of Al can be produced by oral ingestion or infusion during dialysis. However, considering Al as a risk factor in neurodegenerative diseases is not supported by the majority of the scientific community nowadays (Liu et al., 2019). The debate on the role of Al in the human population is still ongoing (Exley, 2017; Inan-Eroglu, 2018; Lidsky, 2014; Perl and Moalem, 2006). The research carried out in the last 50 years has been essential to help us understand its effects on living organisms.

3.1 Interaction of aluminum and β-amyloid protein

The first reports indicated that the lesion produced by Al in the brain was similar but not identical to the neurofibrillary degeneration observed in neurodegenerative diseases like Alzheimer's disease (Krishnan and Dalton, 1973; Miu and Benga, 2006). Alzheimer's disease is characterized by the deposition of senile plaques of β-amyloid protein (Aβ) and the accumulation of neurofibrillary tangles (NFT) constituted by tau protein (Yiannopoulou and Papageorgiou, 2020).

Aβ is a peptide of 36 to 43 amino acids that comes from a membrane glycoprotein known as amyloid precursor protein (APP). Aβ is formed by the sequential cleavage of APP by α-secretase, Aβ, and γ-secretase. In first instance, Aβ is formed by the successive action of β and γ secretases that produce the C-terminal end of the peptide, generally, with 40 or 42 aminoacids in length. Among others, Aβ participates in the activation of kinases (Reichenstein et al., 2021), protection against oxidative stress (Perluigi et al., 2006), and can act as a transcription factor (Leoni and Solomon, 2010). The Aβ42 isoform is more fibrogenic than Aβ40 and, therefore, is associated with the development of neurological disorders (Ye et al., 2007). The etiology of Alzheimer's disease is considered multifactorial and its pathogenesis has not been fully elucidated. However, the formation of senile plaques and NFT by Aβ and tau aggregates may have a critical role in damaging the neuron structure and activating neuroglia (Breijyeh and Karaman, 2020).

Al interacts with Aβ and promotes the formation of aggregates (Folch et al., 2018). The interaction of Al with Aβ induces changes in secondary structure decreasing the α-helix content and increasing the proportion of β-sheet (Exley et al., 1993; Ricchelli et al., 2005). Different Aβ fragments including 1–40, 1–42 and 25–35 amino acids aggregate in the presence of Al (Bondy and Truong, 1999; Exley et al., 1993; Kawahara et al., 1994). Some reports indicated that aggregation induced by Al depended on the presence of phosphate or ATP, probably due to the formation of nucleation sites by Al phosphate or ATP-Al (Bondy and Truong, 1999). Computational studies were carried out to understand how Al interacts with Aβ by studying possible Al-binding sites in Aβ Protein Data Bank (PDB) models. Results suggest that Glu3, Asp7 and Glu11 of Aβ are involved in Al complexation, indicating that negatively charged aminoacids may be essential to Al binding to Aβ (Mujika et al., 2017).

Al can promote the aggregation of Aβ *in vivo* and induce damage to neurons (Breijyeh and Karaman, 2020). Mice expressing a human Aβ protein (APP) with a mutation that promotes the formation of senile plaques (a transgenic line of mice called Tg 2576) were fed on Al in a 2 mg/kg diet. Results showed that the expression of Aβ combined with the Al diet increased the amounts of Aβ aggregates in mice brains and produced high levels of lipid peroxidation (Praticò et al., 2002). The formation of Aβ oligomers induced by Al might be a mechanism to induce long-term toxicity as Aβ oligomers are neurotoxic. This effect may act in synergy with the prooxidant tendency of Al and with its interaction with Fe (Kawahara and Kato-Negishi, 2011). The incubation of liposomes or cultured neurons with Al increases the levels of lipid peroxidation in the presence of Fe (Oteiza, 1994). Alterations in Fe homeostasis can induce oxidative stress, and Fe has been proposed as a mediator in the pathogenesis of Alzheimer's disease (Rolston et al., 2009). Al and Fe compete for the binding regions to ferritin and Tf (Liu et al., 2019). Evidence suggests that both Al or Fe complexation by Aβ may reduce the charge repulsion between different Aβ monomers, inducing aggregation (House et al., 2004). Thus, senile plaques may be an accumulation site for Fe and Al, where reactive oxygen species can be formed. Other metals like copper and zinc may also play a role in modulating oxidative stress in neurological pathologies like Alzheimer's disease (Liu et al., 2019).

3.2 Interaction of aluminum and tau protein

The aggregation of the tau protein has been linked to Al as well. The tau protein is a small protein that stabilizes and promotes the self-assembly of tubulin microtubules in physiological conditions. The tau molecule can be divided into four main domains, which are distinguished by their biochemical properties. The microtubule-binding domain consists of four imperfectly repeated motifs separated by framework regions, and unlike the rest of the protein, the second and third motifs tend to form an ordered β-sheet structure (Mukrasch et al., 2009). Tau is an intrinsically unfolded protein, maintaining a highly flexible conformation. It is subjected to a wide range of post-production modifications but among them, the most common is phosphorylation. Tau contains 85 putative phosphorylation sites. The phosphorylation of tau can induce the organization of structured regions in the protein and has a high impact on its function (Guo et al., 2017). In Alzheimer's disease and tau pathologies, tau is hyperphosphorylated and aggregates forming paired helical filaments (PHF) and straight filaments (Alonso et al., 2001; Pei et al., 2008), which accumulate in NFT. Therefore, studies were performed to understand the effects of Al on tau protein. Al promotes the aggregation of tau but not the formation of fibers (Scott et al., 1993). The co-injection of PHF with Al in rats' brains resulted in PHF accumulations more resistant to *in vivo* proteolysis (Shin et al., 1994). The effect of Al ingestion on tau aggregation *in vivo* was investigated by two studies in transgenic mice expressing human tau. Mice received a chronic administration of Al in drinking water (around 2 mM). One study informed that Al administration produced an increase in NFT deposition (Oshima et al., 2013) where the other indicated that no increase was observed by the Al diet (Akiyama et al., 2012). Thus, it is not clear whether chronic oral administration of Al can induce NFT formation.

3.3 Interaction of aluminum and α-synuclein

α-synuclein is a 15 kDa protein, which is ubiquitously expressed, with the highest expression level in presynaptic nerve terminals (Iwai et al., 1995). It appears in two forms, one bound to the membrane and the other soluble, in the cell cytoplasm. Like the tau protein, α-synuclein is an intrinsically disordered protein but forms regions of secondary structure by interacting with other proteins or ligands, giving rise to a multiplicity of conformations (Fusco et al., 2014). α-synuclein has three regions with different biochemical characteristics. The C-terminal region is rich in acidic residues and the N-terminal region can form an α-helix by interacting with phospholipids, especially acids. The central region is essential for the formation of synuclein sheets and peptides have been identified in Lewy bodies that interact through this region (Fusco et al., 2014). The most common post-translational modification of α-synuclein is phosphorylation, and its alteration lead to altered aggregation behavior (Sato and Kato, 2013). Under physiological conditions, α-synuclein participates in neuronal homeostasis (Devoto et al., 2017).

Al accelerates the aggregation of α-synuclein *in vitro* and its effect is higher than with other metals such as Fe, zinc (Zn), or copper (Cu) (Uversky et al., 2001). Studies in cellular cultures suggest that α-synuclein may mediate Al toxicity (Saberzadeh et al., 2016). Moreover, Al can also induce the formation of tau and α-synuclein co-aggregates and this effect is increased when tau is phosphorylated (Nübling et al., 2012).

The binding of Al with Aβ, tau and α-synuclein motivated many studies to quantify the concentration of Al in the brains of patients suffering from neurodegenerative diseases. Studies in rats showed that daily oral administration of Al for 90 days (100 mg/kg) results in an average Al content of 22 ppm in comparison with 6 ppm in animals where Al was not added to the diet (Bilkei-Gorzó, 1993). Similarly, the normal Al content of a human brain is in the range of 0 to 5 ppm (Priest, 2004). Crapper et al. in 1973 (Crapper et al., 1973) were the first to find Al in Alzheimer's disease patients using atomic absorption spectroscopy. This study showed that Al crosses the blood-brain barrier in humans. Other works measured the brain Al content of patients suffering from Alzheimer's disease by atomic absorption spectrometry (Exley and Clarkson, 2020), scanning electron microscopy with energy-dispersive x-ray spectrometry (Perl et al., 1982), instrumental neutron activation analysis (Marchkesbery et al., 1981), laser microprobe mass spectrometry (Good et al., 1992), inductively coupled plasma mass spectrometry (Beauchemin and Kisilevsky, 1998) and histochemistry (Kasa et al., 1995). The results were not consistent, while some studies reported an increased amount of Al in Alzheimer's disease patients in comparison with controls (Crapper et al., 1976; Perl and Brody, 1980; Exley and Clarkson, 2020), others found no differences (Alexander et al., 1996; Makjanic et al., 1998; McDermott et al., 1977). Some studies have also found increased Al content in patients suffering from Parkinson's disease (Good et al., 1992) and parkinsonism dementia complex of Guam (Garruto et al., 1984). Scanning electron microscopy indicated that the Al deposition is concentrated in groups of neurons containing NFT tangles (Perl and Brody, 1980). Although Al is found in NFT (Perl and Brody, 1980) and senile plaques in human brains (Candy et al., 1986), NFT seems to be the main binding site. In addition, a study informed the co-localization of Fe with Al in NFT (Good et al., 1992).

4. Aluminum and transferrin

4.1 Iron homeostasis

The most common valences of Fe are Fe^{2+} and Fe^{3+}, although Fe^{4+}, Fe^{5+}, and Fe^{6+} states have been described in some Fe-containing enzymes (Hrycay and Bandiera, 2012). It is a strong acid and forms complexes with strong bases such as OH^-. At physiological pH, Fe^{3+} is practically insoluble and precipitates as ferric hydroxide colloids or forms coordination compounds of polyFe (Kawabata, 2022). Fe^{2+} is more soluble in water than Fe^{3+}, however, Fe^{2+} tends to be easily oxidized and deposited in colloids. Fe is an indispensable cofactor of many enzymes and a mediator of redox reactions. Fe concentration is finely regulated because it is involved in the production of highly toxic ROS due to its highly oxidizing capacity (Yamada et al., 2020). Fe is absorbed in the intestine and delivered to all cells in the body through the blood vessels. In a healthy person, about 0.1% of total body Fe (~ 3–4 mg) is in plasma, predominantly bound to transferrin (Tf), which distributes it to cells and tissues.

4.2 Transferrin and receptor-mediated endocytosis

Human Tf is a 76 kDa glycoprotein produced primarily in the liver (Kawabata, 2019). Tf is formed by two domains, located at the N and C ends, connected by a short binding region (Gomme and McCann, 2005). Each domain has a Fe binding site, and in the plasma, Tf can be in the free (apoTf), monoferric (Tf·Fe), or differric (Tf·2Fe or holoTf) forms. Fe is systemically distributed through the holoTf binding to the ubiquitously expressed Tf receptor 1 (TfR1) or the hepatocellular restricted Receptor 2 (TfR2) (Chahine et al., 2012). TfR1 is a homodimer formed by two glycoproteins linked by disulfide bonds, which can bind one molecule of Tf·2Fe each. The binding of Tf·2Fe to TfR on the cell surface (pH 7.4) is followed by endocytosis and the Fe release is produced in the acidic environment (pH 5.6) of the endosome. In the endosome, apoTf and TfR remain bound and then return to the cell membrane, where both proteins dissociate to start a new cycle (Chahine et al., 2012).

Tf does not bind Fe^{2+}, which is the main species in aerobic media, thus Fe^{3+} would be transferred to Tf from a chelator (Harris and Aisen, 1973). By using spectrophotometric techniques and fast kinetics, it has been shown that Tf can rapidly extract Fe^{3+} from low-mass chelators (Cowart et al., 1982). The mechanism by which Tf acquires Fe includes conformational changes in the protein and the presence of a synergistic anion that interacts mainly with the C-domain. In physiological media, HCO_3^- is the usually synergistic anion because its concentration is 25–30 mM (Dhungana et al., 2003). The general mechanism by which Tf acquires Fe begins when the metal is transferred to the C-domain of Tf from a low-mass chelator. This produces a conformational change that leads Fe being occluded, i.e., inaccessible to the surrounding environment. Tf·Fe goes through other conformational changes and a loss of H^+, followed by second Fe binding to the N-domain in case there is enough Fe available. Then, Tf·2Fe (or Tf·Fe) undergoes slight changes until reaching its final equilibrium conformation. These last two events occur in the order of 100s and 1000s, respectively (Chahine et al., 2012).

The binding of Tf·2Fe to RTf occurs through two phases. The first is very fast and is related to the interaction of the C-domain of Tf·2Fe with the helical domain of TFR (Kawabata, 2019). The second phase is slow and involves conformational changes that lead to a stable adduct of both proteins. Finally, endocytosis is in the range of minutes (Sheff et al., 2002). Thus, cellular Fe uptake is a slow and complex process that depends on both the Fe acquisition and the critical conformational changes of Tf.

4.3 Interaction of aluminum and transferrin

The ionic radius of Al is similar to that of Fe (Martin, 1986), hence Al tends to displace Fe from its protein binding sites, and Tf is not an exception. The interaction between Tf and Al was studied for decades and was considered one of the predominant mechanisms for the uptake of Al (Cheng et al., 2018; Chahine et al., 2012; Kim et al., 2011; Roskams and Connor, 1990). However, other studies show that the affinity of Tf for Al is 100 times lower than for Fe, and binding of Al to Tf does not trigger the critical conformational changes for the interaction of Tf·Al with the TfR (Ott et al., 2019; Sakajiri et al., 2010). Currently, the hypothesis of Al influx being mediated by Tf-TfR is controversial in the scientific community.

Under physiological conditions, Al would be transferred to Tf by the citrate-Al complex (Nagasawa et al., 2005). In the presence of HCO_3^-, Al rapidly dissociates from citrate and binds to the C-domain of Tf, therefore, this first step would not be the limiting step in the Al uptake process (Chahine et al., 2012). The metal binding must be followed by a series of conformational changes triggered by H^+ loss and metal occlusion. This leads to the transition from an open to a closed structure of Tf in which Al would be inaccessible to the environmental medium (Peterson et al., 2000). In the case of Al, the conformational changes involving the binding of the second ion to the N-domain occur partially (Sakajiri et al., 2010) and the final equilibrium is reached after hours (Ha-Duong et al., 2008). Thus, Al is partially accessible to the environment in Tf·Al and Tf·2Al has an intermediary structure between the closed form of Tf·2Fe and the open form of apoTf (Sakajiri et al., 2010). In this conformation, Al is not completely occluded explaining why Tf has a 100-fold lower affinity for Al than for Fe (Kubal et al., 1992; Ott et al., 2019; van Veen et al., 2020).

The protein-protein interaction between Tf·2Fe and RTf is fundamental for the endocytosis. The fact that Tf·2Al has a different conformation than that of Tf·2Fe, prevents the interaction with TfR (Chahine et al., 2012) and, consequently, blocks the uptake by the cell. The absence of interaction between Tf·2Fe and RTf has also been described for some abnormal human Tf that produce Fe anemias (Evans et al., 1994).

By combining ultra-high resolution ESI mass spectrometry and CD spectroscopy, Ott et al., studied the competition of Al and Fe for Tf and how the binding of one or both affects the conformation of the protein and its subsequent interaction with TfR1 (Ott et al., 2019). When Tf is incubated with an excess of Al under optimal binding conditions, both Tf·Al and Tf·2Al are formed. However, most of the Tf is found with only one bonded-Al. Similar to Tf·2Al and Tf·2Fe, the structure of Tf·Al is significantly different from that of Tf·Fe. When Fe is added to the solution, most of the protein is in a Tf·Al·Fe hybrid form. Al binds to the N-domain, which initially has less affinity for Fe, but when Fe binds to the C-domain, a second ion gradually displaces Al from the N-domain as well. In agreement with the high affinity that Tf has for Fe, both Tf·2Al and Tf·Al are transformed into TF·2Fe, therefore, Al does not affect the Fe-binding affinity to Tf (Ott et al., 2019). Although the binding of Al to the N-domain affects the structure of the C-domain, the CD spectral properties when Fe is bound to Tf·Al·Fe indicate that the C-domain resembles the Tf·2Al form. The conformational changes that apoTf undergoes on Fe binding are critical for its receptor binding and these do not occur in Tf·2Al. However, a hybrid form Tf·Al·Fe could allow Al to enter the cell through receptor-mediated endocytosis, because the binding of Fe to the C-domain of Tf is the crucial step for the binding to RTf.

Another controversial point for the receptor-mediated Al endocytosis hypothesis refers to the time course of the process (Exley et al., 1993; Exley and Mold, 2015), especially considering that the Tf·2Al formation is even slower than that of Tf·2Fe (Ha-Duong et al., 2008). While kinetic measurements for the Tf·Al·Fe formation have not been described, this complex may explain the Al entry to cells since Fe binding to the C-domain is the rate limiting step of the complex assembly. In this regard, Tenan et al. evaluated Al uptake in Chinese hamster V79 cells in the absence and presence of 10% fetal bovine serum (containing Tf) using the lumogallion probe (Tenan et al., 2021). When Chinese hamster V79 and mammary epithelial cells were incubated with fetal bovine serum and $AlCl_3$ for 3 h, Al was distributed throughout the cytoplasm with a higher concentration in the perinuclear region. When the same experiment was performed in the absence of fetal bovine serum, a higher concentration of Al was observed with a similar cytoplasmic distribution, indicating a lower Al uptake in the presence of Tf. These results question whether the Al uptake pathway is mediated by Tf-RTf endocytosis.

5. Aluminum and Ca^{2+}-ATPases

5.1 Calcium and magnesium homeostasis

Magnesium is the second most abundant cellular cation in mammalian organisms (Romani, 2011). Mg homeostasis depends on the balance between intestinal absorption and renal excretion. In plasma, the free Mg concentration is maintained in a narrow range between 0.75–0.95 mM (Dennis et al., 1990). In the cytoplasm and cell organelles, the total Mg concentration ranges between 17 to 20 mM, and its free concentration is estimated to be 0.8–1.2 mM (Romani, 2011). Thus, the chemical gradient across the plasma and organelle membranes is small. Among others, Mg is essential for the function of many enzymes, especially those with kinase and phosphatase functions, and for reactions that require ATP.

Unlike Mg, the chemical gradient for calcium (Ca) across cell membranes is the largest of all metal ions (Wang et al., 2020). In plasma, the total Ca concentration is about 2.5 mM, and only 1.1 mM is in the free form. Within cells, the most Ca is stored in the endoplasmic reticulum, where its concentration varies from 0.3 to 2 mM. In the resting state, the cytoplasmic concentration of Ca is ~ 0.1 μM, but increases more than 10-fold in response to a stimulus. Ca is the most studied second messenger because it is involved in various signaling pathways, ranging from cell death to immune cell proliferation (Bose et al., 2015). The intracellular Ca concentration is finely regulated and depends on the balance between several mechanisms of Ca influx and efflux of the cytoplasm (Rinaldi et al., 2021).

Mg and Al are similar in size, which is a dominant factor for metal ion competition (Rezabal et al., 2006). Ca is about nine times larger than that Al (Rezabal et al., 2006). Thus, Al is an excellent competitor for Mg but not so much for Ca. However, decades ago it was shown that Al alters intracellular calcium homeostasis (Gandolfi et al., 1998). In principle, this alteration in calcium homeostasis was related to neurodegenerative disorders, such as Alzheimer's disease, where altered levels of Ca and Mg were found in post-mortem studies of patients with this condition. Later it was discovered that Al affected the activity of transport proteins and Ca exchangers (Rengel and Zhang, 2003). In this context, the Ca^{2+}-ATPases play a fundamental role.

5.2 The plasma membrane and sarco(endo)plasmic Ca^{2+}-ATPases

P-ATPases are a family of proteins responsible for the active transport of ions, lipids, amines, and tail-anchored proteins. According to the phylogenetic characteristics and the transported substrate, the P-ATPases are grouped into five subfamilies (I to V), plasma membrane Ca^{2+}-ATPase (PMCA), sarco(endo)plasmic reticulum Ca^{2+}-ATPase (SERCA), H^+, K^+-ATPase and Na^+, K^+-ATPase belong to subgroup II (Zhang and Zhang, 2019).

PMCA and SERCA have an essential role in the regulation of intracellular Ca. These pumps transport Ca against its electrochemical gradient from the cytoplasm to the extracellular medium and the endoplasmic reticulum lumen, respectively. The transmembrane domain is formed by ten α-helices and contains the Ca-binding sites (Dyla et al., 2019). SERCA transports two Ca per hydrolyzed ATP molecule, while PMCA transports only one. The Ca-binding site of PMCA is homologous to the site II of SERCA. In the cytoplasmic region of PMCA and SERCA, the three characteristic domains of P-ATPases: A (actuator), N (nucleotide-binding), and P (phosphorylation) are located.

During the transport cycle, P-ATPases undergo several conformational changes and alternate between phosphorylated and unphosphorylated forms (Figure 3). The transmission of the conformational changes between transmembrane and cytoplasmic domains is crucial to ion transport (Saffioti et al., 2021). The reaction cycle is described by the Post-Albers model (Albers, 1967; Post et al., 1969), which proposes that P-ATPases exist in two main conformations: *E*1 and *E*2. In *E*1, the Ca-binding site is accessible to the cytosol and Ca binds with high affinity. *E*1Ca can be phosphorylated by ATP forming *E*1P, where Ca is occluded in the transmembrane domain. *E*1·P undergoes a conformational transition to *E*2P, and Ca is released from the low-affinity Ca site that is now oriented to the opposite side of the membrane. Then, *E*2P is dephosphorylated to *E*2 and the pump can start a new reaction cycle through a new conformational transition to *E*1. Mg is necessary for optimal catalysis of PMCA (Echarte et al., 2001) and SERCA (Olesen et al., 2007). On the one hand, because the substrate is ATP·Mg, on the other, because Ca^{2+}-ATPases have an Mg-

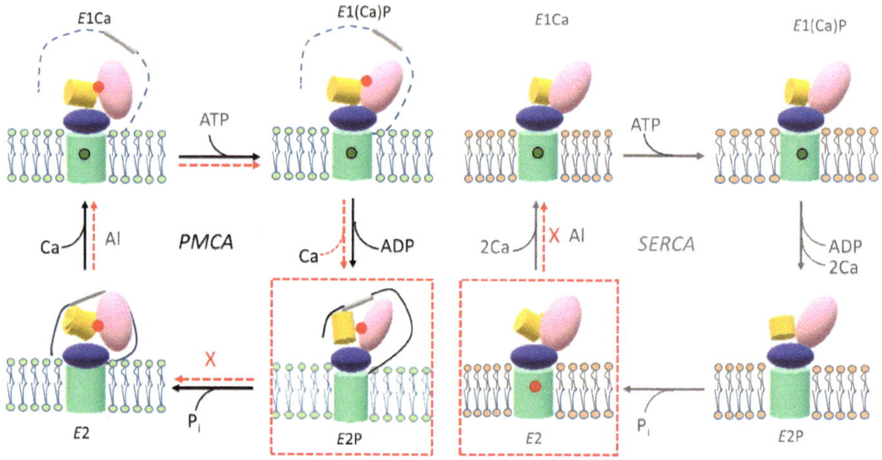

Figure 3. Scheme of the inhibition mechanism of PMCA and SERCA by aluminum. In the figure are shown the main intermediates of the reaction cycle (*E*2; *E*1Ca; *E*1(Ca)P and *E*2P) of PMCA (*right*) and SERCA (*left*). *E*2P corresponds to the product state of the dephosphorylation process, immediately before P_i dissociates (Saffioti et al., 2019). Mg is involved in the *E*1(Ca)P→*E*2P transition and the dephosphorylation step, consequently, binding of Mg to Ca^{2+}-ATPases is essential for the transport of Ca. The binding of Ca (*green circle*) leads to the *E*2→*E*1Ca transition, and in the presence of ATP, the pump is phosphorylated. In the *E*1(Ca)P, Ca is occluded in the transmembrane domain and is inaccessible to both sides of the membrane. Although the binding of Ca and ATP are depicted as sequential steps, ATP can also bind in the absence of Ca (*E*2 conformations) (Mangialavori et al., 2013). However, the binding of Ca to PMCA and SERCA is essential for the phosphorylation from ATP. During the *E*1(Ca) P→*E*2P transition, ADP dissociates and Ca is released to the opposite side of the membrane. Finally, Ca^{2+}-ATPase dephosphorylates (*E*2P→*E*2), and when Ca is present, a new transport cycle starts. PMCA has an autoinhibitory C-terminal tail that regulates the transport of Ca. In *E*2 and *E*2P, the C-terminal tail interacts with the A and N-domains (*closed conformations*) while in the *E*1Ca and *E*1(Ca)P, these autoinhibitory interactions are destabilized and PMCA is in equilibrium between an open and a closed form (*represented as dotted lines*) (Saffioti et al., 2021). The catalytic core of PMCA is slightly different from that of SERCA, particularly in the *E*2P state, where the nucleotide-binding pocket of PMCA is more accessible to the aqueous environment. Al behaves as an irreversible inhibitor of PMCA and as a slowly reversible inhibitor of SERCA. In PMCA (*left*), Al (*red circle*) affects the binding of Mg, but Ca and ATP can bind leading to the correct conformation for the pump to phosphorylate (*red arrows*). The *E*1(Ca)P→*E*2P transition also occurs, but PMCA is not dephosphorylated. Thus, Al stabilizes PMCA in an *E*2P·Al state, where the autoinhibitory interactions between the C-terminal tail and the catalytic core are stable. Unlike PMCA, Al affects the binding of Ca to SERCA (*right*) and stabilizes the *E*2 state. Consequently, SERCA is not phosphorylated from ATP. SERCA has been used as a model for the study of PMCA and other P-ATPases. However, Al inhibits the Ca transport in PMCA and SERCA by stabilizing different reaction intermediates (*red boxes*) with particular structural characteristics.

binding site at the cytoplasmic domain whose occupancy is essential for the maximal activity. Thus, can Al replace Mg and affect Ca homeostasis?

The molecular mechanism by which Al inhibits Ca^{2+}-ATPases was recently studied in PMCA isolated from human erythrocytes and SERCA from rabbit skeletal muscle (de Sautu et al., 2018). Enzymes were purified from natural sources, reconstituted in mixed micelles of phosphatidylcholine and detergent, and exposed to $AlCl_3$ under different experimental conditions. The inhibitory species is Al^{3+} in both cases, but despite the high homology between these pumps, the inhibition mechanism is different.

5.3 Interaction of aluminum and SERCA

In SERCA, Al inhibits the phosphorylation of the enzyme (*E*1P formation) and behaves as a competitive inhibitor of Ca, indicating that Al displaces Ca from its binding site(s). While ATP binds *E*2 and *E*1Ca conformations of SERCA, the phosphorylation of the Asp351 located in the P-domain only occurs after Ca binds to the transmembrane domain (Sautu et al., 2018). Upon ATP binding, the N-domain rotates toward the P-domain. This conformational change leads to the A-domain acquiring a more extended conformation, which is transmitted to the transmembrane domain (Mangialavori et al., 2013b), and Ca is occluded (Ferreira-Gomes et al., 2011). Thus, the γ-phosphate of ATP remains in the correct position for the phosphorylation of the Asp351 and forming *E*1P. When Al displaces Ca, SERCA does not acquire the correct conformation and does not phosphorylate.

The mean time for dissociation of the SERCA·Al complex is ~27 min, indicating that Al is a slowly-reversible inhibitor of SERCA (Sautu et al., 2018). The inhibition constant in terms of $AlCl_3$ was ~ 8 μM. However, as the Al^{3+} concentration in the reaction medium (at pH 7.4) is in the sub-micromolar range (Macdonald and Martin, 1988), Al displaces Ca with a high affinity. SERCA is found in the membranes of the sarcoplasmic reticulum, one of the largest intracellular reservoirs of Ca. SERCA imports Ca into the lumen, playing a crucial role in controlling the concentration of cytoplasmic Ca. Its function is essential in the process of signal-stop after a stimulus. In this context, a high affinity for Al and its slow dissociation constant can have important consequences on Ca homeostasis.

5.4 Interaction of aluminum and PMCA

In PMCA (Figure 3), Al stabilizes the phosphorylated form of the pump (*E*2P) (Saffioti et al., 2021), preventing the continuity of the transport cycle, i.e., does not prevent the phosphorylation as in SERCA (Sautu et al., 2018). The PMCA activity does not recover with increasing Ca concentrations, and the Mg also fails to displace the Al-bound to the pump. The dissociation constant of the PMCA-Al complex is greater than 60 min.

The phosphorylation, the transition between phosphoenzyme conformers, and the dephosphorylation of PMCA depend on the binding of Mg (I) (Mangialavori et al. 2010). In the dephosphorylation of P-ATPases, P_i is released as P_i·Mg complex, thus, the slow dissociation of Al prevents P_i hydrolysis stabilizing the phosphorylated form of PMCA. A similar effect was described for lanthanum (La), which stabilizes PMCA in the phosphorylated form with occluded calcium (*E*1(Ca)P) (Ferreira-Gomes et al., 2011; Ontiveros et al., 2019). The inhibition mechanism involves the competition of ATP·La with ATP·Mg. In the case of Al, it displaces Mg of its binding site in the enzyme and does not compete with ATP·Mg. In fact, 25 μM $AlCl_3$ completely inhibits the PMCA activity in the presence of 2 mM ATP·Mg and 2 mM free Mg (de Sautu et al., 2018).

PMCA is a multi-regulated protein and is considered a high-affinity and low-capacity transporter (Bruce, 2018). Its function is related to the response to changes in Ca microdomains that form near the plasma membrane. PMCA has a long

C-terminal segment regulated by the binding of the Ca-calmodulin complex, which maintains the pump in a partially active state (Mangialavori et al., 2009). The binding of Al to PMCA stabilizes autoinhibitory interactions between the C-terminal domain and the catalytic core (Saffioti et al., 2021). Thus, Al prevents the conformational change that leads to the correct organization of the cytoplasmic domains, keeping the pump in the autoinhibited and phosphorylated state. The inhibition constant for PMCA by $AlCl_3$ is 10 μM (de Sautu et al., 2018), thus, considering the very low dissociation constant of Al and its location in the plasma membrane, PMCA presents as a clear target for the toxicity of this metal. In HEK293T cells, Al inhibits PMCA-mediated Ca transport in a concentration-dependent manner (unpublished results from our group).

While the characterization of the PMCA inhibition mechanism by Al was carried out with the isolated enzyme, the toxicity of this metal in the hematological system was described by several studies (Tenan et al., 2021; Zangeneh et al., 2021; Zhang et al., 2016). Al exposure produces anemia, and the proposed mechanisms include inhibition of hemoglobin biosynthesis (Lin et al., 2013), alteration in membrane integrity, and stimulation of programmed cell death (eryptosis) (Niemoeller et al., 2006). Zhang et al. show that the exposure of erythrocytes to $AlCl_3$ for 24 h produces the activation of Ca-channels, an increase in intracellular Ca, and the consequent eryptosis (Zhang et al., 2016). In human erythrocytes, PMCA plays an essential role in the extrusion of cytoplasmic Ca. Therefore, Al can produce both effects: activate the entry of Ca through the channels and prevent its extrusion by inhibiting PMCA.

5.5 Interaction of aluminum and Na^+, K^+-ATPase

Red blood cells are a simple model to study Al toxicity because they lack intracellular compartments. Furthermore, they are an accessible cell type in terms of interaction with aluminum. Therefore, the effect of Al on human erythrocytes and its physiological implications has been a topic of interest for the scientific community.

The relationship between exposure to Al and arterial hypertension has been proposed early. In rats, intake of $AlCl_3$ in water raised mean and systolic blood pressure and increased osmotic fragility (Zhang et al., 2016). In addition to the effects of oxidative stress, one of the proposed causes involves the Al effect on the disruption of ion homeostasis. During their passage through the blood capillaries, erythrocytes change their rheological properties. The integrity and biophysical properties of the plasma membrane play an essential role in erythrocyte deformability during the blood flux. In the physiological deformations that erythrocytes suffer as they pass the capillaries. In erythrocytes, cation transport between the intracellular and extracellular media is regulated by Na^+, K^+-ATPase, Mg^{2+}-ATPase, and PMCA. Studies in erythrocytes of hypertensive patients showed that Na^+, K^+-ATPase, Mg^{2+}-ATPase, and PMCA have less activity (Ouyz, 1995). The decrease in the activities of these PII-ATPases in the plasmatic membrane alters the ionic balances affecting the function of the intracellular enzymes involved in the function of the membrane erythrocyte, and consequently, they induce hypertension (Zhang et al., 2016).

Silva and Goncalves showed that Al inhibits the Na^+, K^+-ATPase activity of rat brain cortex synaptosomes. Na^+, K^+-ATPase forms oligomers in the plasma membrane, which allow for reaching the maximal enzymatic activity in the presence of high ATP concentrations (Silva et al., 2005). Al affects the monomer interaction in the oligomeric state and reduces activation by high ATP concentrations, leading to cell lysis. Na^+, K^+-ATPase, SERCA, and PMCA are PII-ATPases with high homology, however, the inhibition mechanism by Al seems to be different in the three pumps (Sautu et al., 2018; Silva et al., 2005).

6. Conclusion

The relationship between Al and various pathologies has been extensively documented. However, the causality of Al exposure and human health problems remains a controversial issue in the scientific community. Given the characteristics of Al, its mechanism of toxicity may be related to various cellular targets, which are not mutually exclusive. Proteins have metal-binding sites, are regulated by phosphorylation/dephosphorylation, and alternate between different conformations, all of which are essential for their function and can be affected by Al. In the last years we have acquired valuable information to understand the effect of Al in humans. However, more studies are needed to understand the molecular mechanisms by which Al affects human health.

References

Akiyama, H. Hosokawa, M., Kametani, F., Kondo, H., Chiba, M., Fukushima, M. et al. 2012. Long-term oral intake of aluminium or zinc does not accelerate alzheimer pathology in AβPP and AβPP/Tau transgenic mice. Neuropathology 32(4): 390–97.

Albers, R.W. 1967. Biochemical aspects of active transport. Annual Review of Biochemistry 36: 727–756.

Alexander, J. Gronnesby, J.K., Bakketeig, L.S. and Edwardson, J.A. 1996. Content of brain aluminum is not elevated in Alzheimer disease. Alzheimer Dis. Assoc. Disord 10: 171–74.

Alfrey, A.C., LeGendre, G.R. and Kaehny, W.D. 1976. The dialysis encephalopathy syndrome: possible aluminum intoxication. New England Journal of Medicine 294(4): 184–88.

Alonso, A. Zaidi, T., Novak, M., Grundke-Iqbal, I. and Iqbal, K. 2001. Hyperphosphorylation induces self-assembly of τ into tangles of paired helical filaments/straight filaments. Proceedings of the National Academy of Sciences 98(12): 6923–28.

Beauchemin, D. and Kisilevsky, R. 1998. A method based on ICP-MS for the analysis of Alzheimer's amyloid plaques. Analytical Chemistry 70(5): 1026–29.

Becaria, A., Lahiri, D.K., Bondy, S.C., Chen, D., Hamadeh, A., Li, H. et al. 2006. Aluminum and copper in drinking water enhance inflammatory or oxidative events specifically in the brain. Journal of Neuroimmunology 176(1–2): 16–23.

Bhadauria, M. 2012. Combined treatment of HEDTA and propolis prevents aluminum induced toxicity in rats. Food and Chemical Toxicology 50(7): 2487–95.

Bilkei-Gorzó, A. 1993. Neurotoxic effect of enteral aluminium. Food and Chemical Toxicology 31(5): 357–61.

Bondy, S.C., Liu, D. and Guo-Ross, S. 1998. Aluminum treatment induces nitric oxide synthase in the rat brain. Neurochemistry International 33(1): 51–54.

Bondy, S.C. and Truong, A. 1999. Potentiation of beta-folding of β-amyloid peptide 25–35 by aluminum salts. Neuroscience Letters 267(1): 25–28.

Bose, T., Cie, A. and Wiechec, E. 2015. Role of ion channels in regulating Ca^{2+} homeostasis during the interplay between immune and cancer cells. Cell Death and Disease 6(2): e1648–e1648.

Breijyeh, Z. and Karaman, R. 2020. Comprehensive review on Alzheimer's disease: causes and treatment. Molecules 25(24).

Bruce, J.I.E. 2018. Metabolic regulation of the PMCA: Role in cell death and survival. Cell Calcium 69: 28–36.
Campbell, A., Yang, Y.E., Tsai-Turton, M. and Bondy, S.C. 2002. Pro-inflammatory effects of aluminum in human glioblastoma cells. Brain Research 933(1): 60–65.
Candy, J.M. Oakley, A.E., Klinowski, J., Carpenter, T.A., Perry, R.H., Atack, J.R. et al. 1986. Lancet 327(8477): 354–56.
Cheng, D., Jinlei, T., Xuerui, W., Xinyu, Z. and Shuo, W. 2018. Ecotoxicology and environmental safety effect of aluminum (Al) speciation on erythrocytic antioxidant defense process: correlations between lipid membrane peroxidation and morphological characteristics. Ecotoxicology and Environmental Safety 157(January): 201–6.
Cheng, D., Wang, X., Yu, X., Jiankang, C. and Weibo, J. 2018. Identification of the Al-binding proteins that account for aluminum neurotoxicity and transport: *in vivo*. Toxicology Research 7(1): 127–35.
Connor, D.J., Jope, R.S. and Harrell, L.E. 1988. Chronic, oral aluminum administration to rats: cognition and cholinergic parameters. Pharmacology, Biochemistry and Behavior 31(2): 467–74.
Cowart, R.E., Kojima, N. and Bates, G.W. 1982. The exchange of Fe^{3+} between acetohydroxamic acid and transferrin. Spectrophotometric evidence for a mixed ligand complex. Journal of Biological Chemistry 257(13): 7560–65.
Crapper, D.R., Krishnan, S.S. and Dalton, A.J. 1973. Brain aluminum distribution in Alzheimer's disease and experimental neurofibrillary degeneration. Science 180(4085): 511–13.
Crapper, D.R., Krishnan, S.S. and Quittkat, S. 1976. Aluminium, neurofibrillary degeneration and Alzheimer's disease. Brain: A Journal of Neurology 99(1): 67–80.
Dennis, W.J., Apel, L. and Brierley, G.P. 1990. Matrix free Mg^{2+} changes with metabolic state in isolated heart mitochondria? Biochesmistry 29: 4121–28.
de Sautu, M., Saffioti, N.A., Ferreira-Gomes, M.S., Rossi, R.C., Rossi, J.P.F.C. and Mangialavori, I.C. 2018. Aluminum inhibits the plasma membrane and sarcoplasmic reticulum Ca^{2+}-ATPases by different mechanisms. Biochimica et Biophysica Acta - Biomembranes 1860(8): 1580–88.
de Sautu, M. 2021. The Plasma Membrane Calcium Pump and the Influence of the Lipid Environment: A Comprehensive Study of the Effect of Aluminum. PhD Thesis. Facultad de Farmacia y Bioquímica. Universidad de Buenos Aires. Argentina.
Dhungana, S., Taboy, C.H., Anderson, D.S., Vaughan, K.G., Aisen, P., Mietzner, T.A. et al. 2003. The influence of the synergistic anion on iron chelation by ferric binding protein, a bacterial transferrin. Proceedings of the National Academy of Sciences of the United States of America 100(7): 3659–64.
Dyla, M., Hansen, S.B., Nissen, P. and Kjaergaard, M. 2019. Structural dynamics of P-Type ATPase ion pumps. Biochemical Society Transactions 47(5): 1247–57.
Echarte, M.M., Levi, V., Villamil, A.M., Rossi, R.C. and Rossi, F.C. 2001. Quantitation of plasma membrane calcium pump phosphorylated intermediates by electrophoresis. Analytical Biochemistry 289(2): 267–73.
Evans, R.W., Crawley, J.B., Garratt, R.C., Grossmann, J.G., Neu, M., Aitken, A. et al. 1994. Characterization and structural analysis of a functional human serum transferrin variant and implications for receptor recognition. Biochemistry 33: 12512–20.
Exley, C., Price, N.C., Kelly, S.M. and Birchall, J.D. 1993. An interaction of β-amyloid with aluminium *in vitro*. FEBS Letters 324(3): 293–95.
Exley, C. 2012. The coordination chemistry of aluminium in neurodegenerative disease. Coordination Chemistry Reviews 256(19–20): 2142–46.
Exley, C. and Mold, M.J. 2015. The binding, transport and fate of aluminium in biological cells. Journal of Trace Elements in Medicine and Biology 30: 90–95.
Exley, C. 2017. Aluminum should now be considered a primary etiological factor in Alzheimer's disease. Journal of Alzheimer's Disease Reports 1(1): 23–25.
Exley, C. and Clarkson, E. 2020. Aluminium in Human Brain Tissue from Donors without Neurodegenerative Disease: A Comparison with Alzheimer's Disease, Multiple Sclerosis and Autism. Scientific reports 10(1): 1–7.
Fasman, G.D., Perczel, A. and Moore, C.D. 1995. Solubilization of Beta-Amyloid-(1-42)-peptide: reversing the beta-sheet conformation induced by aluminum with silicates. Proceedings of the National Academy of Sciences 92(2): 369–71.

Ferguson, J.M., Costello, S., Neophytou, A.M., Balmes, J.R., Bradshaw, P.T., Cullen, M.R. et al. 2019. Night and rotational work exposure within the last 12 months and risk of incident hypertension. Scandinavian Journal of Work, Environment and Health 45(3): 256–66.

Ferreira-Gomes, M.S., González-Lebrero, R.M., de la Fuente, M.C., Strehler, E.E., Rossi, R.C. and Rossi, J.P. 2011. Calcium occlusion in plasma membrane Ca^{2+}-ATPase. Journal of Biological Chemistry 286(37): 32018–25.

Flaten, T.P. 2001. Aluminium as a risk factor in Alzheimer's disease, with emphasis on drinking water. Brain Research Bulletin 55(2): 187–96.

Folch, J., Ettcheto, M., Petrov, D., Abad, S., Pedrós, I., Marin, M. et al. 2018. Review of the advances in treatment for Alzheimer disease: Strategies for combating β-amyloid protein. Neurologia 33(1): 35–46.

Fulgenzi, A., Vietti, D. and Ferrero, M.E. 2014. Aluminium involvement in neurotoxicity. BioMed Research International 2014: 758323.

Fusco, G. 2014. Direct observation of the three regions in α-synuclein that determine its membrane-bound behaviour. Nature Communications 5: 1–8.

Gandolfi, L., Stella, M.P., Zambenedetti, P. and Zatta, P. 1998. Aluminum alters intracellular calcium homeostasis *in vitro*. Biochimica et Biophysica Acta - Molecular Basis of Disease 1406(3): 315–20.

Garruto, R.M., Fukatsu, R., Yanagihara, R., Gajdusek, D.C., Hook, G. and Fiori, C.E. 1984. Imaging of calcium and aluminum in neurofibrillary tangle-bearing neurons in parkinsonism-dementia of guam. Proceedings of the National Academy of Sciences 81(6): 1875–79.

Glynn, A.W., Sparen, A., Danielsson, N.G., Sundstrom, B. and Jorhem, L. 2001. The influence of complexing agents on the solubility and absorption of aluminium in rats exposed to aluminium in water. Food Additives and Contaminants 18(6): 515–23.

Gomme, P.T. and McCann, K.B. 2005. Transferrin: Structure, function and potential therapeutic actions. Drug Discovery Today 10(4): 267–73.

Good, P.F., Olanow, C.W. and Perl, D.P. 1992. Neuromelanin-containing neurons of the substantia nigra accumulate iron and aluminum in Parkinson's disease: A LAMMA study. Brain Research 593(2): 343–46.

Good, P.F., Perl, D.P., Bierer, L.B. and Schmeidler, J. 1992. Selective accumulation of aluminum and iron in the neurofibrillary tangles of Alzheimer's disease: A Laser Microprobe (LAMMA) study. Annals of Neurology: Official Journal of the American Neurological Association and the Child Neurology Society 31(3): 286–92.

Guo, T., Noble, W. and Hanger, D.P. 2017. Effect of tau protein in health and disease. Acta Neuropathologica 133(5): 665–704.

Ha-Duong, N.T., Hémadi, M., Chikh, Z. and Chahine, J.M. 2008. Kinetics and thermodynamics of metal-loaded transferrins: transferrin receptor 1 interactions. Biochemical Society Transactions 36(6): 1422–26.

El Hage Chahine, J. Mi., Hémadi, M. and Ha-Duong, N. 2012. Uptake and release of metal ions by transferrin and interaction with receptor 1. Biochimica et Biophysica Acta - General Subjects 1820(3): 334–47.

Harris, D.C. and Aisen, P. 1973. Facilitation of Fe(II) antoxidation by Fe(III) complexing agents. Biochimica et Biophysica Acta (BBA) - General Subjects 329(1): 156–158.

House, E., Collingwood, J., Khan, A., Korchazkina, O., Berthon, G. and Exley, C. 2004. Aluminium, iron, zinc and copper influence the *in vitro* formation of amyloid fibrils of Aβ 42 in a manner which may have consequences for metal chelation therapy in Alzheimer's disease. Journal of Alzheimer's Disease 6(3): 291–301.

Hrycay, E.G. and Bandiera, S.M. 2012. The monooxygenase, peroxidase, and peroxygenase properties of cytochrome P450. Archives of Biochemistry and Biophysics 522(2): 71–89.

Igbokwe, I.O., Igwenagu, E. and Igbokwe, N.A. 2020. Aluminium toxicosis: a review of toxic actions and effects. Interdisciplinary Toxicology 12(2): 45–70.

Inan-Eroglu, E. and Ayaz, A. 2018. Is aluminum exposure a risk factor for neurological disorders? J. Res. Med. Sci. 23–51.

Iwai, A., Masliah, E., Yoshimoto, M., Ge, N., Flanagan, L., de Silva, H.A. et al. 1995. The precursor protein of non-Aβ component of Alzheimer's disease amyloid is a presynaptic protein of the central nervous system. Neuron 14(2): 467–75.

Jeong, C.H., Kwon, H.C., Cheng, W.N., Kim, D.H., Choi, Y. and Han, S.G. 2020. Aluminum exposure promotes the metastatic proclivity of human colorectal cancer cells through matrix metalloproteinases and the TGF-β/smad signaling pathway. Food and Chemical Toxicology 141: 111402.

Jones, D.L. and Kochian, L.V. 1997. Aluminum interaction with plasma membrane lipids and enzyme metal binding sites and its potential role in Al cytotoxicity. FEBS Letters 400(1): 51–57.

Julka, D. and Gill, K.D. 1996. Effect of aluminum on regional brain antioxidant defense status in wistar rats. Research in Experimental Medicine 196(1): 187–94.

Kaiser, L., Schwartz, K.A., Burnatowska-Hledin, M.A. and Mayor, J.H. 1984. Microcytic anemia secondary to intraperitoneal aluminum in normal and uremic rats. Kidney International 26(3): 269–74.

Kar, D., Pradhan, A.A. and Datta, S. 2021. The role of solute transporters in aluminum toxicity and tolerance. Physiologia Plantarum 171(4): 638–52.

Kasa, P., Szerdahelyi, P. and Wisniewski, H.M. 1995. Lack of topographical relationship between sites of aluminum deposition and senile plaques in the Alzheimer's disease brain. Acta Neuropathologica 90(5): 526–31.

Kaur, A. and Gill, K.D. 2005. Disruption of neuronal calcium homeostasis after chronic aluminium toxicity in rats. Basic & Clinical Pharmacology & Toxicology 96(2): 118–22.

Kawabata, H. 2019. Transferrin and transferrin receptors update. Free Radical Biology and Medicine 133: 46–54.

Kawabata, T. 2022. Iron-induced oxidative stress in human diseases. Cells 11(14): 2152.

Kawahara, M. Muramoto, K., Kobayashi, K., Mori, H. and Kuroda, Y. 1994. Aluminum promotes the aggregation of Alzheimer's amyloid β-Protein *in vitro*. Biochemical and Biophysical Research Communications 198(2): 531–35.

Kawahara, M. and Kato-Negishi, M. 2011. Link between aluminum and the pathogenesis of Alzheimer's disease: the integration of the aluminum and amyloid cascade hypotheses. International Journal of Alzheimer's Disease 2011.

Kim, Y. et al. 2011. Bound iron and transferrin bound iron in human. Toxicology Applied Pharmacology 220(3): 349–56.

Klatzo, I., Wiśniewski, H. and Streicher, E. 1965. Experimental production of neurofibrillary degeneration: I. Light microscopic observations. Journal of Neuropathology & Experimental Neurology 24(2): 187–99.

Krewski, D., Yokel, R.A., Nieboer, E., Borchelt, D., Cohen, J., Harry, J. et al. 2007. Human health risk assessment for aluminium, aluminium oxide, and aluminium hydroxide. Journal J. Toxicol. Environ. Health B Crit. Rev. 11(2): 147.

Kruck, T.P., Cui, J.G., Percy, M.E. and Lukiw, W.J. 2004. Molecular shuttle chelation: the use of Ascorbate, Desferrioxamine and Feralex-G in combination to remove nuclear bound aluminum. Cellular and Molecular Neurobiology 24(3): 443–59.

Kubal, G., Mason, A.B, Sadler, P.J., Tucker, A. and Woodworth R.C. 1992. Uptake of Al^{3+} into the N-Lobe of human serum transferrin. Biochemical Journal 285: 711–14.

Kumar, V., Bal, A. and Gill, K.D. 2009. Susceptibility of mitochondrial superoxide dismutase to aluminium induced oxidative damage. Toxicology 255(3): 117–23.

Lai, J.C.K. and Blass, J.P. 1984. Inhibition of brain glycolysis by aluminum. Journal of Neurochemistry 42(2): 438–46.

Leoni, V., Solomon, A. and Kivipelto, M. 2010. Links between ApoE, brain cholesterol metabolism, tau and amyloid beta-peptide in patients with cognitive impairment. Biochem. Soc. Trans 4: 1021–25.

Lidsky, T.I. 2014. Is the aluminum hypothesis dead? Journal of Occupational and Environmental Medicine 56(5 Suppl): S73.

Lin, C., Hsiao, W.C., Huang, C.J., Kao, C.F. and Hsu, G.S. 2013. Heme oxygenase-1 induction by the ROS – JNK pathway plays a role in aluminum-induced anemia. Journal of Inorganic Biochemistry 128: 221–28.

Liu, Y., Nguyen, M., Robert, A. and Meunier, B. 2019. Metal ions in Alzheimer's disease: a key role or not? Accounts of Chemical Research 52(7): 2026–35.

Luque, N.B., Mujika, J.I., Rezabal, E., Ugalde, J.M. and Lopez, X. 2014. Mapping the affinity of aluminum(III) for biophosphates: interaction mode and binding affinity in 1:1 complexes. Physical Chemistry Chemical Physics 16(37): 20107–19.

Macdonald, T.L. and Martin, R.B. 1988. Aluminum ion in biological systems. Trends in Biochemical Sciences 13(1): 15–19.
Maki, H., Sakata, G. and Mizuhata, M. 2017. Quantitative NMR of quadrupolar nucleus as a novel analytical method: hydrolysis behaviour analysis of aluminum ion. Analyst 142(10): 1790–99.
Makjanic, J., McDonald, B., Li-Hsian Chen, P. and Watt, F. 1998. Absence of aluminium in neurofibrillary tangles in Alzheimer's disease. Neuroscience Letters 240(3): 123–26.
Mandriota, S.J., Tenan, M., Ferrari, P. and Sappino, A.P. 2016. Aluminium chloride promotes tumorigenesis and metastasis in normal murine mammary gland epithelial cells. International Journal of Cancer 139(12): 2781–90.
Mangialavori, I., Villamil-Giraldo, A.M., Marino Buslje, C., Ferreira-Gomes, M., Caride, A.J. and Rossi J.P. 2009. A new conformation in sarcoplasmic reticulum calcium pump and plasma membrane Ca^{2+} pumps revealed by a photoactivatable phospholipidic probe. Journal of Biological Chemistry 284(8): 4823–28.
Mangialavori, I., Ferreira-Gomes, M., Pignataro, M.F., Strehler, E.E. and Rossi, J.P. 2010. Determination of the dissociation constants for Ca^{2+} and calmodulin from the plasma membrane Ca^{2+} pump by a lipid probe that senses membrane domain changes. Journal of Biological Chemistry 285(1): 123–30.
Mangialavori, I., Ferreira-Gomes, M.S., Saffioti, N.A., González-Lebrero, R.M., Rossi, R.C. and Rossi, J.P. 2013. Conformational changes produced by ATP binding to the plasma membrane calcium pump. Journal of Biological Chemistry 288(43): 31030–41.
Marchkesbery, W.R. Ehmann, W.D., Hossain, T.I., Alauddin, M. and Goodin, D.T. 1981. Instrumental neutron activation analysis of brain aluminum in Alzheimer disease and aging. Annals of Neurology: Official Journal of the American Neurological Association and the Child Neurology Society 10(6): 511–16.
Martin, R.B. 1986. The chemistry of aluminum as related to biology and medicine. Clinical Chemistry 32(10): 1797–1806.
McDermott, J.R., Smith, A.I., Iqbal, K. and Wisniewski, H.M. 1977. Aluminium and Alzheimer's disease. The Lancet 310(8040): 710–11.
Mirza, A., King, A., Troakes, C. and Exley, C. 2016. The identification of aluminum in human brain tissue using lumogallion and fluorescence microscopy. Journal of Alzheimers Disease 54: 1333–38.
Miu, A.C. and Benga, O. 2006. Aluminum and Alzheimer's disease: a new look. Journal of Alzheimer's Disease 10(2–3): 179–201.
Mousavi, M. and Vaghar, M.I. 2021. The relationship between use of aluminum-containing anti-perspirant and hair color with breast cancer. J. Family Med. Prim. Care 10: 182–6.
Mujika, J.I., Rodríguez-Guerra Pedregal, J., Lopez, X., Ugalde, J.M., Rodríguez-Santiago, L., Sodupe, M. et al. 2017. Elucidating the 3D structures of Al (Iii)–Aβ complexes: a template free strategy based on the pre-organization hypothesis. Chemical Science 8(7): 5041–49.
Mukrasch, M.D., Bibow, S., Korukottu, J., Jeganathan, S., Biernat, J., Griesinger, C. et al. 2009. Structural polymorphism of 441-residue tau at single residue resolution. PLoS Biology 7(2): 0399–0414.
Nagasawa, K., Ito, S., Kakuda, T., Nagai, K., Tamai, I., Tsuji, A. et al. 2005. Transport mechanism for aluminum citrate at the blood-brain barrier: kinetic evidence implies involvement of system Xc - in immortalized rat brain endothelial cells. Toxicology Letters 155(2): 289–96.
Nagasawa, K., Akagi, J., Koma, M., Kakuda, T., Nagai, K., Shimohama, S. et al. 2006. Transport and toxic mechanism for aluminum citrate in human neuroblastoma SH-SY5Y cells. Life Sciences 79(1): 89–97.
Nedzvetsky, V.S., Tuzcu, M., Yasar, A., Tikhomirov, A.A. and Baydas, G. 2006. Effect of Vitamin E against aluminum neurotoxicity in rats. Biochemistry (Moscow) 71(3): 239–44.
Nehru, B. and Anand, P. 2005. Oxidative damage following chronic aluminium exposure in adult and pup rat brains. Journal of Trace Elements in Medicine and Biology 19(2–3): 203–8.
Nie, J. 2018. Exposure to aluminum in daily life and Alzheimer's disease. Advances in Experimental Medicine and Biology 1091: 99–111.
Niemoeller, O.M., Kiedaisch, V., Dreischer, P., Wieder, T. and Lang, F. 2006. Stimulation of eryptosis by aluminium ions. Toxicology Applied Pharmacology 217: 168–75.
Nübling, G., Bader, B., Levin, J., Hildebrandt, J., Kretzschmar, H. and Giese, A. 2012. Synergistic influence of phosphorylation and metal ions on tau oligomer formation and coaggregation with α-synuclein at the single molecule level. Molecular Neurodegeneration 7(1): 1–13.

Olesen, C., Picard, M., Winther, A.M., Gyrup, C., Morth, J.P., Oxvig, C. et al. 2007. The structural basis of calcium transport by the calcium pump. Nature 450(7172): 1036–42.
Ontiveros, M., Rinaldi, D., Marder, M., Espelt, M.V., Mangialavori, I., Vigil, M. et al. 2019. Natural flavonoids inhibit the plasma membrane Ca^{2+}-ATPase. Biochemical Pharmacology 166.
Oshima, E., Ishihara, T., Yokota, O., Nakashima-Yasuda, H., Nagao, S., Ikeda, C. et al. 2013. Accelerated tau aggregation, apoptosis and neurological dysfunction caused by chronic oral administration of aluminum in a mouse model of tauopathies. Brain Pathology 23(6): 633–44.
Oteiza, P.I. 1994. A mechanism for the stimulatory effect of aluminum on iron-induced lipid peroxidation. Archives of Biochemistry and Biophysics 308(2): 374–79.
Ott, D.B., Hartwig, A. and Stillman, M.J. 2019. Competition between Al^{3+} and Fe^{3+} binding to human transferrin and toxicological implications: structural investigations using ultra-high resolution ESI MS and CD spectroscopy. Metallomics 11(5): 968–81.
Ouyz, R.M. and Milne, F.J. 1995. Alterations in intracellular cations and cell membrane ATPase activity in patients with malignant hypertension. Hypertension 13(8): 867–74.
Palop, J.J. and Mucke, L. 2010. Amyloid-β–induced neuronal dysfunction in Alzheimer's Ddisease: From synapses toward neural networks. Nature Neuroscience 13(7): 812–18.
Parkinson, I.S., Ward, M.K. and Kerr, D.Nj. 1981. Dialysis encephalopathy, bone disease and anaemia: the aluminum intoxication syndrome during regular haemodialysis. Journal of Clinical Pathology 34(11): 1285.
Paz, L.N.F., Moura, L.M., Feio, D.C.A., Cardoso, M.S.G., Ximenes, W.L.O., Montenegro, R.C. et al. 2017. Evaluation of *in vivo* and *in vitro* toxicological and genotoxic potential of aluminum chloride. Chemosphere 175: 130–37.
Pei, J., Sjögren, M. and Winblad, B. 2008. Neurofibrillary degeneration in Alzheimer's disease: From molecular mechanisms to identification of drug targets. Current Opinion in Psychiatry 21(6).
Perl, D.P. and Brody, A.R. 1980. Alzheimer's disease: X-Ray spectrometric evidence of aluminum accumulation in neurofibrillary tangle-bearing neurons. Science 208(4441): 297–99.
Perl, D.P., Gajdusek, D.C., Garruto, R.M., Yanagihara, R.T. and Gibbs, C.J. 1982. Intraneuronal aluminum accumulation in amyotrophic lateral sclerosis and Parkinsonism-Dementia of guam. Science 217(4564): 1053–55.
Perl, D.P. and Moalem, S. 2006. Aluminum and Alzheimer's disease, a personal perspective after 25 years. Journal of Alzheimer's Disease 9(s3): 291–300.
Perluigi, M., Joshi, G., Sultana, R., Calabrese, V., De Marco, C., Coccia, R. et al. 2006. In vivo protection by the xanthate tricyclodecan-9-Yl-xanthogenate against amyloid beta-peptide (1-42)-induced oxidative stress. Neuroscience 138(4): 1161–70.
Peterson, N.A., Anderson, B.F., Jameson, G.B., Tweedie, J.W. and Baker, E.N. 2000. Crystal structure and iron-binding properties of the R210K mutant of the n-lobe of human lactoferrin: implications for iron release from transferrins. Biochemistry 39(22): 6625–33.
Platt, B. and Büsselberg, D. 1994. Actions of aluminum on voltage-activated calcium channel currents. Cellular and Molecular Neurobiology 14(6): 819–29.
Post, R.L., Kume, S., Tobin, T., Orcutt, B. and Sen A.K. 1969. Flexibility of an active center in sodium-plus-potassium adenosine triphosphatase. Journal of General Physiology 54(1): 306–326.
Praticò, D., Uryu, K., Sung, S., Tang, S., Trojanowski, J.Q., Lee, V.M. 2002. Aluminum modulates brain amyloidosis through oxidative stress in APP transgenic mice. The FASEB Journal 16(9): 1138–40.
Priest, N.D. Newton, D., Day, J.P., Talbot, R.J. and Warner, A.J. 1995. Human metabolism of Aluminium-26 and Gallium-67 injected as citrates. Human & Experimental Toxicology 14(3): 287–93.
Priest, N.D. 2004. The biological behaviour and bioavailability of aluminium in man, with special reference to studies employing Aluminium-26 as a tracer: review and study update. Journal of Environmental Monitoring 6(5): 375–403.
Pozo Devoto, M.V., Dimopoulos, N., Alloatti, M., Pardi, M.B., Saez, T.M., Otero, M.G. et al. 2017. Asynuclein control of mitochondrial homeostasis in human-derived neurons is disrupted by mutations associated with Parkinson's disease. Scientific Reports 7(1): 1–13.
Rahimzadeh, M.R., Rahimzadeh, M.R., Kazemi, S., Amiri, R.J., Pirzadeh, M. and Moghadamnia A.A. 2022. Aluminum poisoning with emphasis on its mechanism and treatment of intoxication. Emergency Medicine International 2022: 1–13.

Rahman, Md A., Lee, S.H., Ji, H.C., Kabir, A.H., Jones, C.S. and Lee, K.W. 2018. Importance of mineral nutrition for mitigating aluminum toxicity in plants on acidic soils: current status and opportunities. International Journal of Molecular Sciences 19(10).

Reichenstein, M., Borovok, N., Sheinin, A., Brider, T. and Michaelevski, I. 2021. Abelson kinases mediate the depression of spontaneous synaptic activity induced by amyloid beta 1–42 peptides. Cell Mol. Neurobiol. 41(3): 431–48.

Rengel, Z. and Zhang, W.H. 2003. Role of dynamics of intracellular calcium in aluminium-toxicity syndrome. New Phytologist 159(2): 295–314.

Rezabal, E., Mercero, J.M., Lopez, X. and Ugalde, J.M. 2006. A study of the coordination shell of aluminum(III) and magnesium(II) in model protein environments: thermodynamics of the complex formation and metal exchange reactions. Journal of Inorganic Biochemistry 100(3): 374–84.

Ricchelli, F., Drago, D., Filippi, B., Tognon, G. and Zatta, P. 2005. Aluminum-triggered structural modifications and aggregation of β-amyloids. Cellular and Molecular Life Sciences CMLS 62(15): 1724–33.

Rinaldi, E., Ontiveros, M.Q., Saffioti, N.A., Vigil, M.A., Mangialavori, I.C., Rossi, R.C. et al. 2021. Epigallocatechin 3-Gallate inhibits the plasma membrane Ca^{2+}-ATPase: Effects on calcium homeostasis. Heliyon 7: 0–11.

Rolston, R.K., Perry, G., Zhu, X., Castellani, R.J., Dwyer, B.E., Lee, H.G. et al. 2009. Iron: A pathological mediator of Alzheimer disease? Agro Food Industry Hi-tech 19(6): 33.

Romani, A.M. 2011. Cellular magnesium homeostasis. Arch. Biochem. Biophys 512(1): 1–23.

Roskams, A.J. and Connor, J.R. 1990. Aluminum access to the brain: a role for transferrin and its receptor. Proceedings of the National Academy of Sciences of the United States of America 87(22): 9024–27.

Saberzadeh, J., Arabsolghar, R. and Takhshid, M.A. 2016. Alpha synuclein protein is involved in aluminum-induced cell death and oxidative stress in PC12 cells. Brain Research 1635: 153–60.

Saffioti, N.A., de Sautu, M., Ferreira-Gomes, M.S., Rossi, R.C., Berlin, J., Rossi, J.P. et al. 2019. E2P-like states of plasma membrane Ca^{2+}-ATPase characterization of vanadate and fluoride-stabilized phosphoenzyme analogues. Biochimica et Biophysica Acta - Biomembranes 1861(2).

Saffioti, N.A., de Sautu, M., Riesco, A.S., Ferreira-Gomes, M.S., Rossi, J.P.F.C. and Mangialavori, I.C. 2021. Conformational changes during the reaction cycle of plasma membrane Ca^{2+}-ATPase in the autoinhibited and activated states. Biochemical Journal 478(10): 2019–34.

Sakajiri, T.. Yamamura, T., Kikuchi, T., Ichimura, K., Sawada, T. and Yajima, H. 2010. Absence of binding between the human transferrin receptor and the transferrin complex of biological toxic trace element, aluminum, because of an incomplete open/closed form of the complex. Biological Trace Element Research 136(3): 279–86.

Sato, H., Kato, T. and Arawaka, S. 2013. The role of Ser129 phosphorylation of α-synuclein in neurodegeneration of Parkinson's disease: A review of *in vivo* models. Reviews in the Neurosciences 24(2): 115–26.

Schreeder, M.T., Favero, M.S., Hughes, J.R., Petersen, N.J., Bennett, P.H. and Maynard, J.E. 1983. Dialysis encephalopathy and aluminum exposure: an epidemiologic analysis. Journal of Chronic Diseases 36(8): 581–93.

Scott, C.W., Fieles, A., Sygowski, L.A. and Caputo, C. 1993. Aggregation of tau protein by aluminum. Brain Research 628(1–2): 77–84.

Sheff, D., Pelletier. L., O'Connell, C.B., Warren, G. and Mellman, I. 2002. Transferrin receptor recycling in the absence of perinuclear recycling endosomes. Journal of Cell Biology 156(5): 797–804.

Shin, R., Lee, V.M. and Trojanowski, J.Q. 1994. Aluminum modifies the properties of Alzheimer's disease PHF tau proteins *in vivo* and *in vitro*. Journal of Neuroscience 14(11): 7221–33.

Shin, R.W., Kruck, T.P., Murayama, H. and Kitamoto, T. 2003. A novel trivalent cation chelator feralex dissociates binding of aluminum and iron associated with hyperphosphorylated τ of Alzheimer's disease. Brain Research 961(1): 139–46.

Silva, S., Duarte, A.I., Rego, A.C., Oliveira, C.R. and Gonçalves, P.P. 2005. Effect of chronic exposure to aluminium on isoform expression and activity of rat (Na^+/K^+) ATPase. 88(2): 485–94.

Silva, V.S. and Gonçalves, P.P. 2003. The inhibitory effect of aluminium on the (Na^+/K^+) ATPase activity of rat brain cortex synaptosomes. Journal of inorganic biochemistry 97(1): 143–50.

Suarez-Fernandez, M.B., Soldado, A.B., Sanz-Medel, A., Vega, J.A., Novelli, A. and Fernández-Sánchez, M.T. 1999. Aluminum-induced degeneration of astrocytes occurs via apoptosis and results in neuronal death. Brain Research 835(2): 125–36.

Temel, Y., Engü, A.Ş., Akkoyun, H.T., Akkoyun, M. and Ciftci, M. 2017. Effect of astaxanthin and aluminum chloride on erythrocyte G6PD and 6PGD enzyme activities *in vivo* and on erythrocyte G6PD *in vitro* in rats. Journal of Biochemical and Molecular Toxicology 31(10): 30–33.

Tenan, M.R., Nicolle, A., Moralli, D., Verbouwe, E., Jankowska, J.D., Durin, M.A. et al. 2021. Aluminum enters mammalian cells and destabilizes chromosome structure and number. International Journal of Molecular Sciences 22(17): 9515.

Uversky, V. N., Jie, L. and Anthony L. Fink. 2001. Metal-triggered structural transformations, aggregation, and fibrillation of human α-synuclein: a possible molecular link between Parkinson's disease and heavy metal exposure. Journal of Biological Chemistry 276(47): 44284–96.

van Veen, S., Martin, S., Van den Haute, C., Benoy, V., Lyons, J., Vanhoutte, R. et al. 2020. ATP13A2 deficiency disrupts lysosomal polyamine export. Nature 578(7795): 419–24.

Verstraeten, S., Golub, M.S., Keen, C.L. and Oteiza, P.I. 1997. Myelin is a preferential target of aluminum-mediated oxidative damage. Archives of Biochemistry and Biophysics 344(2): 289–94.

Wang, C., Zhang, R., Wei, X., Lv, M. and Jiang, Z. 2020. The metal ion-controlled immunity. Advances in Immunology Metalloimmunology 45: 187–241.

Wang, Z., Wei, X., Yang, J., Suo, J., Chen, J., Liu, X. et al. 2016. Chronic exposure to aluminum and risk of Alzheimer's disease: A meta-analysis. Neuroscience Letters 610: 200–206.

Yamada, N., Karasawa, T., Wakiya, T., Sadatomo, A., Ito, H., Kamata, R. et al. 2020. Iron overload as a risk factor for hepatic ischemia-reperfusion injury in liver transplantation: Potential role of ferroptosis. Am. J. Transplant 20: 1606–1618.

Ye, I., Bassit, B., Zhu, L., Yang, X., Wang, C. and Li, Y.M. 2007. γ-secretase substrate concentration modulates the Aβ42/Aβ40 ratio: Implications for Alzheimer disease. Journal of Biological Chemistry 282(32): 23639–44.

Yiannopoulou, K.G. and Papageorgiou, S.G. 2020. Current and future treatments in Alzheimer disease: An update. Journal of Central Nervous System Disease 12: 1179573520907397.

Yuan, C., Lee, Y. and Wang Hsu, G. 2012. Aluminum overload increases oxidative stress in four functional brain areas of neonatal rats. Journal of Biomedical Science 19(1): 1–9.

Zangeneh, A.R., Takhshid, M.A., Ranjbaran, R., Maleknia, M. and Meshkibaf, M.H. 2021. Diverse effect of Vitamin C and N-acetylcysteine on aluminum-induced eryptosis. Biochemistry Research International 2021: 6670656.

Zatta, P., Lain, E. and Cagnolini, C. 2000. Effects of aluminum on activity of krebs cycle enzymes and glutamate dehydrogenase in rat brain homogenate. European Journal of Biochemistry 267(10): 3049–55.

Zhang, J., Wei, J., Li, D., Kong, X., Rengel, Z., Chen, L. et al. 2017. The role of the plasma membrane H+-ATPase in plant responses to aluminum toxicity. Frontiers in Plant Science 8: 1–9.

Zhang, Q., Cao, Z., Sun, X., Zuang, C., Huang, W. and Li, Y. 2016. Aluminum trichloride induces hypertension and disturbs the function of erythrocyte membrane in male rats. Biological Trace Element Research 171(1): 116–23.

Zhang, X.C. and Zhang, H. 2019. P-Type ATPases use a domain-association mechanism to couple ATP hydrolysis to conformational change. Biophysics Reports 5(4): 167–75.

Chapter 4

Effect of Metals (Hg, Cd, Pb, Zn, Cr and Cu) on Ruminants

Karise Fernanda Nogara,[1,*] *Queila Gouveia Tavares,*[1] *Maila Palmeira,*[2] *Carlos Henrique Milagres Ribeiro*[3] and *Maity Zopollatto*[1]

1. Introduction

Ruminants (cattle, buffaloes, goats, and sheep) are of great importance in the food production chain, providing products with high biological and nutritional value for consumers, such as milk and meat. However, for these foods to reach the consumer's table, several practices are involved on the farm, including animal welfare and balanced nutrition. Nutrition is essential to maintain high animal productivity. Animals have to be healthy, free of disease and toxic effects. However, cases of toxic effects in animals have already been recorded in the literature, especially after the ingestion of food or water contaminated with trace metals. Of particular concern are those that can be bioaccumulate and, subsequently be transferred into products of animal origin. In fact, they may represent a risk to consumers, due to their toxic properties, representing a scenario that involves public health (Tchounwou et al., 2012; Lane et al., 2015; Sharaf et al., 2021).

For instance, some minerals are considered essential, because when supplied via diet they improve animal performance (Herdt and Hoff, 2011), such as cobalt (Co), chromium (Cr), copper (Cu), zinc (Zn), iron (Fe), iodine (I), manganese (Mn), molybdenum (Mo), and selenium (Se) (NASEM, 2021). Accordingly, deficiency

[1] Graduate Program on Animal Science, Department of Animal Science, Federal University of Paraná, Curitiba, Paraná, 80035-050, Brazil.

[2] Graduate Program on Production and Animal Health, Instituto Federal Catarinense, IF-Catarinense, Araquari, Santa Catarina, 89245-000, Brazil.

[3] Graduate Program Agronomia/Fitotecnia, Federal University of Lavras, Lavras, Minas Gerais, 37200-000, Brazil.

* Corresponding author: nogara.karise@gmail.com

of these trace elements can lead to increased risk of infections. In the case of Cu, its deficiency can be associated with increased Mo toxicity, increasing the susceptibility of microbial infections in lambs. This occurs because Cu deficiency generates a dysfunction in the cells of the immune system (Suttle and Jones, 1989). The reproductive performance of ruminants can be affected by several minerals, including Cu, Co, Se, Mn, I, Zn, and Fe. Indeed, failures in reproduction may be associated with enzymatic dysfunctions caused by deficiencies of these minerals (Hidiroglou, 1979).

However, animal toxic effects can also occur if they are exposed to high concentrations of trace or toxic elements such as cadmium (Cd), Cu, lead (Pb), and Zn, either from the soil, water, air, or after intake of unbalanced diets. Cases of Cu intoxication are frequent in sheep and rare in adult cattle (Herdt and Hoff, 2011). However, it has been reported that dietary Cu can influence the optimal productive performance in certain breeds of cattle. For instance, Simmental and Charolais cattle have higher Cu requirements than the Angus breed (Spears and Weiss, 2014), which makes them slightly more resistant to the toxicity of this element. In addition, negative interactions can occur between trace elements. For instance, diets with borderline Cu content and high Mo and/or sulfur (S) content can facilitate Cu^{2+} deficiency and cause toxicity associated with impaired performance (for instance, reduced fertility) (Hidiroglou, 1979). Spears (2003) reported that when there is a high level of calcium (Ca) in the diet, there is a reduction in the concentration of Zn in the serum. Therefore, the mineral requirements of animals must be taken into account to optimize production in cattle and other ruminant species (Spears and Weiss, 2014).

The nutrients required by ruminants can be grouped into the following categories: water, carbohydrates, proteins, lipids, vitamins, and minerals. One of the main nutritional limitations of dairy cattle in tropical regions is associated with mineral deficiency, because some pastures cannot meet the minimal nutrient requirements of the animals (Rotta et al., 2019). And since some of these minerals are critical for adequate growth, physiological functions, and overall productivity of the animals (Herdt and Hoff, 2011), one way to reverse their deficiencies is through supplementation. However, mineral imbalance and intoxications can also cause serious losses in livestock. Nonetheless, the interaction between the elements can cause antagonistic effects, where high levels of S can lead to Cu deficiency, for example. Measuring the blood concentration of the elements is a good option to verify their adequate nutritional intake, but not for all minerals, as is the case of Cu and Zn, which have low blood concentrations. Here, liver biopsy can be indicated (Herdt and Hoff, 2011).

In the report of the Nutrient Requirements of Dairy Cattle (NRC) in 1945 there was an extensive discussion about the symptoms of nutrient deficiency. For example, cows with a deficiency in phosphorus (P) showed a decrease in appetite and developed severe rickets because this was the demand of that moment (to supply the deficiency) (NRC, 1945). However, in the most recent publication (NASEM, 2021) little is discussed about mineral deficiency associated diseases, as the focus is on maximizing productivity in terms of animal health.

Minerals, as well as vitamins, have several functions in protecting tissues and supporting cellular immune function against pathogens (for example, Cr and Cu), improving animal health and productivity (Waldron, 2007; Spears and Weiss, 2014). Deficiencies of these micronutrients and the consequent drop in the animal's immunity, can make it more susceptible to problems such as increased somatic cell count (SCC) in milk, an increase in the incidence of mastitis or metritis, retention of the placenta, worsening of reproductive efficiency, an increased incidence of metabolic diseases, and a decrease in milk production (Suttle and Jones, 1989; Arshad et al., 2021). Therefore, minerals are of great importance in reversing these conditions, as well as contributing to increasing the production of solids in milk and, consequently, increasing its nutritional value.

However, in excess, minerals can also be harmful to the health of the cattle, causing, for instance, from reduction in milk production to blindness; they can cause pollution to the environment, as well as increasing the cost of production and reducing the profitability of the farm. Minerals are divided into macrominerals and microminerals, depending on the amount found in the body (Rotta et al., 2019), usually expressed in mg/kg or ppm (parts per million) (Ortolani, 2002). The macroelements considered essential for ruminant nutrition are Ca, P, Mg, chlorine (Cl), sodium (Na), potassium (K), and S, and the microelements are Cu, Zn, Fe, Co, Mo, Se, I, Mg, and, more recently, Cr (Timm, 2001). The new National Academics of Science, Engineering, Medicine (NASEM) model included nine trace minerals that are considered essential for dairy cattle, including: Co, Cr, Cu, Fe, I, Mg, Mo, Se, and Zn. In this chapter we will only address the toxicology of some essential and non-essential metals for ruminants: mercury, cadmium, lead, copper, chromium, and zinc.

These minerals are also categorized as heavy metals, but not all heavy metals are toxic. For example, Cu, I, Mn, and Zn are called heavy metals, but they are not toxic because they are essential for living of the animals, where in adequate concentrations they contribute to the development of the animal and its metabolic functions. These elements only cause toxic problems when the exposure is much greater than normal, and the body is unable to process them (Costa et al., 2020; Sharaf et al., 2021). For example, heavy metals can alter fatty acid synthesis through oxidative stress, influencing the synthesis of unsaturated fatty acids from bovine milk (Su et al., 2022). On the other hand, Cd, Hg, and Pb are toxic, disturbing the normal physiology by interfering with the function of different organs of the farm animals. Indeed, as they do not have essential physiological functions for living organisms, they have to be avoided even at lower levels (Tchounwou et al., 2012; Costa et al., 2020). The effects of these metals can appear only after long term exposure to contaminated soil, water, and foraging next the industrial location. Their toxicity is usually related to their reactivity and bioaccumulation. Accordingly, toxic metals can have long biological half-life. More specifically, this means that they are slowly eliminated from the body, impairing metabolic functions and causing pathophysiological consequences in the exposed animals (Lane et al., 2015). However, the diagnosis of intoxication is not always clear, as Cu toxicity in dairy cows is less evident than in sheep (Arshad et al., 2021; López-Alonso and Miranda, 2020). There are reports in the literature of eight fatal toxic level cases intoxication due to excess of Cu confirmed by high liver, kidney, and serum Cu concentrations, leading to hepatic necrosis and hepatic and renal abnormalities (Bidewell et al., 2012; López-Alonso and Miranda, 2020).

It is important to assess the impact of toxic metals on domestic animals (López Alonso et al., 2000). Intoxication by toxic metals can occur from exposure to wastes from mining (either via breathing or skin contact) or contaminated food, but there is still little information about these types of contamination in the literature. With the collapse of dams in recent years, it was possible to observe a considerable accumulation of these metals from mining tailings, contaminating water and soil and, consequently, the plant fodder on which animals feed. For example, Cu and Pb contamination can occur in ruminants after grazing in contaminated soil, or consuming contaminated vegetation or water (Johnsen and Aaneby, 2019).

Today, the importance of vitamin nutrition is known, and better supplementation strategies have been developed to improve cow health and production (Weiss, 2017). However, between deficiency and intoxication, it is necessary to seek a balance of these elements, according to the animal's needs. Mineral concentrations in the diet are not always sufficient to meet the nutritional demands of the animal. Animals need minerals to meet high production requirements. However, the risk is exacerbated when production animals are exposed via water supply without any control of physical-chemical quality (regular checking of trace metals in the water), mineral supplements, and agro-industrial by-products in animal feed. In relation to mineral supplements, the knowledge of their composition via bromatological analyses are critical, otherwise these products can be a serious risk factor for the producer, as they may contain high concentrations of heavy metals (Costa et al., 2020).

Therefore, it is extremely important that producers and manufacturers take responsibility for monitoring the levels of heavy metals in the ingredients used in the formulation of ruminant diets. In effect, manufacturers of mineral supplements should be responsible for quality control and the choice of the ingredients used to compose the mineral supplements. An alternative is the use of organic mineral sources to reduce the chances of inclusion of toxic elements that can interfere with the rumen function and health of the animals; however, this still does not guarantee that there will not be high levels of heavy metals if there is no strict quality control. Thus, the objective of this discussion is to address the main active heavy metals that influence the health of ruminants: Cd, Pb, Cu, Cr, Hg, and Zn, both via supplementation and/or diet and by accidental intoxication.

2. Main symptoms of heavy metal contamination in ruminants

The elevation of heavy metals in air, soil and water is usually related to human activities, for instance, industrial and agricultural activities. Accordingly, the application of chemical products (fertilizers, fungicides, pesticides, and herbicides, for example) can contribute to some metals infiltrating soil in the form of impurities, and consequently contaminating vegetation (McLaughlin et al., 1999; He et al., 2005; Balali-Mood et al., 2021). In animals, several pathologies can be caused by the excess of toxic metals or by the deficiency of essential trace elements.

Toxicity and clinical signs depend on the dose, exposure time, exposure route, age, genetic, nutritional status, reproductive and productive phase of the animal (Tchounwou et al., 2012; Nogara et al., 2019; Costa et al., 2020). Heavy metals delay the growth process, induce oxidative stress, apoptosis, changing gene expression and drop in the immune system (Balali-Mood et al., 2021). Furthermore, metal toxicity (like Hg, Pb, Cr and Cd) can also cause lethargy, tremors, convulsions, renal and cardiovascular dysfunction, anemia, liver damage, reduced pulmonary function, dermal diseases, increase the incidence of cancers, degenerative bone disease and disorders in the metabolism of Zn and Cu in humans, and even death (Nogara et al., 2019; Balali-Mood et al., 2021).

Animals, when challenged with high doses of metals, may present alterations in the body tissues, changes in the reproductive system (harming the ovarian health of animals, low birth weight and premature birth), endocrine, neurological (abnormalities in the offspring), and enzymatic systems, DNA damage, neuropsychiatric, and kidney damage (Tchounwou et al., 2012; Costa et al., 2020). Blindness, depression, paralysis, incoordination of movements, excessive salivation, muscle tremors, and convulsions are also some clinical signs of contamination. In cases where there is no clinical manifestation of intoxication, methods of analyses of the blood and milk, and biopsy of animal tissues, such as liver and kidneys, can be performed to identify and determine the concentration of heavy metals (Okada et al., 1977; Herdt and Hoff, 2011; Costa et al., 2020).

The most common symptoms of toxic metal poisoning are apathy, lack of appetite, teeth grinding, abdominal pain, and rumen atony (Radostits et al., 2007). However, blindness, ataxia, increased heart rate, rapid and difficult breathing, tremors, coma, and death can appear within five days when animals are exposed to heavy metals such as lead, causing acute intoxication (Ozmen and Mor, 2004). Elements such as Cd, Hg, and arsenic (As) caused 50% inhibition of the growth of rumen microorganisms even at low concentrations and lead was also shown to inhibit bacterial growth at higher concentrations. The most affected bacteria in the study were: *Bacteroides succinogenes*, *Ruminococcus albus,* and *Eubacterium ruminantium*, which are very important in the digestion of the diet supplied daily to the animals (Forsberg, 1978). In the study by Salem et al. (2011), evaluating the sensitivity of ruminal bacteria from sheep, cattle, and buffalo against some heavy metals (Hg, Cd, Co, Nickel [Ni], Cr, Cu, Silver [Ag], Mo, Zn, Pb, Lithium [Li], and Na), also found that Hg and Cd were the metals that had the most inhibitory effects on ruminal bacteria of the three species. Ruminal bacteria showed 62 and 100% tolerance to Li and Na, respectively, that is, their effects on the rumen were insignificant, whereas Na did not present toxicity to ruminal bacteria. However, cattle tolerated metal toxicity better, while sheep and buffaloes seemed to be more vulnerable to these elements, which demonstrates that the inhibitory effect of these metals can reduce animal productivity.

We will address the symptomatology of metal intoxication in more detail below, according to each element.

3. Main heavy metals that affect ruminants

3.1 Heavy and toxic metals

3.1.1 Mercury (Hg)

Known for being one of the most toxic metals, Hg is characterized by being the only one in the liquid state, and because of its high vapor pressure can be found as vapor at ambient temperature (Clarkson and Magos, 2006). This metal is a non-physiological element and can be found in the environment in three main chemical forms (metallic, organic, and inorganic), all of which cause serious deleterious effects to the living organisms. In mammals the kidneys are the target organs for metallic (Hg^0) and inorganic Hg (Hg^{+2}). For metallic Hg, in addition to the kidneys, lungs and brain can also be the targets for Hg intoxication (Brondani, 2016; Oliveira et al., 2017).

In farm animals, Hg intoxication is almost entirely caused by the supply of grains, pellets, or concentrates treated with mercury-based fungicides, in addition to contaminated water (Radostits et al., 2007). The maximum intake of Hg is 2 mg/kg of the diet, and values above this can cause changes in the gastrointestinal tract caused by the corrosive effect that these compounds have on the mucous membranes. The toxicity of Hg compounds will depend on their interaction with thiol (-SH) groups, and among the symptoms we can mention excessive salivation, nausea, vomiting, bloody diarrhea, and dysentery with fluid loss, which can progress to hypovolemic shock (Oberherr and Rossato, 2011; Oliveira et al., 2017).

Inorganic Hg intoxication occurs more frequently than organic mercury intoxication and is considered more serious, as it causes coagulation of the mucosa of the alimentary tract, favoring the rapid development of gastroenteritis (Oberherr and Rossato, 2011; Costa et al., 2020). When stored in the pituitary gland, Hg affects the production of gonadotropins, thus altering the ovulatory cycle in females and sperm production in males. Furthermore, this metal causes changes in the secretion of progesterone, estradiol, and milk in females, in addition to increasing the risk of miscarriage and death (Costa et al., 2020). Failure in the diagnosis results in animals discarded from the production system, which could have been properly treated, in addition to the lack of implementation of measures that may prevent the emergence of new cases.

Findings by Rowens et al. (1991) point out that acute exposure to Hg vapor can affect the respiratory system. The main symptoms are dyspnea, dry cough, and fever, progressing to interstitial pneumonia, atelectasis, necrotizing bronchiolitis, pulmonary hemorrhage with epistaxis, and pulmonary edema. If ingestion occurs, the gastrointestinal tract must be cleaned by gastric lavage or using laxatives that increase intestinal transit and reduce the mercury absorption. Chelating substances such as ethylenediamine tetraacetic acid (EDTA) can also be used in the treatment of heavy metal toxicity (Pb, Mg, and Cu), removing metals from blood. However, the effects of EDTA vary according to the type of organism studied, the concentration of EDTA, and the metal analyzed (Oviedo and Rodríguez, 2003).

Hg exists in organic and inorganic forms. In the case of inorganic Hg, such as mercuric chloride ($HgCl_2$) and mercurous chloride (Hg_2Cl_2), they are found in aquatic ecosystems Since they are usually poorly absorbed, they have low toxicity.

The inorganic Hg toxicity affects the respiratory, nervous, hepatic, and renal systems, inducing more serious damage, such as locomotor abnormalities, blindness, excitation, abnormal behavior and chewing, incoordination, and seizures (Kahn and Line, 2008; Oliveira et al., 2017).

3.1.2 Cadmium (Cd)

Cd is a highly reactive and toxic element, characterized as a heavy metal, together with As, Hg, and Cr, that have no physiological function, and are thus considered toxic (Underwood and Suttle, 1999; Friberg et al., 2019). Although there is a vast amount of literature on the biochemical responses of animals to Cd, most of them demonstrate little or no nutritional significance (Underwood and Suttle, 1999). However, Cd can be found in soil and fodder, superphosphates, and municipal sewage sludge.

This metal is present in the environment because of various human activities (Rahimzadeh et al., 2017), such as its application in industry as a corrosive agent, stabilizer in products made of PVC, colored pigments, batteries, combustion of fossil fuels, and the use of phosphate fertilizer (Costa et al., 2020). An analysis of the frequency of violations of the suggested maximum limits from 2012 to 2016, showed an incidence of 1.23% violations for Cd and 0.05% for Pb in slaughtered cattle. In these cases, the Ministry of Agriculture, Livestock and Food Supply (MAPA - Brazil) inspects the property to identify the causes of the violation and requires an action plan with corrective and preventive measures to be adopted at the slaughterhouse. Furthermore, these products can be withdrawn from the market and the next batches are analyzed, until five batches present values within the standard. The most common cause of this violation is through toxic metals from animal feed (Tucci, 2017). Regarding toxic metals, 120 were analyzed samples of animal feed, and more outstanding violations were found between 2016 and 2017 of Pb in mineral premix (Pb was found in 82.14% of the tested samples), macromineral (85.71%) and micromineral (10.53%) compared to Cd for the same categories (0.00, 42.86, and 10.00%) (Table 1). The reference limits for premix for Pb and Cd are 200 mg/kg and 15 mg/kg (Tucci, 2017).

Fodder can contain high levels of Cd, with higher concentrations than meat, eggs, milk, and dairy products. This factor is mainly due to atmospheric deposition, mining, and the use of fertilizers containing Cd, causing soil contamination and increased absorption of this metal in plants (Friberg et al., 2019). Small doses of this metal can lead to kidney, liver, cardiovascular, skeletal, and reproductive damage, in addition to deteriorating vision and hearing (Kumar and Sharma, 2019). Cd has attracted public attention due to its toxicity in humans and easy contamination through water and food, as it is considered one of the most important pollutants to public health, with teratogenic, carcinogenic, and mutagenic potential (Papa et al., 2014).

In ruminants, Cd accumulation occurs in the liver and kidneys, signs worsening with increasing doses (Kim et al., 2016). In addition, there is a positive correlation between Cd in the mammary gland and milk production, suggesting that there is more passage of Cd to the milk, contaminating the product and subsequently causing

Table 1. Results of the toxic metals control program in animal feed from 120 samples analyses.

Toxic Metal	Mineral premix supplement	Macromineral	Micromineral	Total
Pb	82.14	85.71	10.53	60.66
Cd	0.00	42.86	10.00	13.11

The data represent the number of samples (in percentage) where the levels of Pb or Cd were higher than 200 or 15 mg/kg, respectively. Source: Tucci (2017).

harm to dairy consumers (Olsson et al., 2001). Deleterious effects have also been observed in the ovaries and uterus of dairy cows, potentially impairing reproductive performance (Kim et al., 2016). Cd can induce apoptosis and interfere the cell cycle of bovine mammary epithelial cells (Chen et al., 2021) and increase markers of oxidative stress in the blood, consequently impairing the animal's health.

Diets containing high Cd concentrations (> 40 mg Cd/kg in dry matter [DM]), induced a loss of appetite, poor growth, retarded testicular development, and parakeratosis in sheep, like signs of zinc deficiency, and were prevented by zinc supplementation (Underwood and Suttle, 1999). Other studies related abnormalities, such as anemia, impaired bone mineralization, loss of wool crimp, abortion, and stillbirths, which are associated with Cu deficiency, beyond teratogenic lesions, and renal failure in animals chronically exposed to Cd (Powell et al., 1964; Underwood and Suttle, 1999).

In view of the pathogenic potential of Cd in both animals and humans, preventive measures must be taken, such as avoiding the rearing of animals close to industries and not using fertilizers contaminated with this metal.

3.1.3 Lead (Pb)

Pb intoxication is associated with environmental contamination, especially when ruminants graze in areas highly contaminated by this element (> 1000 ppm of Pb in the soil). Therefore, Pb intoxications can occur via ingestion of vegetation located in polluted areas (high Pb contents). It can also occur by accidental ingestion of products that contain Pb (oils, greases, paints, or materials related to batteries), in addition to water contaminated by the leaching of these materials (Gava, 2001; Ozmen and Mor, 2004; Roegner et al., 2013). Another risk factor is the grazing of sheep in shooting ranges, due to lead contamination from ammunition (Johnsen and Aaneby, 2019).

The mining-metallurgical industry is also a source of Pb and Cd contamination, as it can affect products obtained from agricultural activities in nearby areas, due to long-term exposure to these elements. In 2018, 40 samples of raw milk were collected from dairy cows, and evaluations found concentrations of Pb (577 μg/kg) and Cd (18.35 μg/kg) above the values recommended by the European Union (20 μg/kg of Pb and 10 μg/kg of Cd in raw milk) (Ayar et al., 2009; European Commission, 2015). In Brazil, of 218 milk samples analyzed, 43 showed Pb levels above (near to 0.20 mg/kg) the maximum limit suggested by Brazilian legislation of 0.05 mg/kg (Okada et al., 1977). This contamination of milk puts human health at risk, especially children and adolescents, due to problems related to the toxicity arising from the bioaccumulation and reactivity of these metals (Costa et al., 2020;

Castro-Bedrinana et al., 2021). These elements (Pb and Cd) are usually related to thyroid changes (Pinheiro and Souza, 2017). Therefore, the permanence of animals near industries and battery storage is a risk factor, as the smoke and its consequent inhalation can cause poisoning and even death of animals (Gava, 2001; Ozmen and Mor, 2004).

Pb intoxication is more common in cattle, mainly in young animals (Kahn and Line, 2008). When absorbed, Pb enters the blood and tissues and is distributed to the bones. Pb influences enzymes that contain free sulfhydryl groups (e.g., cysteine), the thiol group of proteins found in red blood cells can also be altered by Pb (Rocha et al., 2012). In addition, Pb can deplete the antioxidant defenses, particularly in tissues rich in mitochondria (Stohs and Bagchi, 1995; Carmona et al., 2021). Therefore, it acts as an immunosuppressant, which is teratogenic, nephrotoxic, and toxic to the hematopoietic system.

The main clinical signs of Pb intoxication in cattle are muscle tremors, blindness, aggressiveness, or depression, teeth grinding, convulsions, anorexia, rumen atony, and diarrhea. Small, local, momentary, and involuntary muscle contractions of the muscles of the face, neck, and ears may also occur. This intoxication can last from 8 to 30 days, but generally the death of animals occurs within 12 to 24 hours (Gava, 2001).

The age of the animals influences the accumulation of Pb and is greater in cows than in heifers, possibly due to the mobilization of lead present in the bones (López-Alonso et al., 2000). The concentrations of this element are higher in the liver and kidneys of animals, possibly due to the diet received. When the pregnant female has a high concentration of Pb, there is a risk of abortion, fetal malformation, neurological damage, premature birth, and low birth weight and in males drop in sperm motility and pathologies reproductive (Costa et al., 2020).

For prophylaxis and as commented above, it is important to avoid leaving materials that contain Pb within the reach of animals, and not to use pastures close to industries that use this metal. Table 2 describes suggestions for the prevention and treatment of poisoning caused by Pb and other toxic and non-toxic heavy metals. The table also contains information about the main symptoms of poisoning and its acceptable limit in the diet and blood.

3.2 Non-toxic heavy metals

3.2.1 Copper (Cu)

Cu as well as Zn, are important components of some metalloenzymes (enzymes that contain metal as cofactors). Therefore, Cu is a metal of great importance in the functions of red blood cells, immunity, fertility, bones, aerobic respiration, and formation of collagen and elastin and is necessary to produce melanin pigment (Rotta et al., 2019). In addition, copper was listed as an essential nutrient in cattle nutrition in 1971 by the NRC, acting as an important component of several proteins, and cofactor of numerous enzymes, assisting in collagen formation, immune function, gene regulation, erythropoiesis, and iron metabolism (Weiss, 2017; NASEM, 2021). In fact, as early as in 1928, Cu was shown to be important for adequate growth and prevention of disorders in the animals (Underwood and Suttle, 1999). The

Table 2. Main features (acceptable upper limit in blood, maximum acceptable limit in food or diet, symptoms, prevention/treatment) for intoxication by heavy and toxic metals and non-toxic heavy metals.

		Heavy and toxic metals	
Metal	**Hg**	**Cd**	**Pb**
Upper limit acceptable in the blood	-	• 0,01 μg/ml	• 0.30–0.35 ppm
Maximum acceptable limit in food or diet	< 1 ppm Intoxication in cattle 10 mg/kg/d in organic form and 17.4 mg/kg/d in sheep	• Raw materials for animal feed of vegetable origin = 1 mg/kg • Cattle and Sheep: 0.1–0.2 mg/kg DM total diet; • ≥ 1.7 mg Cd/kg DM for sheep = intoxication.	• Raw materials for animal feed (general) = 10 mg/kg; Forage = 30 mg/kg; Yeasts = 5 mg/kg. • Maximum limit in mg/kg of feed for a moisture content of 12%.
Symptoms	• Inorganic Hg toxicity: vomiting, diarrhea, and colic. • Organic Hg toxicity causes: locomotor abnormalities, blindness, excitation, abnormal behavior and chewing, incoordination and seizures	• Loss of appetite; • Poor growth; • Anemia; • Abortion. • Renal failure.	• Accentuated weight loss; • Ataxia; • Salivation; • Eyelid spastic contraction; • Muscle tremors; • Convulsions; • Apathy; • Staggering walk; • Blindness; • Bruxism; • Ruminal atony; • Diarrhea.
Prevention or Treatment	• Chelation therapy with dimercaprol; • Check the amounts of Hg in the products; • Use sodium thiosulphate to contain intoxication; • Use astringents to cleanse the body and control gastroenteritis; • Lavage or using laxatives; • Intake of fluids for animal hydration.	• Avoid raising animals near industries.	• Tissue damage (e.g., nervous system), treatment may not be successful. • Intravenous or subcutaneous administration of Ca disodium ethylenediaminetetraacetic (Ca-EDTA) • Intravenous administration of chelating agents, thiamine, rumenotomy. • Prevent animals from reaching contaminating materials and contaminated areas.

References	Kahn and Line, 2008; Oberherr and Rossato, 2011.	Underwood and Suttle, 1999; Comissão Europeia, 2013; ELIKA, 2015; Kim et al., 2016.	Marçal et al., 1998; Marçal, 2005; Kahn and Line, 2008; Comissão Europeia, 2013.
		Non-toxic heavy metals	
Metal	**Cu**	**Cr**	**Zn**
Upper limit acceptable in the blood	• Toxic ~ 5–20 µg/mL; • Normal ~ 1 µg/mL; • Deficiency < 0.5 µg/mL.	-	• Calf ~ 13 µmol/L • Lamb ~ 18.3 µmol/L
Maximum acceptable limit in food or diet	~ 4–15 mg Cu/kg DM in feedstuffs.	Chromium propionate maximum rate = 0.5 mg Cr/kg of diet DM	500 Zn/kg diet DM.
Symptoms	• Gastroenteritis; • Dehydration; • Apathy and anorexia; • Dyspnea; • Ruminal stasis; • Staggering walk; • Diarrhea; • Pale mucous membranes; • Hemolysis; • Hepatic and Renal Failure.	• Kidney failure; • Skin blisters; • Anemia; • Hemolysis; • Tissue edema; • Liver dysfunction.	• Drop in milk production and food intake.
Prevention or Treatment	• Low treatment efficiency. • For the early stages of the disease: penicillin and Ca versenate; • Tetrathiomolybdate – prevent and treat chronic copper toxicity; • Avoid excessive grazing of legumes, animals staying in places sprayed with Cu-based herbicides, and avoiding mineral supplements that contain amounts of Cu above recommended.	-	-
References	Underwood and Suttle, 1999; Correa, 2001; Kahn and Line, 2008; NASEM, 2021.	NASEM, 2021.	Underwood and Suttle, 1999; NASEM, 2021.

DM = dry matter. Source: Authors (2023).

recommendation aims to circumvent the potential deficiency of this element in the bovine organism, which together with cobalt, represent the most common type of deficiency for microelements in ruminants in Brazil (Moraes et al., 1999). Of economic importance, Cu deficiency is usually involved with pathological conditions in bovines (Tokarnia et al., 2000).

Cu deficiency can be primary or secondary: insufficient intake of the levels required by the animal to maintain its normal metabolic processes, or inadequate absorption and utilization by tissues (even with adequate intake) due to the presence of antagonists in the diet (e.g., Mo and inorganic sulfates), which can interfere with Cu absorption (Mendez, 2001; Vásquez et al., 2001; Suttle, 2012; Clarkson et al., 2020). The reducing ruminal environment metabolizes Mo and sulfate to form thiomolybdate, which interacts with Cu and decreases its absorption by the intestine and its utilization by tissues (Correa, 2001; Clarkson et al., 2020). So, Cu deficiency and toxicity can be dependent on the concentration of Mo in the diet (Johnsen and Aaneby, 2019). In these cases, there is a decrease in the activity of superoxide dismutase (SOD), which is a cupro-enzyme that catalyzes the dismutation of superoxide radical ($O_2{\bullet}^-$) into molecular oxygen (O_2) and hydrogen peroxide (H_2O_2), the oxidant group that participates in neutrophil defense reactions (Babior et al., 1973). The enzyme lysyl oxidase is important for cross-linking connective tissue. With Cu deficiency, and the consequent absence of lysyl oxidase, there is a failure in collagen formation, resulting in structural rigidity and loss of collagen elasticity (Mcdowell, 1992). In addition, Cu deficiency can lead to low growth rate, increased prevalence of disease, reduced reproductive efficiency, diarrhea, anemia, osteoporosis, and reduced ability of phagocytic cells (NASEM, 2021).

The main symptoms of Cu deficiency are marked hemosiderosis in the spleen, lymph nodes, and in some cases deposits in the liver (Tokarnia et al., 2000; Oliveira et al., 2022). It can also cause apathy, anemia, thirst, anorexia, depression, staggering gait, osteoporosis, hemoglobinuria, pale mucous membranes, lateral recumbency, changes in hair pigmentation, lower animal performance (lower body growth and reproductive performance), diarrhea, and sudden death (Mendez, 2001; Oliveira et al., 2022). In sheep, Cu deficiency during gestation causes an ataxic disease in newborn lambs (Underwood and Suttle, 1999). Furthermore, in sheep, it can lead to brittleness, curling, and depigmentation, causing losses to the textile industry. Deficiency is a greater risk to grazing animals, due to the low availability of this element in pastures, particularly lacking in soils that are sandy, poor in organic matter, and eroded (Correa, 2001). However, to be sure of the deficiency it is necessary to diagnose via chemical dosages.

The research of Kincaid et al. (1986) pioneered the use of organic minerals in animal nutrition since organic Cu was less affected by antagonists than Cu sulfate. Additionally, the benefit of supplementation of organic sources of Cu (but also of Zn and Mn) during the transition period and at the beginning of lactation of dairy cows was also verified, modulating plasma markers against oxidative stress, and reducing the concentration of variables that represent the antioxidant status and lipid oxidation products. This was possible because cows that were fed organic mineral sources had a lower demand for antioxidant activity due to the lower plasma concentration of lipid oxidation products (less lipid mobilization, non-esterified fatty acids [NEFA], and β-hydroxybutyrate [BHBA]) (Yasui et al., 2014).

Poisoning occurs through the ingestion of food and mineral supplements with high levels of Cu (e.g., ~ 4–15 mg Cu/kg DM in feedstuffs), by grazing on legumes (such as white or red clover) and in areas that have received some chemical treatment based on Cu (Rosa and Gomes, 1982; Kahn and Line, 2008; Oliveira et al., 2022), which is characterized as primary chronic intoxication. Single doses of 20–110 mg of Cu per kg of live weight (LW) or daily intake of 3.5 mg of Cu/kg of LW produce intoxication in sheep (Radostits, 2007). Secondary chronic intoxication comes from the ingestion of foods with normal amounts of Cu, but the element accumulates in the liver due to the ingestion of vegetation with a low concentration of molybdenum and/or the ingestion of toxic plants, causing liver damage (Rodríguez-Marín et al., 2019).

Toxicity can occur accidentally, through an excessive amount of soluble Cu salts, present in anthelmintics, mineral mixtures, or poorly formulated feeds. Secondary poisoning can also occur after ingestion of plants, such as subterranean clover (*Trifolium subterranenum*), through mineral imbalance, retaining Cu. Similarly, prolonged ingestion of *Heliotropium europaeum* or *Senecio* spp., can cause chronic hepatic intoxication by Cu due to its retention (Kahn and Line, 2008).

However, severe intoxication is exposed when the maximum limit of Cu is accumulated in the liver. In addition, stress may result in liberation of large amounts of Cu into the bloodstream, causing a hemolytic crisis (anemia, hemolysis, and hemoglobinuria), possibly leading to the animal's death (NASEM, 2021). Until then, the animals remain apparently healthy, but when they get sick, they die quickly, due to the anemia caused (Mendez, 2001). Acute intoxications appear with the ingestion of 20–100 mg Cu/kg in sheep and 200–800 mg/kg in adult cattle. Chronic intoxications are derived from the ingestion of 3.5 mg of Cu/kg in pastures containing 15–20 ppm Cu (DM basis) and low levels of Mo (Kahn and Line, 2008).

The most affected species is sheep, because of the lower tolerance of this species, however the breed factor is subject to variation regarding the concentration of Cu in the liver of ewes, since the Texel breed is more vulnerable to toxicity when they receive too much Cu (Underwood and Suttle, 1999; Radostits, 2007). Excess Cu in a mineral mix caused the death of more than 15 adult sheep (Rosa and Gomes, 1982). Another study also showed the death of 146 cattle due to the excessive supply of Cu mixed with the offered mineral salt (Oliveira et al., 2022). However, to receive the correct diagnosis, it is necessary to determine the levels of Cu in the liver and/or kidneys, where levels above 500 ppm and 80 ppm, respectively, already characterize a condition of Cu intoxication (Mendez, 2001).

Regarding the Cu requirement of cattle and sheep, the use of 5 to 9 ppm of Cu in the dry matter of the diet is recommended, although tolerance levels differ between species. Beef cattle are more tolerant (100 ppm) than sheep (20–25 ppm) (Rosa and Gomes, 1982; Radostits, 2007). It is important to provide the animals with nutrients as required, avoiding the risk of nutritional deficiency (underfeeding) and possible intoxication (overfeeding), as this can generate accumulation in the body and cause lethal intoxication, especially for sheep (Babior et al., 1973; Pinheiro and Souza,

2017). Concentrates with levels of 15–20 mg/kg of Cu in confined sheep can lead to primary Cu intoxication, as sheep are very sensitive, due to their propensity to accumulate Cu in the body (Pereira and Rivero, 1993; Mendez, 2001). Advances in understanding the physiology of intracellular Cu transport from fundamental biology have not effectively entered the field of ruminant nutrition, leading to widespread misunderstanding and consequently widespread Cu imbalance in practice (Clarkson et al., 2020).

3.2.2 Chromium (Cr)

Cr is a micronutrient of growing interest to be used as a source of supplementation for the animal feed, as it presents benefits such as yield, growth rate, immune response and metabolic changes. Apparently, Cr can be critical in metabolic processes, for instance, it impacts carbohydrates, lipids, ammonia, and nucleic acids metabolism, which ultimately influence the quality of colostrum. Thus, Cr can have additional beneficial effects in pre-ruminant calves, for instance, potentiating insulin release and glucose uptake in the calf's first moments of life, and consequently reducing the risk of metabolic disorders, such as incidence and duration of diarrhea (Oliveira and Soares Filho, 2005; Bompadre et al., 2020; Khare et al., 2022). In feedstuff, values range from 0.01 to 4.2 mg Cr/kg DM, with cereals relatively poor and legumes relatively rich in chromium (Underwood and Suttle, 1999).

Cr supplementation facilitates glucose uptake in early life stages of calves and may have a beneficial effect on immunity in susceptible animals by improving the immune system (Lashkari et al., 2018). In addition, Cr is a component of a small peptide called chromodula, responsible for potentiating the action of insulin, but studies are needed to confirm its antioxidant activity and its safety as a supplement for ruminants (Lashkari et al., 2018; NASEM, 2021). Ruminants use acetate rather than glucose as a carbon source for lipogenesis and this may lead to a decreased insulin sensitivity. Supplementation of Cr in the diet of an adult sheep with 1 mg/kg DM (as amino acid chelate) increases the potential for glucose to be used for fat synthesis by 30%, by increasing the activity of ATP-citrate lyase, a marker of the glucose-insulin axis (Underwood and Suttle, 1999).

However, inorganic sources of Cr in ruminant diets are not recommended, despite being poorly absorbed, since the high level of these forms in the diet can lead to intoxication (Chang et al., 1995; Underwood and Suttle, 1999) with organic sources being the most recommended form.

Cr toxicity is rare, as even soluble sources such as chloride and chromate are tolerated at concentrations of > 1000 mg/kg DM by cattle (Underwood and Suttle, 1999). However, there are some cases reported in the literature as in the study by Kerr and Edwards (1981), the authors observed chromate toxicity effect on livestock from oil field residues. Research has shown that Cr picolinate toxicity is associated with kidney failure, skin blisters, anemia, hemolysis, tissue edema, liver dysfunction, neuronal damage, increased production of hydroxyl radicals, chromosomal aberrations, and DNA damage (Bagchi et al., 2002).

However, its requirements have not been quantified for cattle because concentrations are very low (μg/kg DM) in feeds. It is very difficult to check the concentration of Cr in food because the concentration is very low. Therefore, what is known for lactating dairy cows is that supplementation of 0.01 mg Cr/kg body weight (~ 6 mg/d) appears to increase milk production in early lactation and may promote benefits in the animal's immune function (NASEM, 2021). Nevertheless, supplementation with Cr is recommended in the following cases: animals in high production, under stress, a diet with low protein content and rich in silage, low fiber content (0.5 mg/kg of organic Cr source), three weeks before slaughter (from 0.2 to 0.3 mg/kg of organic source), in the case of early weaning, and before and after delivery (Chang and Mowat, 1992; Yang et al., 1996; Bin-Jumag et al., 2020).

3.2.3 Zinc (Zn)

Zn acts as a component of some metalloenzymes, such as superoxide dismutase, carbonic anhydrase, alcohol dehydrogenase, carboxypeptidase, alkaline phosphate, DNA and RNA polymerases (NRC, 2001). It is involved with the metabolism of carbohydrates, lipids, synthesis of proteins and nucleic acids, in addition to the use of vitamin A. It plays an important role in animal fertility, immune function, hormone, and gene regulation, in skin health, in addition to cell signaling, as it is the key component of more than 200 enzymes (Mir et al., 2020; NASEM, 2021). It acts as an enzymatic activator, mainly in the processes of bone formation, vision process, and reproductive systems, in addition to growth and other functions for the development of the animal (Kahn and Line, 2008; Rotta et al., 2019). Accordingly, Zn supplementation has been reported to ameliorate the growth of small ruminants (Angeles-Hernandez et al., 2021).

Livestock exhibit considerable tolerance to high intakes of Zn. However, it was observed that the growth rate of calves was reduced when the Zn concentration was greater than 700 mg/kg DM, suggesting that calves are less tolerant to high levels of this metal. Effects were also observed in lambs, with a decrease in weight gain and feed efficiency of the animals when fed diets containing 1 g of Zn/kg DM. In ruminants, high concentrations of Zn end up generating a toxic effect on ruminal microorganisms, reducing volatile fatty acids production, and the acetate: propionate ratio (Underwood and Suttle, 1999).

The main problems related to Zn deficiency are the decrease in animal consumption and their lower growth rate. In cases of prolonged deficiency, the hoof can be affected (pododermatitis), since Zn acts on the integrity of the hoof horn, accelerating wound healing, increasing the speed of epithelial tissue repair, and maintaining cellular integrity (Pardo et al., 2004). Therefore, Zn-based therapeutic supplements are currently used for the treatment and control of interdigital necrobacillosis and interdigital diseases in cattle. In addition, Zn deficiency was associated with parakeratosis in the skin of the legs, head, and neck of animals, resulting in brittle hair. An increased risk of mastitis and other infectious diseases has also been reported (NRC, 2001; NASEM, 2021).

The adequate concentration of Zn in the serum is between 0.7 and 1.3 μg/mL, and in the diet the minimum requirement is around 500 mg of Zn/kg of dry matter. Generally, animals are tolerant to high Zn concentrations, but clinical signs of toxicity may appear when the diet contains 900 mg of Zn/kg DM, with a drop in milk production and intake. For dry cows, 25 mg/kg of Zn is recommended and for lactating cows 60 mg/kg of Zn in total diet (NASEM, 2021). However, Zn absorption can be affected by its antagonists, such as Ca, Cu, and Fe. But its absorption is favored by Mn, phosphate, and vitamin D (Rosa and Gomes, 1982). In addition, to be absorbed, Zn (as well as Cu) must be in free or ionic form, that is, the metal must be released near the absorption site in the intestine so that it is captured by the enterocytes (NASEM, 2021).

4. Conclusion

This chapter can serve as a guide for new researchers working with the pharmacology and toxicology of essential and xenobiotic metals, in particular for those who work with production animals, which are used for human food (milk and meat).

Due to the high exposure to these substances, heavy metal intoxication can lead to the development of several serious diseases ranging from anemia, blindness, depression, paralysis, to death. This clinical condition can lead to a reduction in animal production and contribute to an increase in animals discarded from the system. Adequate treatment can save such animals and the implementation of appropriate measures can prevent the emergence of new cases.

As Cd, Hg, and Pb do not have essential physiological functions for living organisms, and are toxic, affecting the animal organism, they should be avoided. However, the effect of these metals may appear in the long term, even at low doses because they are reactive and bioaccumulative elements, that is, they are difficult to eliminate from the body, impairing metabolic functions. Furthermore, from the point of view of human health, the increase in the concentration of both toxic and essential metals in food derived from ruminants (for instance, milk and meat) must be avoided.

References

Angeles-Hernandez, J.C., Miranda, M., Munoz-Benitez, A.L., Vieyra-Alberto, R., Morales-Aguilar, N., Paz, E.A. et al. 2021. Zinc supplementation improves growth performance in small ruminants: A systematic review and meta-regression analysis. Anim. Prod. Sci. 61: 621–629.

Arshad, M.A., Ebeid, H.M. and Hassan, F.U. 2021. Revisiting the effects of different dietary sources of selenium on the health and performance of dairy animals: a review. Biol. Trace Elem. Res. 199: 3319–3337.

Ayar, A., Sert, D. and Akin, N. 2009. The trace metal levels in milk and dairy products consumed in middle Anatolia-Turkey. Environ. Monit. Assess. 152: 1–12.

Babior, B.M., Kipnes, R.S. and Curnutte, J.T. 1973. Biological defense mechanisms. The production by leukocytes of superoxide, a potential bacterial agent. J. Clin. Investig. 52: 741–744.

Bagchi, D., Stohs, S.J., Downs, B.W., Bagchi, M. and Preuss, H.G. 2002. Citotoxicidade e mecanismos oxidativos de diferentes formas de cromo. Toxicol. 180: 5–22.

Balali-Mood, M., Naseri, K., Tahergorabi, Z., Khazdair, M.R. and Sadeghi, M. 2021. Toxic mechanisms of five heavy metals: mercury, lead, chromium, cadmium, and arsenic. Front. Pharmacol. 12: 1–19.

Bidewell, C.A., Drew, J.R., Payne, J.H., Sayers, A.R., Higgins, R.J. and Livesey, C.T. 2012. Case study of copper poisoning in a British dairy herd. Vet. Rec. Case Rep. 1: e100267.

Bin-Jumah, M., Abd El-Hack, M.E., Abdelnour, S.A., Hendy, Y.A., Ghanem, H.A., Alsafy, S.A. et al. 2020. Potential use of chromium to combat thermal stress in animals: A review. Sci. Total Environ. 707: 1–11.

Bompadre, T.F.V., Moretti, D.B., Sakita, G.Z., Ieda, E.H., Martinez, M.I.V., Fernandes, E.A.N. et al. 2020. Long-term chromium picolinate supplementation improves colostrum profile of Santa Ines ewe. Biol. Trace Elem. Res. 193: 414–421.

Brondani, J.C. 2016. Composição fitoquímica, avaliação do perfil toxicológico e atividade protetora frente a intoxicação por mercúrio das folhas de *Dolichandra unguis-cati* L. MSc. Dissertation, Federal University of Santa Maria, Santa Maria, Rio Grande do Sul, BRA.

Carmona, A., Roudeau, S. and Ortega, R. 2021. Molecular mechanisms of environmental metal neurotoxicity: A focus on the interactions of metals with synapse structure and function. Toxics. 9: 1–15.

Castro-Bedrinana, J., Chirinos-Peinado, D., Ríos-Ríos, E., Machuca-Campuzano, M. and Gomez-Ventura, E. 2021. Dietary risk of milk contaminated with lead and cadmium in areas near mining-metallurgical industries in the Central Andes of Peru. Ecotoxicol. Environ. Saf. 200: 112382.

Chang, X. and Mowat, D.N. 1992. Supplemental chromium for stressed and growing feeder calves. J. Anim. Sci. 70: 559–565.

Chang, X., Mowatand, D.N. and Mallard, B.A. 1995. Supplemental chromium and niacin for stressed feeder calves. Can. J. Anim. Sci. 75: 351–358.

Chen, Z., Liang, Y., Lu, Q.Y., Nazar, M., Mao, Y., Aboragah, A. et al. 2021. Cadmium promotes apoptosis and inflammation via thecirc08409/miR-133a/TGFB2 axis in bovine mammary epithelial cells and mouse mammary gland. Ecotoxicol. Environ. Saf. 222: 112477.

Clarkson, A., Paine, S., Martín-Tereso, J. and Kendall, N. 2020. Copper physiology in ruminants: Trafficking of systemic copper, adaptations to variation in nutritional supply and thiomolybdate challenge. Nutr. Res. Rev. 33: 3–49.

Clarkson, T.W. and Magos, L. 2006. The toxicology of mercury and its chemical compounds. Crit. Rev. Toxicol. 36: 609–62.

Comissão Europeia. 2013. Regulamento nº 1275/2013 limites máximos de arsénio, cádmio, chumbo, nitrito, essência volátil de mostarda e impurezas botânicas prejudiciais. Jornal Oficial da União Europeia.

Correa, F.R. 2001. Deficiência de cobre. pp. 312–320. *In*: Correa, F.R., Schild, A.L., Mendez, M.D.C. and Lemos, R.A. (eds.). Doenças de ruminantes e equinos. Varela, São Paulo, SP, BRA.

Costa, A.G., Borges, A.M. and Soto-Blanco, B. 2020. Metais tóxicos e seus efeitos sobre a reprodução dos animais: Revisão. RBHSA. 14: 108–124.

ELIKA Fundación Vasca para la Seguridad Agroalimentaria. 2015. Sustancias indeseables | Alimentación animal. 2: 1–5.

European Commission. 2015. Regulation (EC) N° 1881/2006 as regards maximum levels of lead in certain food stuffs. Official Journal of the European Union.

Forsberg, C.W. 1978. Effects of heavy metals and other trace elements on the fermentative activity of the rumen microflora and growth of functionally important rumen bacteria. Can. J. Microbiol. 24: 298–306.

Friberg, L.T., Elinder, C.G., Kjellstromand, T. and Nordberg, G.F. 2019. Cadmium and Health: A Toxicological and Epidemiological Appraisal. CRC Press, Boca Raton.

Gava, A. 2001. Intoxicação por chumbo. pp. 181–186. *In*: Correa, F.R., Schild, A.L., Mendez, M.D.C. and Lemos, R.A. (eds.). Doenças de ruminantes e equinos. Varela, São Paulo, SP, BRA.

He, Z.L., Yang, X.E. and Stoffella, P.J. 2005. Trace elements in agroecosystems and impacts on the environment. J. Trace Elem. Med. Biol. 19: 125–140.

Herdt, T.H. and Hoff, B. 2011. The use of blood analysis to evaluate trace mineral status in ruminant livestock. Vet. Clin. North Am. Food Anim. Pract. 28: 255–83.

Hidiroglou, M. 1979. Trace element deficiencies and fertility in ruminants: a review. J. Dairy Sci. 62: 1195–1206.

Johnsen, I.V. and Aaneby, J. 2019. Soil intake in ruminants grazing on heavy-metal contaminated shooting ranges. Sci. Total Environ. 687: 41–49.

Kahn, C.M. and Line, S. 2008. Manual Merck de Veterinária. Roca, São Paulo.

Kerr, L.A. and Edwards, W.C. 1981. Chromate poisoning in livestock from oil field wastes. Vet. Hum. Toxicol. 23: 401–402.

Khare, S., Kumar, M., Kumar, V., Kushwaha, R., Vaswani, S., Kumar, A. et al. 2022. Dietary chromium picolinate supplementation improves glucose utilization in transition calf by ameliorating insulin response. Biol. Trace Elem. Res. 1–16.
Kim, D.G., Kim, M., Shin, J.Y. and Son, S.W. 2016. Cadmium and lead in animal tissue (muscle, liver and kidney), cow milk and dairy products in Korea. Food Addit. Contam. Part B Surveill. 9: 33–37.
Kincaid, R.L., Blauwiekel, R.M. and Cronrath, J.D. 1986. Supplementation of copper as copper sulfate or copper proteinate for growing calves fed forages containing molybdenum. J. Dairy Sci. 69: 160–163.
Kumar, S. and Sharma, A. 2019. Cadmium toxicity: effects on human reproduction and fertility. Rev. Environ. Health. 34: 327–338.
Lane, E.A., Canty, M.J. and More, S.J. 2015. Cadmium exposure and consequence for the health and productivity of farmed ruminants. Res. Vet. Sci. 101: 132–139.
Lashkari, S., Habibian, M. and Jensen, S.K. 2018. A review on the role of chromium supplementation in ruminant nutrition—effects on productive performance, blood metabolites, antioxidant status, and immunocompetence. Biol. Trace Elem. Res. 186: 305–321.
López Alonso, M., Benedito, J.L., Miranda, M., Castillo, C., Hernandez, J. and Shore, R.F. 2000. Arsenic, cadmium, lead, copper and zinc in cattle from Galicia, NW Spain. Sci. Total Environ. 246: 237–248.
López-Alonso, M. and Miranda, M. 2020. Copper supplementation, a challenge in cattle. Animals 10: 1–21.
Marçal, W.S., Campos Neto, O. and Nascimento, M.R. 1998. Valores sanguíneos de chumbo em bovinos Nelore suplementados com sal mineral naturalmente contaminado por chumbo. Cien. Rural. 28: 53–57.
Marçal, W.S. 2005. Intoxicação por chumbo em gado bovino em zona rural próxima a indústria metalífera. Vet. Notícias. 11: 87–93.
McDowell, L.R. 1992. Minerals in Animal and Human Nutrition. Academic Press, San Diego.
McLaughlin, M.J., Parker, D.R. and Clarke, J.M. 1999. Metals and micronutrients - Food safety issues. Field Crops Res. 60: 143–163.
Mendez, M.D.C. 2001. Intoxicação crônica por cobre. pp. 181–186. *In*: Correa, F.R., Schild, A.L., Mendez, M.D.C. and Lemos, R.A. (eds.). Doenças de ruminantes e equinos. Varela, São Paulo, SP, BRA.
Mir, S.H., Mani, V., Pal, R.P., Malik, T.A. and Sharma, H. 2020. Zinc in ruminants: Metabolism and homeostasis. Proc. Natl. Acad. Sci. India Sect. B Biol. Sci. 90: 9–19.
Moraes, S.S., Tokarnia, C.H. and Döbereiner, J. 1999. Deficiências e desequilíbrios de microelementos em bovinos e ovinos em algumas regiões do Brasil. Pesqui. Vet. Bras. 19: 19–33.
National Academies of Sciences, Engineering, and Medicine (NASEM). 2021. Nutrient Requirements of Dairy Cattle. National Academy Press, Washington, D.C.
National Research Council (NRC). 1945. Recommended Nutrient Allowances for Domestic Animals: Recommended Nutrient Allowances for Dairy Cattle. National Academy Press, Washington, D.C.
National Research Council (NRC). 2001. Nutrient Requirements of Dairy Cattle. National Academy Press, Washington, D.C.
Nogara, P.A., Oliveira, C.S., Schmitz, G.L., Piquini, P.C., Farina, M., Aschner, M. et al. 2019. Methylmercury's chemistry: From the environment to the mammalian brain. Biochim. Biophys. Acta. Gen. Subj. 1863: 1–23.
Oberherr, J. and Rossato, C.K. 2011. Intoxicação por mercúrio em animais domésticos: revisão bibliográfica. XVI Seminário Interinstitucional de Ensino, Pesquisa e Extensão. BRA.
Okada, I.A., Sakuma, A.M., Maio, F.D., Dovidauskas, S. and Zenebon, O. 1977. Avaliação dos níveis de chumbo e cádmio em leite em decorrência de contaminação ambiental na região do Vale do Paraíba, Sudeste do Brasil. Rev. Saúde Pública. 31: 140–143.
Oliveira, C.S., Piccoli, B.C., Aschner, M. and Rocha, J.B.T. 2017. Chemical speciation of selenium and mercury as determinant of their neurotoxicity. pp. 53–83. *In*: Aschner, M. and Costa, L. (eds.). Neurotoxicity of Metals. Advances in Neurobiology, Springer, Cham.
Oliveira, D.J. and Soares Filho, C.V. 2005. Suplementação com cromo para ruminantes. Arq. cienc. vet. zool. UNIPAR. 8: 71–77.
Oliveira, M.C., Barbosa, J.D., Oliveira, C.M.C., Malafaia, P., Pires, A.P.C. and Bomjardim, H.A. et al. 2022. Agentes tóxicos em ruminantes diagnosticados no Serviço de Anatomia Patológica da UFRRJ, Rio de Janeiro - Brasil. Acta Sci. Vet. 50: 1–8.

Olsson, I.M., Jonsson, S. and Oskarsson, A. 2001. Cadmium and zinc in kidney, liver, muscle and mammary tissue from dairy cows in conventional and organic farming. J. Environ. Monit. 3: 531–538.

Ortolani, E.L. 2002. Macro e microelementos. pp. 641–651. *In*: Spinosa, H.S., Górniak, S.L. and Bernardi, M.M. (eds.). Farmacologia aplicada à Medicina Veterinária. Guanabara Koogan, Rio de Janeiro, RJ, BRA.

Oviedo, C. and Rodríguez, J. 2003. EDTA: the chelating agent under environmental scrutiny. Quím. Nova. 26: 901–905.

Ozmen, O. and Mor, F. 2004. Acute lead intoxication in cattle housed in an old battery factory. Vet. Hum. Toxicol. 46: 255–256.

Papa, V., Wannenes, F., Crescioli, C., Caporossi, D., Lenzi, A., Migliaccio, S. et al. 2014. The environmental pollutant cadmium induces homeostasis alteration in muscle cells *in vitro*. J. Endocrinol. Investig. 37: 1073–1080.

Pardo, P.E., Bremer Neto, H., Chiacchio, S.B., Nagoshi, M. and Padilha, P.M. 2004. Determinação de zinco da sola do casco de bovinos leiteiros com ou sem lesões podais, suplementados ou não com levedura seca de cana-de-açúcar. Cienc. Rural. 34: 1501–1504.

Pereira, D. and Rivero, 1993. Intoxicação crônica fitógena por cobre. pp. 279–307. *In*: Correa, F.R., Méndez, M.C. and Schild, A.L. (eds.). Intoxicação por plantas e micotoxicoses em animais domésticos. Hemisfério Sul do Brasil, Pelotas, RS, BRA.

Pinheiro, M.O. and Souza, C.B. 2017. Efeitos teratogênicos dos metais pesados sobre a infertilidade humana e malformações congênitas. UNILUS Ens. Pesq. 14: 47–58.

Powell, G.W., Miller, W.J., Morton, J.D. and Clifton, C.M. 1964. Influence of dietary cadmium level and supplemental zinc on cadmium toxicity in the bovine. J. Nutr. 84: 205–213.

Radostits, O.M., Gay, C.C., Hinchcliff, K.W. and Constable, P.D. 2007. Veterinary Medicine: A Text Book of the Diseases of Cattle, Horses, Sheep, Pigs, and Goats. Elsevier Health Sciences, Saunders, USA.

Rahimzadeh, M.R., Rahimzadeh, M.R., Kazemi, S. and Moghadamnia, A. 2017. Cadmium toxicity and treatment: An update. Caspian J. Intern. Med. 8: 135–145.

Rocha, J.B.T., Saraiva, R.A., Garcia, S.C., Gravina, F.S. and Nogueira, C.W. 2012. Aminolevulinate dehydratase (ALA-D) as a marker protein of intoxication with metals and other pro-oxidant situations. Tox. Res. 1: 85–102.

Rodríguez-Marín, N., Hardisson, A., Gutiérrez, A.J., Luis-González, G., González-Weller, D., Rubio, C. et al. 2019. Toxic (Al, Cd, and Pb) and trace metal (B, Ba, Cu, Fe, Mn, Sr, and Zn) levels in tissues of slaughtered steers: risk assessment for the consumers. Environ. Sci. Pollut. Res. 26: 28787–28795.

Roegner, A.F., Giannitti, Woods, L.W., Mete, A. and Puschner, B. 2013. Public health implications of lead poisoning in backyard chickens and cattle: four cases. Vet. Med: Res. Rep. 4: 11–20.

Rosa, I.V. and Gomes, R.F. 1982. Intoxicação crônica por cobre em ovinos. EMBRAPA. Campo Grande, MS.

Rotta, P.P., Marcondes, M.I. and Pereira, B.M. 2019. Nutrição e manejo de vacas leiteiras. UFV, Viçosa.

Rowens, B., Guerrero-Betancourt, D., Gottlieb, C.A., Boyes, R.J. and Eichenhorn, M.S. 1991. Respiratory failure and death following acute inhalation of mercury vapor: a clinical and histologic perspective. Chest. 99: 185–190.

Salem, A.Z.M., Ammar, H., Lopez, S., Gohar, Y.M. and González, J.S. 2011.Sensitivity of ruminal bacteria isolates of sheep, cattle and buffalo to some heavy metals. Anim. Feed Sci. Technol. 163: 143–149.

Sharaf, S., Khan, M.U.R., Aslam, A., Rabbani, M., Sharf, A., Ijaz, M. et al. 2021. Toxico-pathological effects of heavy metals from industrial drainage wastewater on vital organs of small ruminants in Lahore. Environ. Sci. Pollut. Res. 28: 3533–3543.

Spears, J.W. 2003. Trace mineral bioavailability in ruminants. J. Nutr. 133: 1506–1509.

Spears, J.W. and Weiss, W.P. 2014. Mineral and vitamin nutrition in ruminants. Prof. Anim. Sci. 30: 180–191.

Stohs, S.J. and Bagchi, D. 1995. Oxidative mechanisms in the toxicity of metal ions. Free Radic. Biol. Med. 18: 321–336.

Su, C., Qu, X., Gao, Y., Zhou, X., Yang, X. and Zheng, N. 2022. Effects of heavy metal exposure from leather processing plants on serum oxidative stress and the milk fatty acid composition of dairy cows: a preliminary study. Animals 12: 1–10.

Suttle, N.F. and Jones, D.G. 1989. Recent developments in trace element metabolism and function: trace elements, disease resistance and immune responsiveness in ruminants. J. Nutr. 119: 1055–61.
Suttle, N.F. 2012. Copper imbalances in ruminants and humans unexpected common ground. Adv. Nutr. 3: 666–674.
Tchounwou, P.B., Yedjou, C.G., Patlolla, A.K. and Sutton, D.J. 2012. Heavy metal toxicity and the environment. *In*: Lunch, A. (ed.). Molecular, Clinical and Environmental Toxicology. Springer, Basel.
Timm, C.D. 2001.Carências minerais. pp. 301–309. *In*: Correa, F.R., Schild, A.L., Mendez, M.D.C. and Lemos, R.A. Doenças de ruminantes e equinos. Varela, São Paulo, SP, BRA.
Tokarnia, C.H., Döbereiner, J. and Peixoto, P.V. 2000. Deficiências minerais em animais de fazenda, principalmente bovinos em regime de campo. Pesq. Vet. Bras. 20: 127–138.
Tucci, F.M. 2017. Metais tóxicos: Controles oficiais. Ministério da Agricultura, Pecuária e Abastecimento. Brasília.
Underwood, E.J. and Suttle, N.F. 1999. The Mineral Nutrition of Livestock. CABI Digital Library, Wallingford, Oxon.
Vásquez, E.FA., Herrera, A.D.P.N. and Santiago, G.S. 2001. Interação cobre, molibdênio e enxofre em ruminantes. Cienc. Rural. 31: 1101–1106.
Waldron, M.R. 2007. Nutritional strategies to enhance immunity during the transition period of dairy cows. Fl. Rumin. Nutr. Symp. Gainesville, 11p.
Weiss, W.P.A. 2017. 100-Year Review: From ascorbic acid to zinc—Mineral and vitamin nutrition of dairy cows. J. Dairy Sci. 100: 10045–10060.
Yang, WZ., Mowatl, D.N., Subiyatnol, A. and Liptrap, R.M. 1996. Effects of chromium supplementation on early lactation performance of Holstein cows. Canadian J. Anim. Sci. 76: 221–230.
Yasui, T., Ryan, C.M., Gilbert, R.O., Perryman, K.R. and Overton, T.R. 2014. Effects of hydroxy trace minerals on oxidative metabolism, cytological endometritis, and performance of transition dairy cows. J. Dairy Sci. 97: 3728–3738.

CHAPTER 5

Mercury Toxicity

A Brief Overview

Juliane Hostert,[1,2,†] *Maria Eduarda A. Galiciolli,*[1,2,†] *Luíza Siqueira Lima,*[1,2,†] *Júlia Vicentin Souza*[1,2,†] and *Cláudia Sirlene de Oliveira*[1,2,*]

1. Introduction

Mercury is a heavy metal known for its toxicity. Inorganic mercury comprises mercurous or mercuric salts (Hg_2^{2+} and Hg^{2+}, respectively) and elemental mercury/vapor (Hg^0) (Mutter et al., 2010; Bernhoft, 2012; Oliveira et al., 2017a). Organic mercury, in turn, is mainly represented by species such as methylmercury ($MeHg^+$) and ethylmercury ($EtHg^+$) (Dórea, 2017; Oliveira et al., 2018; Nogara et al., 2019).

According to the World Health Organization (WHO), food is the main source of mercury for people not occupationally exposed, noting that the average intake of this element in the diet of the population of several countries is around 2 to 20 mg/day per person. Furthermore, according to the WHO, almost all the mercury present in drinking water is in the form of Hg^{2+}; its guidelines for the quality of drinking water, establishes a reference value of 6 µg/L Hg^{2+}. However, it is extremely important to point out that despite the existence of an established limit, exposure to any level of mercury should be a matter of concern, since it causes several toxic effects, mainly renal and neuronal, and has no biological function (WHO, 2017a; Oliveira et al., 2018; Nogara et al., 2019).

[1] Instituto de Pesquisa Pelé Pequeno Príncipe, Avenida Silva Jardim, 1632, Água Verde, 80.250-200, Curitiba, PR, Brazil.
[2] Faculdades Pequeno Príncipe, Curitiba, PR, Brazil.
† These authors equally contributed to this work.
* Corresponding author: claudia.bioquimica@yahoo.com.br

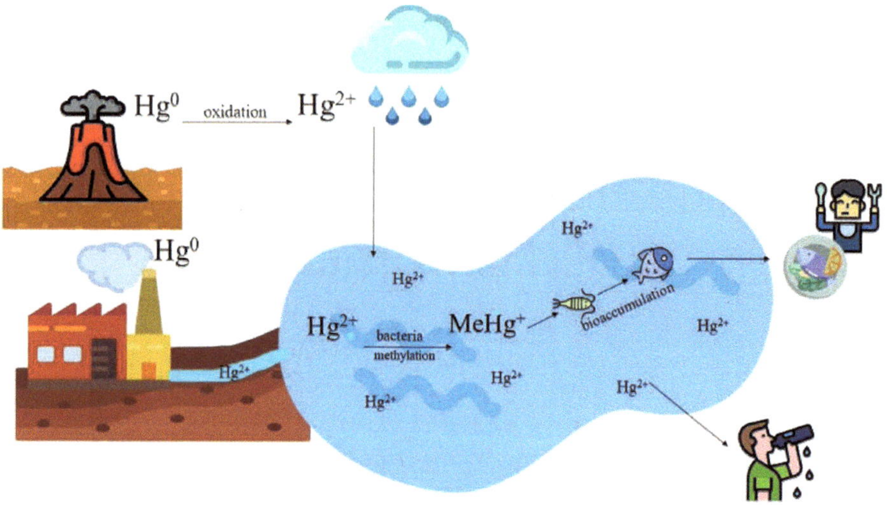

Figure 1. The cycle of mercury in the environment and sources of human contamination.

In the air, elemental mercury (in the form of vapor, Hg^0) is emitted by industries or by natural geomorphological processes, such as erosion, and transformed into oxidized mercury by biological or physicochemical processes. The formed Hg^{2+} can be transported by rain and deposited in soil and water. Depending on the level of Hg contamination of these environments, they can significantly contribute to human exposure to inorganic Hg. Furthermore, the methylation of Hg^{2+} in aquatic environments by the action of bacteria results in human exposure to organic mercury ($MeHg^+$) (Figure 1) (WHO, 1991; Clarkson, 1993; Branco et al., 2017; Farina and Aschner, 2017; Oliveira et al., 2018).

Therefore, $MeHg^+$ is mainly found as a pollutant in the aquatic environment, so the main route of human exposure to $MeHg^+$ is the ingestion of contaminated fish. $MeHg^+$ is bioaccumulated in high concentrations in predatory fish, mollusks, and marine mammals. Thus, people whose diet consists mainly of fish and shellfish may be exposed to high levels of $MeHg^+$ (Fujimura and Usuki, 2015; Antunes et al., 2016; Oliveira et al., 2018; Nogara et al., 2019; de Leeuw et al., 2022). In relation to $EtHg^+$, the human-relevant source of contamination is in the form of thimerosal (ethylmercury thiosalicylate) and can be found in topical antiseptics and as a preservative in multidose vaccines. Both $MeHg^+$ and $EtHg^+$ are of concern for human health, as they are potent neurotoxic agents (WHO, 2017a; Dórea, 2017; Oliveira et al., 2018).

Thus, this chapter will seek to bring an overview of mercury and its involvement with human health and current disasters, addressing mainly $MeHg^+$, the most environmentally inimical organic species for human exposure.

2. Inorganic mercury

In Ancient China, one of the first records of inorganic mercury use was in production of a red dye. Later, cinnabar (mercuric sulfide) has been used in traditional Chinese medicine (Liu et al., 2008; Zhao et al., 2022). In western countries, inorganic mercury was most used in dental amalgams (elemental mercury) (Mackert et al., 1991; Mitchell et al., 2005) and in calomel (Hg_2Cl_2)-containing products, which were used to treat several illnesses, namely, syphilis, seborrheic eczema, psoriasis, childhood teething and constipation, among others (Cornbleet et al., 1939; Fleming and Wolf, 1946; Davis, 2000). The use of mercury for these purposes was gradually disappearing, as the toxicological effects of mercury came to light, as well as the cases of toxicity (Holzel and James, 1952; Mutter et al., 2010; Park and Zheng, 2012). Inorganic mercury is still used in the formulation of whitening skin cosmetics and germicidal solutions. Several products contain inorganic mercury, such as thermometers, batteries, and lamps. Moreover, of great concern is the use of inorganic mercury in artisanal and small-scale mining (Oliveira et al., 2017a, 2018; Esdaile and Chalker, 2018).

Despite all the inorganic mercury salts being water soluble, mercuric (Hg^{2+}) compounds tends to be more toxic than mercurous Hg^+ compounds since their water solubility is higher. Thus, oxidized mercury ions are more corrosive and can cause more acute effects, especially when ingested orally, than elemental mercury (Von Burg, 1995; Langford and Ferner, 1999; Park and Zheng, 2012; Oliveira et al., 2017a).

According to the WHO, human Hg exposure occurs mainly through diet since food can be contaminated when there is mercury concentration in water and soil (WHO, 1991). In addition, some skin care (skin lighteners and anti-aging treatments) products have inorganic mercury in their formulation, increasing human exposure to this toxicant (Sun et al., 2022; Gao et al., 2022). Inorganic mercury's behavior in humans is not fully understood, i.e., the primary toxic target is still unclear (Oliveira et al., 2020). It is known that elemental mercury is frequently inhaled, causing lung toxicity and neurotoxicity (Rowens et al., 1991; Lorscheider et al., 1995; Fields et al., 2017; Oliveira et al., 2017a). Hg^0 can be oxidized intracellularly to Hg^{2+}, which will deposit, mainly, in the kidneys (Lorscheider et al., 1995; Morgan et al., 2002; Oliveira et al., 2017a). On the other hand, oxidized inorganic mercury, when ingested in diet, is poorly absorbed in the gastrointestinal tract (Lorscheider et al., 1995). Inorganic mercury excretion occurs mostly through urine and feces (Kazantzis, 1970; Aschner and Aschner, 1990; Syversen and Kaur, 2012) (Figure 2).

Although poorly absorbed in the gastrointestinal tract, rats exposed to Hg^{2+} in drinking water presented an increase in the levels of metallothionein in the kidneys and liver and in the renal weight and levels of Hg, as well as a decreased intake of water and food (Galiciolli et al., 2022) and increase in the levels of Hg in the feces (Oliveira et al., 2016). Once in the body, the oxidized mercury will bind to thiol- or selenol-containing proteins and will be distributed to the organs, mainly to the kidney (Aschner and Aschner, 1989; Azevedo et al., 2012; Dias et al., 2016; Oliveira et al., 2017a, 2019). Rodents exposed to inorganic mercury have several signs of nephrotoxicity, such as an increase of serum urea and creatinine levels, proteinuria, an increase of renal metallothionein, accumulation of Hg in the kidneys, and several histopathological alterations (Favero et al., 2014; Oliveira et al., 2015; Mesquita et al., 2016; Oliveira et al., 2017, 2020; Moraes-Silva et al., 2018; Galiciolli et al., 2022).

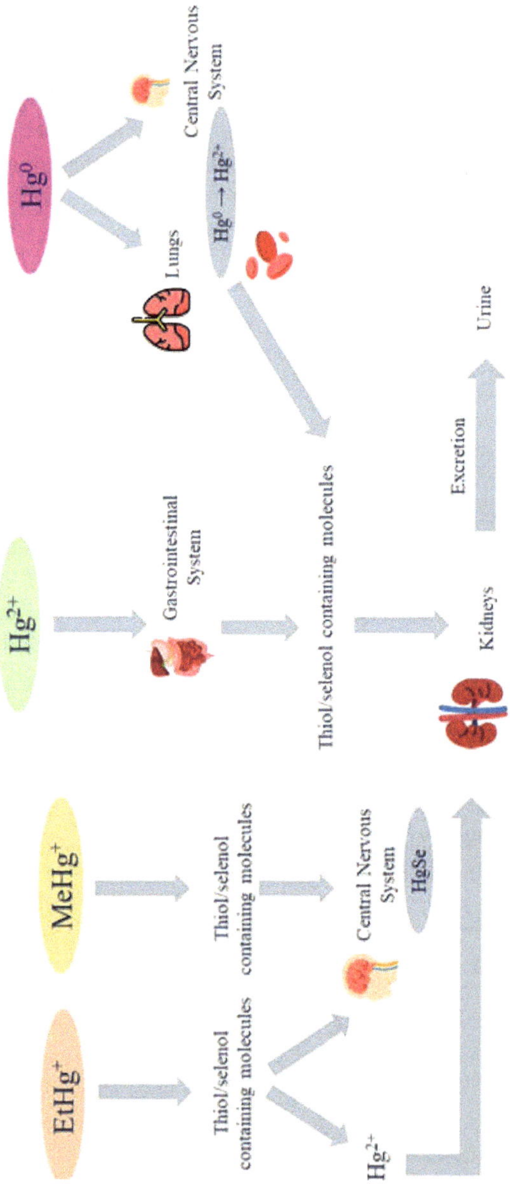

Figure 2. The entry of mercury into the human body and its absorption, distribution, and excretion.

Oxidized inorganic mercury has low lipophilicity, i.e., low ability to cross organic barriers; however, studies with animals in development demonstrated that exposure to inorganic mercury alters the pup's behavior and the cerebral activity of acetylcholinesterase (Franciscato et al., 2009; Moraes-Silva et al., 2014). In addition, Oliveira et al. (2015) observed inorganic mercury accumulation in the brain of pups exposed to Hg^{2+} *in utero*. Moreover, Teixeira et al. (2018) observed that adult rats chronically exposed to inorganic mercury had several alterations in the motor cortex. The neurotoxicity of Hg^{2+} can be mediated by glutamate, the most abundant amino acid in the central nervous system, which acts as an excitatory neurotransmitter. Inorganic mercury can inhibit the uptake and stimulate the release of glutamate, which considerably increases extracellular glutamate levels leading to Ca^{2+} influx in the cytosol, an increase of free radicals, and consequent oxidative stress (Albrecht and Matyja, 1996; Meador-Woodruff and Healy, 2000; Faro et al., 2001; Valli, 2014).

Selenium plays a role in protecting against Hg toxicity (Oliveira et al., 2017a). Thus, studies have evaluated this mechanism *in vitro* (Hossain et al., 2021) and *in vivo* (Orr et al., 2020) and have found that Hg-Se complexes formed in the body have few toxic effects (Oliveira et al., 2017a). Furthermore, the Hg-Se complex has been observed in large marine animals from Hg-contaminated areas (Ikemoto et al., 2004; López-Berenguer et al., 2020). However, this high affinity of mercurial ions for this essential trace element can disturb the redox balance, i.e., trapping all selenium available to selenoprotein synthesis (relevant antioxidant enzymes) (Walach et al., 2015; Oliveira et al., 2017a).

The clinical features of inorganic mercury exposure vary according to the route and intensity of exposure. As mercury salts can be deposited in all tissues, clinical symptoms can vary considerably (Cappelletti et al., 2019). As previously mentioned, the kidneys are the organs most affected by this metal, so in acute intoxication, the amount of mercury excreted through the urine is directly linked to the severity of the kidney injury. In fact, Hg^{2+} has a renal tropism which can lead to acute tubular necrosis (Chan, 2011). It has been estimated that the average lethal dose ranges from 1 to 4 g of mercuric chloride in human adults. Oral exposure to high doses of Hg^{2+} usually was associated with gastrointestinal damage, which was followed by renal failure (Park and Zheng, 2012; Cappelletti et al., 2019). The main symptoms of subchronic and chronic inorganic mercury (Hg^{2+}) exposure are nephrotic syndrome, fever, and gastrointestinal alterations, such as a metallic taste in the mouth, gum inflammation, excess saliva production, diarrhea, and nausea (Rosenman et al., 1986; Chan, 2011). Symptoms also include excessive urination and proteinuria, which may progress to the hematuria (Park and Zheng, 2012). Despite the low permeability of the blood-brain barrier to Hg^{2+}, some symptoms are attributed to the nervous system toxicity of Hg^{2+}, such as tremors, insomnia, memory loss, and depression (Rosenman et al., 1986). Harada et al. (1999) found polyneuropathy, neurasthenia, and tremor in gold miners diagnosed with mild inorganic mercury poisoning.

3. Organic mercury

The world's population has trace amounts of mercury in body tissues, the amount of organic mercury, especially $MeHg^{+}$, being of concern to human health (WHO,

2017b). Another known organic mercury molecule is $EtHg^{+}$, which is found as a component of thimerosal (ethylmercury thiosalicylate) and is marketed as a topical antiseptic and preservative in multidose vaccines (Dórea, 2017; Oliveira et al., 2018).

$MeHg^{+}$ is found mainly as a pollutant in the aquatic environment (Fujimura and Usuki, 2015). The methylation of Hg^{2+} by methanogenic bacteria leads to its release into water and bioaccumulation in the food chain. Thus, the main route of human exposure to $MeHg^{+}$ is the intake of contaminated fish (Figure 1). $MeHg^{+}$ is bioaccumulated in high concentrations in predatory fish, shellfish, and marine mammals. Consequently, people whose diet consists primarily of fish and shellfish may be exposed to high levels of $MeHg^{+}$ (Antunes et al., 2016; Nogara et al., 2019; de Leeuw et al., 2022). The bioaccumulation of mercury in marine fish is highly variable and its concentration is affected by species-specific differences in physiological and ecological characteristics (Taylor et al., 2014; Sinkus et al., 2016; Nogara et al., 2019). Moreover, the half-life of mercury in the human body takes around 74 days, the exposure to contaminated shellfish and fish, even in low amounts, can cause an accumulation of mercury in the body and affect human health (Björkman et al., 2007).

Due to $MeHg^{+}$ liposolubility, it is rapidly absorbed in the gastrointestinal tract (~ 99%) and redistributed vascularly to other parts of the body, mainly the kidneys, liver, and brain (Figure 2) (Madejczyk et al., 2007). Besides, it easily crosses the placental and blood-brain barriers (Lorscheider et al., 1995; Fujimura and Usuki, 2015; Oliveira et al., 2017a,b, 2018; Nogara et al., 2019).

The electrophilic properties of the organic mercury molecules mediate their toxicity. Kanai et al. (1998) report that interactions of $MeHg^{+}$ or Hg^{2+} with thiols dictate their fate and toxicity. $MeHg^{+}$ is a soft electrophile and preferentially reacts with soft nucleophiles (Nogara et al., 2019), which justifies the difficulty to detect free $MeHg^{+}$ in human cells (Onyido et al., 2004). In the body, $MeHg^{+}$ reacts with selenol and thiol groups, the two types of soft nucleophiles found in mammal proteins (Aschner et al., 2000; Oliveira et al., 2017a; Nogara et al., 2019). The interaction between $MeHg^{+}$ and thiol or selenol groups forms stable complexes, depleting important antioxidant molecules, such as reduced glutathione. Moreover, $MeHg^{+}$ can cause the inhibition of several enzymes, namely, thioredoxin reductase, glutathione peroxidase, and porphobilinogen synthase, among others (Rocha et al., 1993; Wagner et al., 2010; Branco et al., 2011; Meinerz et al., 2017; Nogara et al., 2019), impairing their catalytic function. Shreds of evidence indicate that oxidative stress is one of the factors responsible for the neurotoxic effects of $MeHg^{+}$; however, the molecular mechanisms that regulate the neurotoxicity generated by $MeHg^{+}$ are not completely understood (Farina et al., 2011; Oliveira et al., 2020).

Since conjugates of $MeHg^{+}$ with thiol compounds are easily formed *in vivo*, the metabolism and transport of glutathione, cysteine, and their derivatives are determinants of tissue distribution and elimination of $MeHg^{+}$ (Antunes et al., 2016; Oliveira et al., 2017a; Nogara et al., 2019; Bjørklund et al., 2022). When ingested, $MeHg^{+}$ readily binds to cysteine, forming the methylmercury-cysteinyl complex (MeHg-Cys) (Nogara et al., 2019). This complex is recognized by some amino acid transporters, such as the neutral amino acid transporter LAT1, system B(0,+), which allows its transport across the plasma membrane as a mimic of the amino

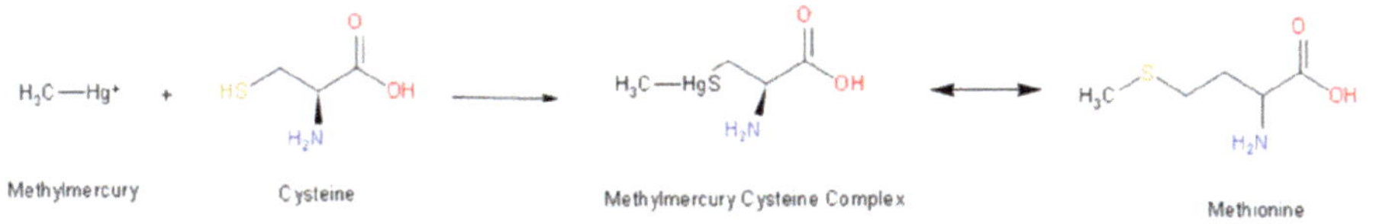

Figure 3. Formation of the cysteine-methylmercury complex and the similarity between methionine and the methylmercury-cysteine-complex.

acid methionine (Figure 3) (Bridges and Zalups, 2005, 2006a,b; Bridges et al., 2004; Ganapathy et al., 2021). The mobility of the MeHg-Cys complex allows it to cross not only the blood-brain barrier but also the placental barrier (Bridges and Zalups, 2005, 2006a,b). The chemistry of $MeHg^+$ is also dictated by exchange reactions (Rabensten's reactions) among a $MeHg^+$ bound with a thiol or selenol group with a free thiol or selenol group, as elegantly reviewed by Nogara et al. (2019).

3.1 Organic mercury and the central nervous system

The major toxicological target of $MeHg^+$ is the central nervous system (Ni et al., 2012; Oliveira et al., 2017a, 2018; Nogara et al., 2019). In adults, in the case of the mature central nervous system, before the onset of symptoms, the toxic effects of $MeHg^+$ are characterized by a long latency period (Clarkson et al., 2003). Initial symptoms include weight loss, blurred vision, and paresthesia of the circumoral area, feet, and hands, followed by visual field constriction, psychiatric symptomatology, and ataxia (Clarkson et al., 2003; Bakir et al., 1973). Furthermore, in adults, symptoms and signs of $MeHg^+$ poisoning are associated with the loss of neuronal cells in specific regions of the brain, such as the cerebellum and visual cortex (Aschner and Syversen, 2005). A significant loss of cerebellar granule cells with considerable preservation of the adjacent layer of Purkinje cells was found in *post-mortem* histopathological analysis of individuals chronically exposed to high levels of $MeHg^+$ in the Minamata outbreak (Takeuchi, 1982).

Furthermore, the developing brain is highly sensitive to $MeHg^+$ poisoning (Oliveira et al., 2018). Indeed, infants exposed to $MeHg^+$ may exhibit impaired movement, decreased IQ, disruption of visual-spatial perception and speech, and medical symptomatology known as fetal Minamata disease (Harada, 1995; Grandjean et al., 1997; Castoldi et al., 2008). In animal studies, the developmental effects of exposure to $MeHg^+$ include, among others, memory impairment (Sakamoto et al., 2002; Dare et al., 2003; Carratu et al., 2006), reduced motor activity (Bjorklund et al., 2007) and learning impairments (Paletz et al., 2006).

4. Cases of Hg contamination

Cases of poisoning resulting from exposure to mercury have occurred for decades, the most notable being that of Minamata, Japan (Yorifuji and Tsuda, 2014), as well as others that worry the world, such as cases in Iraq (Clarkson et al., 1976; Skerfving and Copplestone, 1976), and, recently, in Brazil (Lima, 2023). Mercury is a widespread environmental contaminant, which causes public health problems, with Minamata disease being the most remarkable case of contamination by this metal. The outbreak of Minamata Bay contamination with mercury was the first example of how the industrial exploitation of toxic agents can have neurodevelopmental and long-lasting brain toxicological effects. In fact, the recurrent cases of human mercury intoxication culminated, in 2017, with the creation of the Minamata Convention (Bank, 2020). The main objective of the Minamata Convention is to reduce the emission of mercury in the environment, both in the air and in the water, being a global effort to minimize the damage caused by mercury (Wang et al., 2019). One hundred and forty countries currently adhere to the convention, joining efforts to reduce the emission of mercury, among them Brazil, the United States, Canada, France, Italy, China, and Japan (UN, 2023). Even with worldwide efforts to avoid damage from mercury exposure, some cases still take place in great proportions, for instance, the recent contamination of indigenous Yanomami in Brazil due to illegal mining practiced on their lands (Ramos et al., 2020).

4.1 The Minamata disease

The Minamata disaster is the most striking and famous case related to $MeHg^+$ contamination in food, which occurred in Japan. The contaminant, mercury, a by-product of the production of acetaldehyde, was deposited in the environment by the Chisso factory between 1932 to 1968. The contamination of the Minamata Bay waters with mercury resulted in the bioaccumulation of $MeHg^+$ in the marine fauna, and through the trophic chain, reached the human population (Yorifuji and Tsuda, 2014; Silva et al., 2017). The first human case of intoxication was reported in 1956 (Eto et al., 2010; Yorifuji and Tsuda, 2014). A child was admitted to a hospital with unusual neurological signs, presenting with seizures, speech, and walking difficulties (Hachiya, 2006).

The $MeHg^+$ poisoning was designated as Minamata disease, established by daily consumption of large amounts of mercury. The disease presents clinical symptoms that vary from the $MeHg^+$ exposure level and individual susceptibility (Hachiya, 2006; Eto et al., 2010; Silva et al., 2017). The Minamata population, especially individuals exposed to $MeHg^+$ *in utero*, had several irreversible neurological symptoms, such as paresthesia, ataxia, and visual and auditory alterations. Psychiatric symptoms, including cognitive and mood changes, behavioral dysfunction, neurological signs, and motor disturbances, were also observed in the Minamata disease (Yorifuji et al., 2011; Yorifuji and Tsuda, 2014). Changes in other organs and tissues were generally light and included erosive lesions inflammation of the digestive tract, hypoplasia of the bone marrow, and changes in the liver, kidney, and lymph nodes, in addition to the change in pancreatic islet cells (Eto et al., 2010).

4.2 Methylmercury poisoning outbreak in Iraq

In the 1970s in Iraq, there were cases of intoxication resulting from homemade bread made with wheat treated with $MeHg^+$. This wheat was intended exclusively for planting; however, an unknown amount was used to produce bread. Around

459 people died and approximately 6,900 were admitted to the hospital (Clarkson et al., 1976; Skerfving and Copplestone, 1976).

At the beginning of the intoxication cases, little was known about the wheat with $MeHg^+$ used in bread production. Over time, the sample of this wheat was analyzed through the techniques of selective atomic absorption spectrometry and gas chromatography. About 2 to 6 mg $MeHg^+$/kg were found in wheat samples, and about 5 to 10 mg $MeHg^+$/kg in bread made with this wheat. It should be noted that these levels did not decrease during bread baking (Clarkson et al., 1976; Skerfving and Copplestone, 1976).

Signs and symptoms presented by intoxicated individuals mainly included sensory, motor, and visual disturbances, showing the central nervous system as the main site of action of $MeHg^+$. In some cases, individuals also had gastrointestinal symptoms. Furthermore, there were cases of prenatal intoxication whose diagnosis was difficult since, in mild cases, the signs and symptoms only manifested sometime after birth. Cases have been reported of children born blind and deaf and even with severe motor dysfunctions (Clarkson et al., 1976; Skerfving and Copplestone, 1976).

4.3 Yanomami contamination and illegal artisanal and small-scale mining

In Brazil, there were several cases, of mercury contamination in soil and water by mining activity, such as the contamination of the Madeira River, which crosses the states of Roraima and Amazonas, during the 1970s and 1980s (Bastos, 2004). It is estimated that between 1550 and 1880, about 200,000 tons of mercury were released into the environment in South America (Malm, 1998). Barbosa et al. (1995) and Malm (1998) demonstrated an increase in Hg levels in the trophic chain of the Madeira, Tapajós, and Negro rivers. Another case of environmental contamination was the disaster of the rupture of the Fundão barrage, in Mariana - Minas Gerais, in 2015, releasing more than 50 million m^3 of mud, rich in toxic metals, including mercury (de Faria, 2019; Paulelli et al., 2022). The most recent case of contamination in Brazil was of the Yanomami group (Lima, 2023).

The Yanomami are a group of indigenous people that inhabit the largest indigenous territory covered by forest and carry out hunting and agriculture activities. They occupy a territory that extends from the Guyana massif, on both sides of the border between Brazil (Basins of the Upper Rio Branco and left bank of the Rio Negro) and Venezuela (Basins of the Upper Orinoco and Cassiquiare), within the Amazon rainforest, with a population of approximately 28,000 individuals, covering 9.6 million hectares (Orellana et al., 2019; Tupy et al., 2022).

The defense of indigenous peoples and their territories is ensured in Brazil by the Fundação Nacional dos Povos Indígenas (FUNAI), protecting and ensuring their rights, promoting identification and delimitation studies, demarcation, land regularization and registration of lands traditionally occupied by indigenous people, in addition to monitoring and supervising (Brasil, 2020). Areas such as the Uraricora River, demarcated by the Brazilian government in 1992 as the Yanomani Indigenous Land, suffer from the consequences of illegal artisanal and small-scale mining, with worrying levels of mercury contamination (Ramos et al., 2020).

The use of mercury is allowed if it is authorized by the Agência Nacional de Mineração (ANM) and regulated and monitored by the Instituto Brasileiro do Meio Ambiente e dos Recursos Naturais Renováveis (IBAMA); however, its use in indigenous lands is illegal, due to the right guaranteed by the Federal Constitution, the non-use of these by other people, as well as in environmental conservation units (Marques et al., 2022; Alves, 2023). Activities such as deforestation and illegal mining carried out in the Amazon Forest are responsible for increasing the concentration of mercury in the soil and water, factors that expose riverside and indigenous populations to high levels of mercury (Ramos et al., 2020). Due to the lack of authorization, the amount of metal used is not supervised, with high amounts being deposited in the environment that impacts the health, environment, sociocultural, and economy of the riverside communities (Alves, 2023). In addition to mercury contamination from mining, the Amazonian soils have the highest concentrations of natural mercury in the world (Ramos et al., 2020), thus, when the soil is deforested by logging companies, agriculture, livestock, and mining activities, erosion occurs, leading to the deposition of the metal in the aquatic environment. This is methylated and incorporated into the trophic network, leading to bioaccumulation so that organic mercury reaches humans (Bastos et al., 2015; Ramos et al., 2020; Queiroz et al., 2022; Tupy et al., 2022).

In 2017, Brazil joined the Minamata Convention, which aims to reduce and control the use of mercury, to avoid damage to human health and the environment, rights which are also guaranteed by the Federal Constitution of 1988 (Brasil, 1988). The 169 Convention of the International Labor Organization – ILO, was enacted through Decree No. 5,051, of April 19, 2004 (Brasil, 2004), recognizing the rights of indigenous people to assume control over the use of their lands, with the State having to assure the conservation of resources. Article 18 of this Convention further provides that sanctions must be applied against the unauthorized intrusion of indigenous lands and any unauthorized use by non-indigenous individuals (Ramos et al., 2020; Tupy et al., 2022). The ores present on Brazilian soil are the property of the Union, so for their use, there must be free, prior, and informed consultation with the people to determine whether there is an authorization for the exploration of resources present in the territory, subject to compensation for damages that may occur due to the extraction (Ramos et al., 2020). Between 2018 and 2019, Roraima exported 194 kg of gold to India, even though there are no legally registered mines (Tupy et al., 2022).

In 2020, the Yanomami lost about 414 hectares of their territory to mining (Tupy et al., 2022). Also reporting that complaints about illegal exploitation were already being made since 2018 to various spheres of Public Power, which were ignored by the rulers, further aggravating the situation of the Yanomami population. Furthermore, in 2021, a Yanomami leader complained about mercury contamination in rivers that flow through indigenous lands, reporting health problems resulting from exposure to the metal, causing deformities in children and women (Lima, 2021). In fact, for some years, there have been complaints about the illegal exploitation of gold in indigenous territories, which have not been investigated due to the negligence of the government in question, which took media proportions in 2023 (Tupy et al., 2022).

In January 2023, it was declared a state of health emergency regarding the situation of the Yanomami in Brazil due to the illegal mining activity, which has brought significant damage to the health of the inhabitants (Lima, 2023). In artisanal and small-scale mining practices, mercury is used to extract gold, separating it from other sediments as it adheres to the gold, culminating in an amalgam. Then, this amalgam is heated to separate the gold from the Hg so that, with heating, the Hg evaporates into the atmosphere. Vapors from burning amalgam can be inhaled by humans. In addition, usually, the Hg that does not amalgamate with gold is dumped into rivers and is transformed by microorganisms into $MeHg^{+}$, which is incorporated by aquatic organisms, mainly fish, and transmitted to humans through the trophic chain (Bastos et al., 2015; Oliveira et al., 2017a; Esdaile and Chalker, 2018; Bakker, 2021; Alves, 2023).

Although the state of emergency has now been declared, the exposure of the Yanomami indigenous tribes to mercury has been going on for many years. In 2014, a study carried out by the National School of Public Health (Ensp/Fiocruz) in partnership with the Socioenvironmental Institute (ISA) analyzed 239 hair samples from 19 indigenous villages in the region where the Yanomami and Ye'kwana ethnic groups reside. Samples were collected from indigenous people belonging to groups more vulnerable to mercury contamination, i.e., women of childbearing age, children, and adults with direct contact with the mining. The result obtained by analyzing the hair samples revealed that in the Yanomami community of the Waikás region, 92% of the samples had a high level of Hg contamination (Vega et al., 2018). This community has a mine close to its location (ISA, 2016).

Simultaneously to mercury exposure, malnutrition in Yanomami children is another important factor characterizing the state of health emergency declared in 2023 for this community, currently considered one of the most serious in the world (Carvalho, 2023). Since malnutrition in developing children can also disrupt the development of the brain, the association of mercury exposure with malnutrition can have synergistic and unpredictable long-term effects on cognition and other brain-related functions. In the report produced by the Hutukara Associação Yanomami and the Associação Wanasseduume Ye'Kwana published in 2022, several factors are highlighted that demonstrate the association of illegal mining with malnutrition. Among these factors are the contamination of water and fish with Hg and the reduced availability of land for harvesting due to environmental destruction associated with diseases transmitted by miners' activity (for instance, such as malaria). The entire catastrophic situation prevented the indigenous community from taking proper care of its children (Ye'Kwan, 2022). As mentioned above, malnutrition together with the poor social care may exacerbate the long-term MeHg toxic effects in the Yanomami children.

Due to the lack of public policies related to the inspection of illegal gold mining, the socioeconomic damage to the Yanomami population has become increasingly worrying. Thus, studies have been carried out to find solutions and estimate values for repairing the damage caused to this ethnic group (Bakker et al., 2021; Queiroz et al., 2022). Bakker et al. (2021) pointed out that the standardization of the assessment of mercury impacts is essential to containing illegal gold mining in the Brazilian

Amazon. Furthermore, these same authors showed that the economic impact of illegal mining on health can reach about 69 million dollars only in relation to the release of Hg in the environment in the year 2020. Also, the Urihi Associação Yanomami, responsible for representing around 150 Yanomami communities, estimated around R$6.6 billion for repairing social, moral, and environmental damage in Yanomami indigenous lands (Ribeiro, 2023).

5. Conclusion and remarks

Here we presented a brief overview of mercury toxicity, mainly pointing to mercury interaction with endogenous thiol and selenol groups. Since the Minamata outbreak, the mechanism of mercury toxicity has been extensively explored by researchers, but we are still far from figuring out primary toxicity target of mercury. In addition, we dedicated half of this chapter to presenting some cases of populations suffering from the toxic effects of mercury. Although two of the cohorts with mercury toxicity (the Minamata and Iraq outbreaks) are widely explored in the scientific literature, the last one, the intoxication of a Brazilian indigenous cohort was recently discovered. Based on that, gaps in the understanding of mercury's mechanism of toxicity and the recurrence of cases of populations contaminated with mercury, it is necessary for several areas of research (chemistry, biology, epidemiology, and medicine, among others) to collaborate in the understanding this environmental contaminant's behavior in the human body. In addition, areas such as biotechnology will be pivotal to develop new technologies to decrease the mercury use in medical devices, and even in the gold mining.

References

Albrecht, J. and Matyja, E. 1996. Glutamate: a potential mediator of inorganic mercury neurotoxicity. Metab. Brain Dis. 11: 175–184.

Alves, B. 2023. Os riscos à saúde causados pelo uso de mercúrio no garimpo. BBC News Brasil.

Antunes Dos Santos, A., Hort, M.A., Culbreth, M., Lòpez – Graneo, C., Farina, M., Rocha, J.B.T. et al. 2016. Methylmercury and brain development: A review of recent literature. J. Trace Elem. Med. Biol. 38: 99–107.

Aschner, M. and Aschner, J.L. 1990. Mercury neurotoxicity: mechanisms of blood-brain barrier transport. Neurosci. Biobehav. Rev. 14: 169–176.

Aschner, M., Yao, C.P., Allen, J.W. and Tan, K.H. 2000. Methylmercury alters glutamate transport in astrocytes. Neurochem. Int. 37: 199–206.

Aschner, M. and Syversen, T. 2005. Methylmercury: recent advances in the understanding of its neurotoxicity. Ther. Drug Monit. 27: 278–283.

Azevedo, F.B., Furieri, L.B., Peçanha, F.M., Wiggers, G.A., Vassallo, P.F., Simões, M.R. et al. 2012. Toxic effects of mercury on the cardiovascular and central nervous systems. J. Biomed. Biotechnol. 2012: 949048.

Bakir, F., Damluji, S.F., Amin-Zaki, L., Murtadha, M., Khalidi, A., al-Rawi, N.Y. et al. 1973. Methylmercury poisoning in Iraq. Science 181: 230–241.

Bakker, L.B., Gasparinetti, P., de Queiroz, J.M. and de Vasconcellos, A.C.S. 2021. Economic impacts on human health resulting from the use of mercury in the illegal gold mining in the brazilian amazon: A methodological assessment. Int. J. Environ. Res. Public Health. 18.

Bank, M.S. 2020. The mercury science-policy interface: History, evolution and progress of the Minamata Convention. Sci. Total Environ. 722: 137832.

Barbosa, A.C., Boischio, A.A., East, G.A., Ferrari, I., Gonçalves, A., Silva, P.R.M. et al. 1995. Mercury contamination in the Brazilian Amazon. Environmental and occupational aspects. Wat Air and Soil Poll. 80: 109–121.

Bastos, W.R. 2004. A contaminação por mercúrio na Bacia do Rio Madeira: uma breve revisão. Geochim. Bras. 18.

Bastos, W.R., Dórea, J.G., Bernardi, J.V.E., Lauthartte, L.C., Mussy, M.H., Lacerda, L.D. et al. 2015. Mercury in fish of the Madeira river (temporal and spatial assessment), Brazilian Amazon. Environ. Res. 140: 191–197.

Bernhoft, R.A. 2012. Mercury toxicity and treatment: a review of the literature. J. Environ. Health. 1–10.

Bjorklund, O., Kahlstrom, J., Salmi, P., Ogren, S.O., Vahter, M., Chen, J.F. et al. 2007. The effects of methylmercury on motor activity are sex- and age dependent, and modulated by genetic deletion of adenosine receptors and caffeine administration. Toxicol. 241: 119–133.

Björkman, L., Lundekvam, B.F., Laegreid, T., Bertelsen, B.I., Morild, I., Lilleng, P. et al. 2007. Mercury in human brain, blood, muscle and toenails in relation to exposure: an autopsy study. Environ. Health. 6: 1–14.

Bjørklund, G., Antonyak, H., Polishchuk, A., Semenova, Y., Lesiv, M., Lysiuk, R. et al. 2022. Effect of methylmercury on fetal neurobehavioral development: an overview of the possible mechanisms of toxicity and the neuroprotective effect of phytochemicals. Arch. Toxicol. 96: 3175–3199.

Branco, V., Canário, J., Holmgren, A. and Carvalho, C. 2011. Inhibition of the thioredoxin system in the brain and liver of zebra–seabreams exposed to waterborne methylmercury. Toxicol. Apll. Pharmacol. 251: 95–103.

Branco, V., Caito, S., Farina, M., Teixeira da Rocha, J., Aschner, M. and Carvalho, C. 2017. Biomarkers of mercury toxicity: past, present, and future trends. J. Toxicol. Environ. 20: 119–154.

Brasil. 2004. Decreto No. 5.051 de 19 de abril de 2004. PR, Brasilia, DF, Brazil.

Brasil. 2020. Fundação Nacional dos Povos Indígenas. Ministério dos Povos Indígenas.

Brasil. Constituição. 1988. Constituição da República Federativa do Brasil. Brasília, DF: Senado Federal.

Bridges, C.C., Bauch, C., Verrey, F. and Zalups, R.K. 2004. Mercuric conjugates of cysteine are transported by the amino acid transporter system b(0,+): implications of molecular mimicry. J. Am. Soc. Nephrol. 15: 663–673.

Bridges, C.C. and Zalups, R.K. 2005. Molecular and ionic mimicry and the transport of toxic metals. Toxicol. Appl. Pharmacol. 204: 274–308.

Bridges, C.C. and Zalups, R.K. 2006a. System b0,+ and the transport of thiol-s-conjugates of methylmercury. J. Pharmacol. Exp. Ther. 319: 948–956.

Bridges, CC. and Zalups, R.K. 2006b. Molecular mimicry as a mechanism for the uptake of cysteine S-conjugates of methylmercury and inorganic mercury. Chem. Res. Toxicol. 19: 1117–1118.

Cappelletti, S., Piacentino, D., Fineschi, V., Frati, P., D'Errico, S. and Aromatario, M. 2019. Mercuric chloride poisoning: symptoms, analysis, therapies, and autoptic findings. A review of the literature. Crit. Rev. Toxicol. 49: 329–341.

Carratu, M.R., Borracci, P., Coluccia, A., Giustino, A., Renna, G., Tomasini, M.C. et al. 2006. Acute exposure to methylmercury at two developmental windows: focus on neurobehavioral and neurochemical effects in rat offspring. Neuroscience 141: 1619–1629.

Carvalho, G.M. 2023. Pesquisadora do IFF/Fiocruz analisa o quadro de desnutrição das crianças yanomami. Fundação Oswaldo Cruz (Fiocruz).

Castoldi, A.F.C., Johansson, C., Onishchenko, N., Coccini, T., Roda, E., Vahter, M. et al. 2008. Human developmental neurotoxicity of methylmercury: impact of variables and risk modifiers. Regul. Toxicol. Pharmacol. 51: 201–214.

Chan, T.Y. 2011. Inorganic mercury poisoning associated with skin-lightening cosmetic products. Clin. Toxicol. 49: 886–891.

Clarkson, T.W., Amin-Zaki, L. and Al-Tikriti, S.K. 1976. An outbreak of methylmercury poisoning due to consumption of contaminated grain. Fed. Proc. 35: 2395–2399.

Clarkson, T.W. 1993. Mercury: major issues in environmental health. Environ. Health Perspect. 100: 31–38.

Clarkson, T.W., Magos, L. and Myers, G.J. 2003. The toxicology of mercury—current exposures and clinical manifestations. N. Engl. J. Med. 349: 1731–1737.

Cornbleet, T., Slepyan, A.H. and Ebert, M.H. 1939. The use of colloidal calomel ointment in dermatology. J. Am. Med. Assoc. 113: 1804–1806.
Dare, E., Fetissov, S., Hokfelt, T., Hall, H., Ogren, S.O. and Ceccatelli, S. 2003. Effects of prenatal exposure to methylmercury on dopamine-mediated locomotor activity and dopamine D2 receptor binding. Naunyn Schmiedebergs Arch. Pharmacol. 367: 500–508.
Davis, L.E. 2000. Unregulated potions still cause mercury poisoning. West. J. Med. 173: 19.
de Faria, M.P. 2019. Mariana e Brumadinho: a repercussão dos desastres do setor de mineração na saúde ambiental. In 17° Congresso Da Associação Nacional de Medicina Do Trabalho. Presented at the 17° Congresso da Associação Nacional de Medicina do Trabalho (16–17). Superintendência Regional do Trabalho Minas Gerais.
De Leeuw, V.C., Oostrom, C.T.M.V., Wackers, P.F.K., Pennings, J.L.A., Hodemaekers, H.M., Piersma, A.H. et al. 2022. Neuronal differentiation pathways and compound-induced developmental neurotoxicity in the human neural progenitor cell test (hNPT) revealed by RNA-seq. Chemosphere 304.
Dias, D., Bessa, J., Guimarães, S., Soares, M.E., de Lourdes, M.B. and Teixeira, H.M. 2016. Inorganic mercury intoxication: A case report. Forensic Sci. Int. 259: 20–24.
Dórea, J.G. 2017. Low-dose Thimerosal in pediatric vaccines: Adverse effects in perspective. Environ. Res. 152: 280–293.
Esdaile, L.J. and Chalker, J.M. 2018. The mercury problem in artisanal and small-scale gold mining. Chem. Eur. J. 24: 6905–6916.
Eto, K., Marumoto, M. and Takeya, M. 2010. The pathology of methylmercury poisoning (Minamata disease) the 50th anniversary of Japanese society of neuropathology. Neuropathology 30: 471–479.
Farina, M., Rocha, J.B.T. and Aschner, M. 2011. Mechanisms of methylmercury-induced neurotoxicity: evidence from experimental studies. Life Sci. 89: 555–563.
Farina, M. and Aschner, M. 2017. Methylmercury-induced neurotoxicity: focus on pro oxidative events and related consequences. Adv. Neurobiol. 18: 267–286.
Faro, L.R.F., Do Nascimento, J.L.M., Alfonso, M. and Duran, R. 2001. *In vivo* effects of inorganic mercury ($HgCl_2$) on striatal dopaminergic system. Ecotoxicol. Environ. Saf. 48: 263–267.
Favero, A.M., Oliveira, C.S., Franciscato, C., Oliveira, V.A., Pereira, J.S.F., Bertoncheli, C.M. et al. 2014. Lactating and nonlactating rats differ to renal toxicity induced by mercuric chloride: the preventive effect of zinc chloride. Cell Biochem. Funct. 32: 420–428.
Fields, C.A., Borak, J. and Louis, E.D. 2017. Mercury-induced motor and sensory neurotoxicity: systematic review of workers currently exposed to mercury vapor. Crit. Rev. Toxicol. 47: 811–844.
Fleming, W.L. and Wolf, M.H. 1946. The relative prophylactic effectiveness against syphilis of ointments, containing calomel in different particle size. Am. J. Syph. 30: 47–53.
Franciscato, C., Goulart, F.R., Lovatto, N.M., Duarte, F.A., Flores, E.M.M., Dressler, V.L. et al. 2009. $ZnCl_2$ exposure protects against behavioral and acetylcholinesterase changes induced by $HgCl_2$. Int. J. Dev. Neurosci. 27: 459–468.
Fujimura, M. and Usuki, F. 2015. Methylmercury causes neuronal cell death through the suppression of the TrkA pathway: *in vitro* and *in vivo* effects of TrkA pathway activators. Toxicol. Appl. Pharmacol. 282: 259–266.
Galiciolli, M.E.A., Pedroso, T.F., Mesquita, M., Oliveira, V.A., Pereira, M.E. and Oliveira, C.S. 2022. Biochemical parameters of female wistar rats and their offspring exposed to inorganic mercury in drinking water during the gestational and lactational periods. Toxics 10: 664.
Ganapathy, S., Farrel, E.R., Vaghela, S., Joshee, L., Ford, E.G., Uchakina, O. et al. 2021. Transport and toxicity of methylmercury-cysteine in cultured BeWo cells. Int. J. Mol. Sci. 23: 394.
Gao, H., Liu, G., He, Y. and Chen, J. 2022. Nephrotic syndrome of minimal change disease following exposure to mercury-containing skin lightening cream: A case report and literature review. Clin. Nephrol. 98: 107–112.
Grandjean, P., Weihe, P., Whit, R.F., Debes, F., Araki, S., Yokoyama, K. et al. 1997. Cognitive deficit in 7-year-old children with prenatal exposure to methylmercury. Neurotoxicol. Teratol. 19: 417–428.
Hachiya, N. 2006. The history and the present of Minamata disease. Japan Med. Assoc. J. 49: 112–118.
Harada, M. 1995. Minamata disease: methylmercury poisoning in Japan caused by environmental pollution. Crit. Rev. Toxicol. 25: 1–24.

Harada, M.S., Nakachi, S., Cheu, T., Hamada, H., Ono, Y., Tsuda, T. et al. 1999. Monitoring of mercury pollution in Tanzania: relation between head hair mercury and health. Sci. Total Environ. 227: 249–256.

Holzel, A. and James, T. 1952. Mercury and pink disease. Lancet. 259: 441–443.

Hossain, K.F.B., Rahman, M.M., Sikder, M.T., Hosokawa, T., Saito, T. and Kurasaki, M. 2021. Selenium modulates inorganic mercury induced cytotoxicity and intrinsic apoptosis in PC12 cells. Ecotoxicol. Environ. Saf. 207: 111262.

Ikemoto, T.T., Kunito, T., Anan, Y., Tanaka, H., Baba, N., Miyazaki, N. et al. 2004. Association of heavy metals with metallothionein and other proteins in hepatic cytosol of marine mammals and seabirds. Environ. Toxicol. Chem. 23: 2008–2016.

Instituto Socioambiental (ISA). 2016. O povo Yanomami está contaminado por mercúrio do garimpo. Available at: https://medium.com/hist%C3%B3rias-socioambientais/o-povo-yanomami-est%C3%A1-contaminado-por-merc%C3%BArio-do-garimpo-fa0876819312#.4qbjrn4u8. Accessed on March 4, 2023.

Kanai, Y., Segawa, H., Miyamoto, K., Uchino, H., Takeda, E. and Endou, H. 1998. Expression cloning and characterization of a transporter for large neutral amino acids activated by the heavy chain of 4F2 antigen (CD98). J. Biol. Chem. 273: 29–32.

Kazantzis, G. 1970. Mercury and the kidney. Occup. Med. 20: 54–59.

Langford, N.J. and Ferner, R.E. 1999. Toxicity of mercury. J. Hum. Hypertens. 13: 651–656.

Lima, I. 2021. Líder Yanomami denuncia contaminação de rios, deformidades e doenças causadas por mercúrio. Fundação Oswaldo Cruz (Fiocruz).

Lima, L. 2023. Yanomami: "É um cenário de guerra", diz Sesai. Fundação Oswaldo Cruz (Fiocruz).

Liu, J., Shi, J.Z., Yu, L.M., Goyer, R.A. and Waalkes, M.P. 2008. Mercury in traditional medicines: is cinnabar toxicologically similar to common mercurials? Exp. Biol. Med. 233: 810–817.

López-Berenguer, G., Peñalver, J. and Martínez-López, E. 2020. A critical review about neurotoxic effects in marine mammals of mercury and other trace elements. Chemosphere 246: 125688.

Lorscheider, F.L., Vimy, M.J. and Summers, A.O. 1995. Mercury exposure from "silver" tooth fillings: emerging evidence questions a traditional dental paradigm. FASEB J. 9: 504–508.

Mackert Jr, J.R., Leffell, M.S., Wagner, D.A. and Powell, B.J. 1991. Lymphocyte levels in subjects with and without amalgam restorations. J. Am. Dent. Assoc. 22: 49–53.

Madejczyk, M.S., Aremu, D.A., Simmons-Willis, T.A., Clarkson, T.W. and Ballatori, N. 2007. Accelerated urinary excretion of methylmercury following administration of its antidote N-Acetylcysteine requires Mrp2/Abcc2, the apical multidrug resistance-associated protein. J. Pharmacol. Exp. Ther. 322: 378–84.

Malm, O. 1998. Gold mining as a source of mercury exposure in the Brazilian Amazon. Environ. Res. 77: 73–78.

Marques, R.L.S., Pozzeti, V.C., Lopes, M.T.G. and das Chagas Seixas, C. 2022. Use of Mercury in the Brazilian Amazon: Contamination, problems and current legislation. Rev. Catalana Dret. Ambient. 13.

Meador-Woodruff, J.H. and Healy, D.J. 2000. Glutamate receptor expression in schizophrenic brain. Brain Res. Rev. 31: 288–294.

Meinerz, D.F., Branco, V., Aschner, M., Carvalho, C. and Rocha, J.B.T. 2017. Diphenyl diselenide protects against methylmercury-induced inhibition of thioredoxin reductase and glutathione peroxidase in human neuroblastoma cells: a comparison with ebselen. J. Appl. Toxicol. 37: 1073–1081.

Mesquita, M., Pedroso, T.F., Oliveira, C.S., Oliveira, V.A., do Santos, R.F., Bizzi, C.A. et al. 2016. Effects of zinc against mercury toxicity in female rats 12 and 48 hours after $HgCl_2$ exposure. EXCLI J. 15: 256.

Mitchell, R.J., Osborne, P.B. and Haubenreich, J.E. 2005. Dental amalgam restorations: daily mercury dose and biocompatibility. J. Long-Term Eff. Med. Implants. 15.

Moraes-Silva, L., Siqueira, L.F., Oliveira, V.A., Oliveira, C.S., Ineu, R.P., Pedroso, T.F. et al. 2014. Preventive effect of $CuCl_2$ on behavioral alterations and mercury accumulation in central nervous system induced by $HgCl_2$ in newborn rats. J. Biochem. Mol. Toxicol. 28: 328–335.

Moraes-Silva, L., Oliveira, C.S., Peixoto, N.C. and Pereira, M.E. 2018. Copper attenuates early and late biochemical alterations induced by inorganic mercury in young rats. J. Toxicol. Environ. Health. Part A 81: 633–644.

Morgan, D.L., Chanda, S.M., Price, H.C., Fernando, R., Liu, J., Brambila, E. et al. 2002. Disposition of inhaled mercury vapor in pregnant rats: maternal toxicity and effects on developmental outcome. Toxicol. Sci. 66: 261–273.

Mutter, J., Curth, A., Naumann, J., Deth, R. and Walach, H. 2010. Does inorganic mercury play a role in Alzheimer's disease? A systematic review and an integrated molecular mechanism. J. Alzheimer's Dis. 22: 357–374.

Ni, M., Li, X., Rocha, J.B., Farina, M. and Aschner, M. 2012. Glia and methylmercury neurotoxicity. J. Toxicol. Environ. 75: 1091–1101.

Nogara, P.A., Oliveira, C.S., Schmitz, G.L., Piquini, P.C., Farina, M., Aschner, M. et al. 2019. Methylmercury's chemistry: From the environment to the mammalian brain. Biochim. Biophys. Acta Gen. Subj. 12: 129284.

Oliveira, C.S., Oliveira, V.A., Ineu, R.P., Moraes-Silva, L. and Pereira, M.E. 2012. Biochemical parameters of pregnant rats and their offspring exposed to different doses of inorganic mercury in drinking water. Food Chem. Toxicol. 50: 2382–2387.

Oliveira, C.S., Joshee, L., Zalups, R.K., Pereira, M.E. and Bridges, C.C. 2015. Disposition of inorganic mercury in pregnant rats and their offspring. Toxicology 335: 62–71.

Oliveira, C.S., Oliveira, V.A., Costa, L.M., Pedroso, T.F., Fonseca, M.M., Bernardi, J.S. et al. 2016. Inorganic mercury exposure in drinking water alters essential metal homeostasis in pregnant rats without altering rat pup behavior. Reprod. Toxicol. 65: 18–23.

Oliveira, C.S., Piccoli, B.C., Aschner, M. and Rocha, J.B.T. 2017a. Chemical speciation of selenium and mercury as determinant of their neurotoxicity. Neurotox. Metals. 53–83.

Oliveira, C.S., Joshee, L., George, H., Nijhara, S. and Bridges, C. 2017b. Oral exposure of pregnant rats to toxic doses of methylmercury alters fetal accumulation. Reprod. Toxicol. 69: 265–275.

Oliveira, C.S., Nogara, P.A., Ardisson – Araújo, D.M.P., Aschner, M., Rocha, J.B.T. and Dórea, J.G. 2018. Neurodevelopmental effects of mercury. Adv. Neurotoxicol. 2: 27–86.

Oliveira, C.S., Nogara, P.A., Garlet, Q.I., Rieder, G.S. and Rocha, J.B.T. 2019. Biological thiols and their interaction with mercury. *In*: Carlos C. McAlpine. (Org.). Biological Thiols and Their Interaction with Mercury. 1ed. Hauppauge, New York, USA: Nova Science Publishers. 1: 1–60.

Oliveira, C.S., Segatto, A.L., Nogara, P.A., Piccoli, B.C., Loreto, E.L.S., Aschner, M. et al. 2020. Transcriptomic and proteomic tools in the study of Hg toxicity: what is missing? Front. Genet. 11: 425.

Oliveira, V.A., Favero, G., Stacchiotti, A., Giugno, L., Buffoli, B., Oliveira, C.S. et al. 2017. Acute mercury exposition of virgin, pregnant, and lactating rats: Histopathological kidney and liver evaluations. Environ. Toxicol. 32: 1500–1512.

Oliveira, V.A., da Costa, N.S., Mesquita, M., Pedroso, T.F., Fiuza, T.L., Peixoto, N.C. et al. 2020. Mercury toxicity in pregnant and lactating rats: zinc and N-acetylcysteine as alternative of prevention. Environ. Sci. Poll. Res. 27: 40563–40572.

Onyido, I., Norris, A.R. and Buncel, E. 2004. Biomolecule—mercury interactions: modalities of DNA base—mercury binding mechanisms. Remediation strategies. Chem Rev. 104: 5911–5929.

Orellana, J.D.Y., Marrero, L., Alves, C.L.M., Ruiz, C.M.V., Hacon, S.S., Oliveira, M.W. et al. 2019. Associação de baixa estatura severa em crianças indígenas Yanomami com baixa estatura materna: indícios de transmissão intergeracional. Ciênc. Saúde Colet. 24: 1875–1883.

Orr, S.E., George, H.S., Barnes, M.C., Mathis, T.N., Joshee, L., Barkin, J. et al. 2020. Co-administration of selenium with inorganic mercury alters the disposition of mercuric ions in rats. Biol. Trace Elem. Res. 195: 187–195.

Paletz, E.M., Craig-Schmidt, M.C. and Newland, M.C. 2006. Gestational exposure to methylmercury and n-3 fatty acids: effects on high- and low-rate operant behavior in adulthood. Neurotoxicol. Teratol. 28: 59–73.

Park, J.D. and Zheng, W. 2012. Human exposure and health effects of inorganic and elemental mercury. J. Prev. Med. Public Health. 45: 344.

Paulelli, A.C.C., Cesila, C.A., Devóz, P.P., Oliveira, S.R., Ximenez, J.P.B., Filho, W.R.P. et al. 2022. Fundão tailings dam failure in Brazil: Evidence of a population exposed to high levels of Al, As, Hg, and Ni after a human biomonitoring study. Environ. Res. 205.

Queiroz, J., Gasparinetti, P., Bakker, L.B., Lobo, F. and Nagel, G. 2022. Socioeconomic cost of dredge boat gold mining in the Tapajós basin, eastern Amazon. Resour. Policy. 79.

Ramos, A.R.A., Oliveira, K.A.D. and Rodrigues, F.D.S. 2020. Mercury-based mining in Yanomami indigenous lands and accountabilities. Ambient. Soc. 23: 1–22.
Ribeiro, R. 2023. Associação Yanomami quer reparação de danos causados pelo garimpo. Globo. Available at: https://umsoplaneta.globo.com/sociedade/noticia/2023/03/09/associacao-yanomami-quer-reparacao-de-danos-causados-pelo-garimpo.ghtml. Accessed on March 4, 2023.
Rocha, J.B.T., Freitas, A.J., Marques, M.B., Pereira, M.E., Emanuelli, T. and Souza, D.O. 1993. Effects of methylmercury exposure during the second stage of rapid postnatal brain growth on negative geotaxis and on delta-aminolevulinate dehydratase of suckling rats. Braz. J. Med. Biol. 26: 1077–1083.
Rosenman, K.D., Valciukas, J.A., Glickman, L., Meyers, B.R. and Cinotti, A. 1986. Sensitive indicators of inorganic mercury toxicity. Arch. Environ. Health. 41: 208–215.
Rowens, B., Guerrero-Betancourt, D., Gottlieb, C.A., Boyes, R.J. and Eichenhorn, M.S. 1991. Respiratory failure and death following acute inhalation of mercury vapor. A clinical and histologic perspective. Chest. 99: 185–190.
Sakamoto, M., Kakita, A., Wakabayashi, K., Takahashi, H., Nakano, A. and Akagi, H. 2002. Evaluation of changes in methylmercury accumulation in the developing rat brain and its effects: a study with consecutive and moderate dose exposure throughout gestation and lactation periods. Brain Research. 949: 51–59.
Santos-Sacramento, L., Arrifano, G.P., Lopes - Araújo, A., Augusto-Oliveira, M., Albuquerque-Santos, R., Takeda, P.Y. et al. 2021. Human neurotoxicity of mercury in the Amazon: A scoping review with insights and critical considerations. Ecotoxicol. Environ. Saf. 208.
Silva, R.R.D., Branco, J.C., Thomaz, S.M.T. and Cesar, A. 2017. Convenção de Minamata: análise dos impactos socioambientais de uma solução em longo prazo. Saúde debate. 41: 50–62.
Sinkus, W., Shervette, V., Ballenger, J., Reed, L.A., Plante, C. and White, B. 2016. Mercury bioaccumulation in offshore reef fishes from waters of the Southeastern USA. Environ. Pollut. 228: 222–233.
Skerfving, S.B. and Copplestone, J.F. 1976. Poisoning caused by the consumption of organomercury-dressed seed in Iraq. Bull. World Health Organ. 54: 101–112
Sun, S., Chang, J., Jiang, X. and Gu, H. 2022. Irritant contact dermatitis caused by cosmetics containing excessive mercury. J. Cosmet. Dermatol. 21: 6688–6690.
Syversen, T. and Kaur, P. 2012. The toxicology of mercury and its compounds. J. Trace Elem. Med. Biol. 26: 215–226.
Takeuchi, T. 1982. Pathology of Minamata disease. With special reference to its pathogenesis. Acta Pathol. Jpn. 1: 73–99.
Taylor, D.L., Kutil, N.J., Malek, A.J. and Collie, J.S. 2014. Mercury bioaccumulation in cartilaginous fishes from Southern New England coastal waters: contamination from a trophic ecology and human health perspective. Mar. Environ. Res. 99: 20–33.
Teixeira, F.B., de Oliveira, A.C.A., Leão, L.K.R., Fagundes, N.C.F., Fernandes, R.M., Fernandes, L.M.P. et al. 2018. Exposure to inorganic mercury causes oxidative stress, cell death, and functional deficits in the motor cortex. Front. Mol. Neurosci. 11: 125.
Tupy, G.S., Santos, A.A., de Brito Silva, R.T., de Carvalho, K.M. and dos Santos Morato, R.B. 2023. Governança policêntrica no combate ao uso do mercúrio na mineração ilegal em território Yanomami. Rev. Políticas Públicas. 26: 760–781.
United Nations, Environment Programme- UN. Minamata Convention on Mercury. Party Profiles.
Valli, L.G. 2014. Mecanismo de ação do glutamato no sistema nervoso central e a relação com doenças neurodegenerativas. Rev. Bras. de Neurol. e Psiquiatr. 18.
Vega, C.M., Orellana, J.D.Y., Oliveira, M.W., Hacon, S.S. and Basta, P.C. 2018. Human mercury exposure in yanomami indigenous villages from the Brazilian Amazon. Int. J. Environ. Res. Public Health. 15.
Von Burg, R. 1995. Inorganic mercury. J. Appl. Toxicol. 15: 483–493.
Wagner, C., Sudati, J.H., Nogueira, C.W. and Rocha, J.B. 2010. *In vivo* and *in vitro* inhibition of mice thioredoxin reductase by methylmercury. Biometals. 23: 1171–1177.
Walach, H., Mutter, J. and Deth, R. 2015. Inorganic mercury and Alzheimer's disease results of a review and a molecular mechanism. Diet and Nutrition in Dementia and Cognitive Decline 593–601.
Wang, F., Outridge, P.M., Feng, X., Meng, B., Heimbürger-Boavida, L.E. and Mason, R.P. 2019. How closely do mercury trends in fish and other aquatic wildlife track those in the atmosphere?–

Implications for evaluating the effectiveness of the Minamata Convention. Sci. Total Environ. 674: 58–70.
World Health Organization (WHO). 1991. Inorganic mercury.
World Health Organization (WHO). 2017a. Guidelines for Drinking-Water Quality: First Addendum to The Fourth Edition.
World Health Organization (WHO). 2017b. Mercury and Health.
Ye'kwana, A.W. 2022. Hutukara Associação Yanomami; Instituto Socioambiental (ISA). Yanomami sob ataque: garimpo ilegal na Terra Indígena Yanomami e propostas para combatê-lo. Boa vista: 120.
Yorifuji, T., Tsuda, T., Inoue, S., Takao, S. and Harada, M. 2011. Long-term exposure to methylmercury and psychiatric symptoms in residents of Minamata, Japan. Environ. Int. 37: 907–913.
Yorifuji, T. and Tsuda, T. 2014. "Minamata." Encyclopedia of Toxicology: Third Edition. Elsevier. 340–344.

Chapter 6

Lead Neurotoxicity

Interactions with Key Ethanol Metabolizing Enzymes

Miriam B. Virgolini[1,*] and *Michael Aschner*[2]

1. Introduction

Trace elements as essential components of living organisms have different metabolic properties and take part in several chemical processes, including mitochondrial functionality and enzymatic reactions. Several of the essential elements for humans are metals. Accordingly, in addition to sodium (Na), potassium (K), magnesium (Mg), and calcium (Ca), essential metallic elements include manganese (Mn), iron (Fe), cobalt (Co), magnesium (Mg), copper (Cu), zinc (Zn), selenium (Se) and molybdenum (Mo), which are collectively known as trace metals (Jomova et al., 2022). Remarkably, several of these elements are essential components of enzymes that attract or subtract molecules and facilitate their conversion to specific end products. They also have the potential to be substituted by other metals or metalloids, an event that may be a determinant of toxicity.

Metalloenzymes are proteins containing tightly bound metal ions for structural or functional purposes. Structural metal ions are required to properly fold a protein that can also serve as regulatory capacity, for example, Zn finger proteins. Functional metal ions are found at the active site of metalloenzymes, carrying out a diverse range of processes, such as electron transfer, substrate recognition/binding, and catalysis, that altogether serve a wide variety of biological functions. Thus, the metal ions form coordinated covalent bonds with the enzyme's amino acids or a prosthetic group to

[1] IFEC-CONICET. Departamento de Farmacología Otto Orsingher. Facultad de Ciencias Químicas. Universidad Nacional de Córdoba.

[2] Department of Molecular Pharmacology, Albert Einstein College of Medicine, Bronx, NY 10461.

* Corresponding author: miriam.virgolini@unc.edu.ar

act as a coenzyme and carry out the enzyme's activity. They are located in a specific region on the enzyme surface that does not disrupt the substrate's binding with the active site. Besides the enzymes that are associated with catalytic activity in other metalloproteins, the metal is a temporary component involved in non-enzyme electron transfer reactions (e.g., cytochromes) or may act as transport (e.g., transferrin) or storage (e.g., ferritin) proteins. The metalloenzymes containing Fe, Zn, Cu, and Mn as metal centers are widespread in nature including superoxide dismutase (Zn and Cu), carboxypeptidase A (Zn), carbonic anhydrase (Zn), cytochrome oxidase (Fe and Cu), and xanthine oxidase (Co and Fe) (Eom and Song, 2019).

In relation to the metalloenzymes participating in ethanol disposition, besides the non-oxidative pathways that metabolize a small portion of the drug, the enzymes alcohol dehydrogenase (ADH), cytochrome P450 2E1 (CYP2E1), catalase, and aldehyde dehydrogenase (ALDH), all contribute to ethanol elimination through oxidative metabolism. In this respect, in addition to enzyme polymorphisms which affect ethanol metabolism and thereby the risk of alcohol use disorders (AUD) (Vaswani, 2019), we propose that potential substitution of the enzymes' metal-bound site could be also a mechanism which will determine the ensuing ethanol-induced toxicity.

Accordingly, the main four enzymes involved in ethanol oxidative metabolism are discussed in detail with an emphasis in their metalloenzyme properties. Thus, predominantly, the ADH family (in particular ADH1 and to a lesser extent ADH3) accounts for the majority of ethanol oxidation to acetaldehyde in the liver, with nicotinamide adenine dinucleotide (NAD^+) being the limiting factor in this reversible reaction. Thereby, at low alcohol concentrations, the inducible CYP2E1 enzyme may account for only 10% of the total alcohol oxidizing capacity of the liver, whereas at high ethanol levels, the relevance of CYP2E1 in the oxidative metabolism increases. Finally, catalase can oxidize ethanol in addition to its well-known antioxidant properties to catalyze H_2O_2 removal. Although its role in the liver is insignificant, it acquires relevance in the brain, in particular, related to acetaldehyde's positive reinforcing properties, either by itself or via the interaction with catecholamines to produce isoquinoline-derived condensation products such as salsolinol. Finally, independently of the organ, acetaldehyde oxidation to acetate is mediated by the mitochondrial ALDH2 isoform, a reaction also dependent on NAD^+ as a limiting factor (Cederbaum, 2012; Israel et al., 2015; Cavallaro et al., 2017; Acquas et al., 2019).

Based on these statements, the present chapter focuses on the metal-ligand properties of the main enzymes involved in ethanol metabolism to provide evidence on the importance of the metallic core in the differential effects of ethanol in response to lead (Pb) exposure in several experimental models (see Figure 1). Thus, it is plausible that Pb as a nonessential heavy metal may gain access to these target molecules and upon binding to them, and modify their biological activity by ligand substitution reactions (Virgolini and Aschner, 2021). Competition between essential and toxic metal ions for physiologically important metal ion binding sites is thought to comprise a major category of reactions leading to cell injury (Zalups and Koropatnick, 2004). Thus, a detailed presentation of each enzyme is provided below

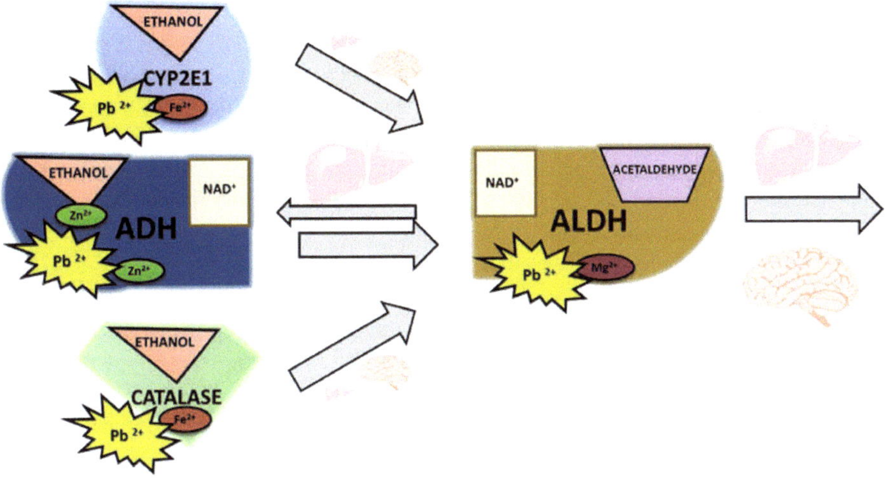

Figure 1. Effects of ethanol in response to Pb exposure in several experimental models.

followed by experimental evidence on Pb effects on the enzymes in the context of biochemical and behavioral effects of ethanol.

2. Enzymes and oxidative ethanol metabolism

2.1 Alcohol dehydrogenases

The ADHs enzymes (EC 1.1.1.1) represent a group of dehydrogenases present in many living organisms, which facilitate the reversible interconversion between alcohols and aldehydes or ketones accompanied with the reduction of nicotinamide adenine dinucleotide (NAD to NADH) and *vice versa*. The direction of the reaction depends on the relative cofactor concentrations (NADH/NAD^+ ratio) with preponderance to the aldehyde/ketone formation *in vivo* (Cederbaum, 2012).

Human ADHs are metalloenzymes that can be classified into 5 classes, while there are 7 different human ADH genes (*ADH1A, ADH1B, ADH1C, ADH4, ADH5, ADH6,* and *ADH7*) aligned along a small region of chromosome 4q21 that encode medium-chain ADHs. The three class I genes, *ADH1A, ADH1B,* and *ADH1C*, are very closely related; they encode the α, β, and γ subunits, which can form homodimers or heterodimers that account for most of the ethanol-oxidizing capacity in the liver. The ADH4 isoform encodes π-ADH, which contributes significantly to ethanol oxidation at higher concentrations. ADH5 encodes χ-ADH, a ubiquitously expressed formaldehyde dehydrogenase with very low affinity for ethanol, while ADH6 mRNA is present in fetal and adult liver, although the enzyme has not been isolated from tissue and little is known about its nature. Finally, ADH7 encodes σ-ADH, which contributes to both ethanol and retinol oxidation (Edenberg, 2007; Edenberg and McClintick, 2018; Vaswani, 2019). Notably, polymorphisms in these enzymes, such as ADH1B*2 are protective against the development of AUD (Chen et al., 2009; Tawa et al., 2016).

In relation to the subunit arrangement, ADHs are dimeric cytosolic metalloproteins in which each 40 kDa subunit binds to two Zn^{2+} ions, although only one of them is catalytically active. The Zn^{2+} ion located at the active site of an enzyme has a distorted tetrahedral geometry, coordinated to one histidine and two cysteine residues, as well as one water molecule, directly participating in the bond-making or -breaking step (Vallee and Hoch, 1955). The non-catalytic Zn^{2+} ion plays a structural role accounting for ADH enzyme stability and forming a tetrahedral structure with four cysteine residues. Unlike the ions of other first-row of transition metals, Zn^{2+} contains a complete *d* orbit (d10), and therefore does not participate in redox reactions but rather functions as a Lewis acid to accept a pair of electrons. Therefore, all ligand geometries can be used by Zn-metalloenzymes to alter the reactivity of the metal conforming an important factor in the ability of the ion to catalyze chemical reactions (Hambidge et al., 2000; Baj et al., 2020).

Each subunit of the dimeric enzyme is divided into a cofactor binding domain and a catalytic domain separated by a cleft containing a deep pocket. The cleft is open when NAD^+ is not bound to the enzyme. When NAD^+ binds to ADH, the cleft between the catalytic and NAD^+ binding domains closes around the cofactor, an event that allows for the displacement of the water molecule by an alcohol substrate. The conformational changes of the enzyme also orientate the Zn^{2+}-bound substrate in the right direction for effective hydride transfer from the alcohol substrate to NAD^+. Subsequently, two hydrogens are stripped off the ethanol by Zn^{2+}, also involving a conformational change, a 10^0 rotation that makes the protein move from its apo "open" form to the halo "closed", hence, demonstrating the importance of protein dynamics in catalysis. Deprotonation of the coordinated alcohol yields a Zn alkoxide intermediate, which then undergoes hydride transfer to NAD^+ to give the Zn^{2+}-bound aldehyde and NADH. Next, a water molecule displaces the aldehyde to regenerate the original catalytic Zn^{2+} center, and finally NADH is released to complete the cycle. Thus, the role of Zn^{2+} in the dehydrogenation reaction is to promote deprotonation of the alcohol to enhance hydride transfer from the Zn^{2+} alkoxide intermediate. Conversely, in the reverse hydrogenation reaction, its role is to enhance the electrophilicity of the carbonyl carbon atom. NAD^+ assist in the removal of a hydride ion from the -CH group of the metal-bound alcohol substrate by forming NADH (Plapp et al., 2017; Di et al., 2021). Importantly, the binding to NAD^+ or NADH is a requisite for the subsequent binding of any of the substrates (alcohols or aldehydes/ketones) to the ADH enzymes (Di et al., 2021).

2.1.1 ADH inhibitors and activators

Despite the importance of ADH in oxidative ethanol metabolism, there was not major interest in either activating or inhibiting this enzyme, other than the efforts directed to treat methanol or ethylene glycol intoxication. As mentioned, ADH activation can lead to generation of a highly toxic acetaldehyde metabolite, whereas its inhibition prolongs the half-life of ethanol and can lead to elevated blood alcohol concentration (BAC) after excessive ethanol consumption. As Zn^{2+} removal from ADH leads to a complete loss of its catalytic activity, the ethanol-induced Zn^{2+} depletion observed in alcoholic patients is potentially linked to an altered metabolic pathway, promoting a shift from ADH to CYP2E1 in the liver of alcoholic individuals. Accordingly, the Zn^{2+} ion site has been the target for the development of ADH inhibitors, emphasizing the utility of inhibiting metalloenzymes with metal-binding compounds (Chen et al., 2019). For example, imidazole was first identified as a weak inhibitor due to its ability to coordinate Zn^{2+}. This drug lead later to the development of the FDA-approved 4-methylpyrazole (commercialized as fomepizole), a metalloenzyme inhibitor that also acts by directly binding to the catalytic Zn^{2+} site. In addition, a formamide-based transition state ADH mimetics has been developed to bind the Zn^{2+} ion. Interestingly, by coordinating with the catalytic Zn^{2+} ion through the aldehyde this compound can distinguish between different human ADH isoenzymes (Gibbons and Hurley, 2004).

2.2 Cytochromes (CYP2E1)

The cytochrome superfamily P450 is characterized by inducible hemoproteins that possess a protein motif (apoprotein) and a heme group (prosthetic group) where a Fe atom is located. The active CYP2E1 (CYP2E1- EC 1.14.13.n7) is localized in both the endoplasmic reticulum (erCYP2E1) and the mitochondria (mtCYP2E1) and expressed in high levels in the liver and in some extrahepatic tissues. They function as a monooxygenase that conforms the microsomal ethanol oxidizing system (MEOS) catalyzing reactions in which only one of the oxygen atoms is incorporated and the other is reduced to H_2O. In addition, the enzyme requires two substrates; the main substrate accepts one of the O_2, while a co-substrate provides H atoms to reduce the other to H_2O. However, in many cases, CYP2E1 oxidation results in the formation of reactive metabolites, such as epoxides or aldehydes. Furthermore, in addition to the oxidation of several substrates, CYP2E1 undergoes "uncoupling" of its catalytic cycle wherein electrons are consumed to generate reactive oxygen species (Hartman et al., 2017; Teschke, 2019).

2.2.1 CYP inhibitors and activators

Diallyl sulfide, a major chemical constituent of garlic (*Allium sativum*) (Rao et al., 2015) or chlormethiazole (Gebhardt et al., 1997), which binds to CYP 2E1, is not a substrate, yet it competes for the common metabolic site of the enzyme leading to impaired metabolism. Several mechanisms of CYP2E1 induction have been described with focus on the prevention of its degradation by the proteolytic proteasome pathway, which results in its increased CYP2E1 half-life and consequently protein levels. Specifically, ethanol induction of CYP2E1 in the liver occurs by a 2-step mechanism; the first step of the induction is associated with low BACs and appears to be posttranscriptional, whereas high BACs observed in step two induction are associated with increased CYP2E1 gene transcription (Teschke, 2019).

2.3 Catalases

Mammalian catalases (EC 1.11.1.6) are highly conserved antioxidant enzymes that belong to a family of Fe-containing proteins capable of H_2O_2 dismutation. Crystallography data show that the 22 catalases are tetramers with each of the four active sites consisting of a pentacoordinate-Fe protoporphyrin IX prosthetic group with a tyrosinated axial ligand. They also contain NADPH as a cofactor, although the enzyme lacks a known nucleotide binding motif. Heme groups are deeply buried inside a central cavity of the catalase tetramers, with the Fe atoms at more than 20 Å from the nearest molecular surface connected to several channels (Díaz et al., 2012). It has been described recently that these enzymes also exhibit oxidase activity in addition to their well-known ability to convert H_2O_2 into H_2O and O_2 (catalatic activity), as well as their ability to oxidize low molecular-weight alcohols (peroxidatic activity) in the presence of low H_2O_2 concentrations (Hansberg, 2022) as follows: (a) the *catalatic* reaction pathway: the conversion of H_2O_2 to H_2O and O_2 by catalase is a two-step process whereby catalase heme Fe^{3+} reduces one molecule of H_2O_2 to H_2O, generating a high-valent Fe intermediate covalent $Fe^{4+} = O$ oxyferryl specie and a porphyrin cation radical. This reaction intermediate, referred to as Compound I, which, in contrast to other hydro peroxidases is reduced back to the resting state by further reactions with H_2O_2, with one oxygen separating as H_2O and the other remaining in the heme Fe; (b) the *peroxidatic* activity of catalase results from the ability of compound I to oxidize H donors, e.g., methanol, ethanol, formic acid, and phenols to aldehydes and H_2O. Each catalase monomer binds one heme molecule, while the holoenzyme binds two NADPH molecules, although the precise role of this cofactor in enzymatic activity is unclear because H_2O_2 provides both oxidative and reductive potential during catalysis. Recent studies suggest that NADPH may be important in maintaining catalase in an active state (Oshino et al., 1973; Vetrano et al., 2005; Zamocky et al., 2008).

Catalase kinetics do not abide by this normal pattern. On the one hand it is not possible to saturate the enzyme with "substrate" within the feasible concentration range (up to 5 M H_2O_2), and on the other there is a rapid inactivation at H_2O_2 concentrations above 0.1 M, when the active enzyme-H_2O_2 complex I is converted to the inactive complexes II or III (Aebi, 1984). Finally, it is noteworthy that although the enzyme's localization is not under discussion, recent findings provide evidence for dynamic and highly regulated dual subcellular localization of catalase both in peroxisomes and the cytosol, which has important implications for the cellular redox balance (Fujiki and Bassik, 2021).

2.3.1 Catalase activators and inhibitors

Various classes of inhibitors have been investigated (Goyal and Basak, 2010; Sepasi Tehrani and Moosavi-Movahedi, 2018). The oxidation by H_2O_2 does not directly affect the active site but determines conformational changes to the enzyme that are necessary for catalysis by oxidation of amino acid residues. Both cyanide and azide bind to heme, thus inhibiting the catalysis irreversibly by blocking the heme access to other potential iron ligands. The latter inhibits the peroxidatic activity by preventing the formation of compound II from compound I in the slow loop intermediate. Inhibitors such as 3-aminotriazole and derivates are those that act indirectly, reacting with compounds I and II in the near vicinity of the active site, either changing the protein configuration or sterically affecting the binding (Margoliash et al., 1959). Although no direct catalase activators have been described, conformational changes induced by various substances may facilitate the accessibility of substrate to the heme catalytic center activating the enzyme. These include sodium n-dodecyl sulphate, curcumin, and signaling pathways such as calcium/calmodulin (Sepasi Tehrani and Moosavi-Movahedi, 2018). There are also reports on catalase activation after 3-nitropropionic acid administration, a reactive oxygen species (ROS) generator that elicits a compensatory antioxidant response against a pro oxidant environment (Binienda et al., 1998; Manrique et al., 2006; Mattalloni et al., 2013). Additionally, superoxide dismutase/catalase mimetics such as compounds, known as EUKs, are also considered catalase activators, although the exact mechanisms involved have yet to be characterized (Baker et al., 1998; Peng et al., 2005).

2.4 Aldehyde dehydrogenases (ALDHs)

The mammalian ALDH gene superfamily encodes a group of evolutionarily-related sequences whose protein products have pyridine nucleotide-dependent oxidation activity catalyzing the irreversible oxidation of aldehydic substrates to their corresponding carboxylic acids. Although many ALDH enzymes display broad substrate specificity and oxidize a variety of aliphatic and aromatic aldehydes, others retain unique substrate preferences. To date, 19 putatively functional ALDH genes have been identified in the human genome with allelic variations in genes coding aldehyde-metabolizing enzymes that induce variations in alcohol absorption and metabolism which, in turn, contribute to AUD. In this regard, the ALDH1 family consists of six human ALDH genes: ALDH1A1, ALDH1A2, ALDH1A3, ALDH1B1, ALDH1L1 and ALDH1L2. The ALDH2, is a mitochondrial enzyme involved in the oxidation of acetaldehyde and the metabolites of dopamine and norepinephrine, DOPAL and DOPEGAL, respectively, whereas the ALDH3 group is formed by ALDH3A1, ALDH3A2, ALDH3B1 and ALDH3B2; ALDH4A1; ALDH5A1; ALDH6A1; ALDH7A1; ALDH8A1; ALDH9A1; ALDH16A1 and ALDH18A1 (Black et al., 2012; Jackson et al., 2011; Sophos and Vasiliou, 2003; Marchitti et al., 2007). A detailed ALDH database resource is available at www.aldh.org. Human ALDH1A1 and ALDH2 exhibit K_m for acetaldehyde in the micromolar to sub-micromolar range at a near physiological pH and cytoplasmic NAD^+ concentration. Human ALDH2 is expressed ubiquitously in all tissues but is most abundant in the liver; it is also found in high amounts in organs that require high mitochondrial oxidative phosphorylation, such as the heart and brain. In addition, ALDH2 plays a key role in oxidizing endogenous aldehydic products that arise from lipid peroxidation upon oxidative stress conditions, such as 4-hydroxy-2-nonenal (4-HNE) and malondialdehyde (MDA), as well as environmental aldehydes, such as acrolein. This enzyme is a 517-amino acid polypeptide encoded by a nuclear gene located at chromosome 12q24 and transported to the mitochondrial matrix in a process that is dependent on its NH_2 terminus 17-amino acid mitochondrial targeting sequence, which is cleaved as part of the complete folding and maturation of the enzyme inside the mitochondria. It is a tetrameric enzyme with approximately 56 kDa identical subunits. The tetramer is a dimer of dimers with only two of the catalytic sites on each enzyme complex maintaining the activity. Each subunit consists of three main domains: the catalytic domain, the coenzyme- or NAD^+-

binding domain, and the oligomerization domain (bridging). The active site is located at the base of the hydrophobic tunnel, in vicinity to the tetrameric interface and opposite the cofactor binding site. The residues lining this tunnel are believed to confer the substrate specificity of each isoenzyme (Koppaka et al., 2012). The NAD^+ cofactor is held to the enzyme through 10 hydrogen bonds, four of which involve two residues, an amino acid lysine at position 192 (Lys192) and Glu399, which are highly conserved across all ALDH isoenzymes (Sheikh et al., 1997). Two conformations of the NAD^+ binding pocket has been identified: the hydride transfer (closed or extended) and the hydrolysis (open and contracted). In general, divalent metals such as Mg^{2+} enhance the enzymatic activity by stabilizing the binding of the cofactor which precedes the substrate binding. It is the tight binding between the Mg^{2+} ions and NAD^+ that enhances the rate of conversion from the closed to open conformation state after NAD^+ reduction (Morrison, 2020). Moreover, Mg^{2+} is coupled to enhancement in the number of functional units rather than the modulation of the catalytic properties of the active site (Takahashi and Weiner, 1980; Takahashi et al., 1981; Vallari and Pietruszko, 1984). Interestingly, the Mg^{2+} ion has been shown to have opposite effects on liver cytosolic and mitochondrial human ALDHs (Rawles et al., 1987). First, cytosolic ALDH1A1 is inhibited by Mg^{2+} because these ions are responsible for the binding between the enzyme and cofactor in the binary complex. In this scenario the NADH cofactor release represents the rate-limiting step of the reaction, and the dissociation constant for both NAD^+ and NADH cofactors are decreased in the presence of Mg^{2+} (Dickinson and Hart, 1982; Bennett et al., 1983; Weiner and Takahashi, 1983; Gonnella et al., 2013), an effect that is independent on the cofactor concentration (Pequerul et al., 2020). In contrast, mitochondrial ALDH2 is activated by Mg^{2+} because the rate-determining step of the reaction is the diacylation, in which the carboxylic acid generated is released from the ternary complex enzyme-NADH-carboxylic acid (Takahashi and Weiner, 1980; Takahashi et al., 1981). Other data suggest that the presence of Mg^{2+} may lead to selection of particular conformations and speed isomerization of the reduced cofactor following hydride transfer promoting ALDH2 activation (Perez-Miller and Hurley, 2003).

In terms of the catalysis mechanism, it consists of five distinct steps as follows: first, the catalytic site (being a cysteine at position 302 (Cys302) the main amino acid) is activated through a proton abstraction of H^+ mediated by a H_2O molecule provided by a glutamic acid in position 268 (Glu268). The consequent nucleophilic attack on the carbonyl of the aldehyde substrate is then carried out by the thiolate group of Cys302. Subsequently, a tetrahedral thio-hemiacetal intermediate is formed by diacylation, with concomitant hydride transfer to the pyridine ring of the NAD(P) cofactor. Hydrolysis of the resulting thioester intermediate then occurs followed by a final dissociation of the reduced cofactor and subsequent regeneration of the enzyme (Sheikh et al., 1997; Liu and Tanner, 2019).

2.4.1 ALDH inhibitors and activators

As noted before, the enzymes belonging to the ALDH family metabolize toxic biogenic and environmental aldehydes, including 4-HNE and the environmental pollutant, acrolein. ALDH2 also bioactivates nitroglycerin, but it is best known for its role in ethanol metabolism. The most relevant variant is the ALDH2*2 allele, which is found in as many as 35–45% of East Asians (i.e., Chinese, Japanese, Korean, and Taiwanese). The ALDH2*2 carriers have lower enzymatic activity, and this deficiency is manifested by the characteristic facial flushing, headaches, nausea, dizziness, and cardiac palpitations after consumption of alcoholic beverages. This ethanol-induced flushing syndrome in ALDH2*2-carrying individuals is caused by a single G to A nucleotide change, which leads to a substitution of glutamate to lysine at position 487 (E487K, or E504K in some literature). The E487K mutation exerts a dominant effect over the wild-type monomer encoding the ALDH2*1 allele. Structural analysis of the apo- and holoenzyme forms of ALDH2*2 reveals structural deficits in both the coenzyme-binding and active sites that are consistent with the observed kinetic properties. Therefore, heterozygotic individuals (ALDH2*1/*2) have less than 50% of the wild-type's enzymatic activity, and ALDH2*2/*2 homozygotes have 1–4% of the wild-type activity (Chen et al., 2014). On the basis of these features, several ALDH inhibitors were developed (Koppaka et al., 2012). The discovery of alcohol deterrent substances was led by ALDH inhibitors, in particular disulfiram (Barth and Malcolm, 2010). Disulfiram has been used in the United States from 1951; it is a unique medication that relies on "psychological threat" to avoid disulfiram-ethanol reactions, although the consumption of higher amounts of alcohol can lead to a more severe disulfiram–ethanol reaction with potentially fatal toxic effects which prevents its unlimited prescription. It is a potent inhibitor of cytosolic ALDH1 enzyme in vitro, whereas the formation of the metabolite methanethiol is responsible for mitochondrial ALDH2 irreversible inhibition *in vivo*, ALDH2 being the main isozyme involved in acetaldehyde metabolism (Sanny and Rymas, 1993). Thus, disulfiram reacts with thiols to form diethyl ammonium diethyldithiocarbamates, carbon disulfide and disulfide derived from thiol groups (Hu et al., 1997), this is a broadly acting, non-specific inhibitor of many physiologically important sulfhydryl-containing compounds including enzymes. A similar mechanism is exerted by cyanamide, a calcium carbamide that potently inhibits liver ALDH, but is less effective in blocking the brain enzyme, raising questions on whether the drug or its metabolites are able to cross the blood-brain barrier. This drug exerts preferential ALDH2 inhibition which peaks 1–2 h after drug administration, with 80% restoration of the activity occurring within 24 h, a feature that has limited its clinical use because of its short duration compared to disulfiram (Deitrich et al., 1976). In addition, based on ancient therapeutics, several plant derivatives have been developed for the treatment of alcohol dependency (Rezvani et al., 2003; Overstreet et al., 2003). Among them, daidzin is a potent ALDH2 inhibitor because the phenolic moiety interacts directly with two essential active site residues, Cys302 and Glu268, inhibiting the enzyme by restricting substrate binding and catalysis (Keung et al., 1997; Keung, 2003) and related to a more favorable relationship between enthalpic and entropic features (da Silva Cunha et al., 2020). Interestingly, although both, daidzin and Alda-1 share overlapping binding sites, the very different effects of daidzin and Alda-1 on ALDH2 activity can be explained, in part, given their unique crystal structures. In the daidzin-bound structure, the phenolic moiety interacts directly with two essential active-site residues, Cys302 and Glu268, inhibiting the enzyme by restricting substrate binding and catalysis. In contrast, Alda-1 binds at the entrance to the active site but does not sterically interfere with the catalytic residues (Perez-Miller et al., 2010).

There has also been great interest in the developing of ALDHs agonists to prevent toxic aldehydes build up. It is known that NAD(H) binding ameliorates in large measure the structural disorder in the coenzyme-binding site, but does not improve the positioning of critical residues involved in catalysis. In this sense, the small N-benzyl benzamide derivative Alda-1 activates the wild-type enzyme and restores the activity of ALDH2*2 by acting as a structural chaperone, binding at the entrance to the active binding site without sterically interfere with the catalytic residues (Chen et al., 2008; Perez-Miller et al., 2010). However, given the toxic aldehydes-removal properties of ALDH2 that goes beyond ethanol-derived acetaldehyde metabolism, it was recently demonstrated that Alda-1 administration led to beneficial changes in the expression of genes and proteins related to neuroplasticity and mitochondrial function (Stachowicz et al., 2017). Moreover, *in vivo* data demonstrated that the activation of ALDH2 by Alda-1 can inhibit both the acquisition and the maintenance of chronic ethanol intake by alcohol-preferring rats. Thus, the activation of brain ALDH2 may constitute a novel approach in the treatment of alcohol use disorders (Rivera-Meza et al., 2019). More recently, and based on Alda-1 skeleton, three new classes of compounds were designed and synthesized to improve their water solubility and drug-like properties (Hu et al., 2019). In addition, benzyl isothiocyanate, an organosulfur compound from cruciferous vegetables, was shown to reduce toxic aldehydes buildup by increasing the activity and gene expression of the cytosolic/microsomal and mitochondrial fractions of total ALDH in hepatoma-derived cells (Liu et al., 2017). More recently, omeprazole was proposed as a potent activator of human ALDH1A1 after docking experiments (Calleja et al., 2020).

3. Lead and enzymes

Among the main mechanisms of actions associated with Pb^{2+} toxicity is its ability to substitute for diverse essential divalent cations such as Ca^{2+}, Fe^{2+}, and Zn^{2+}, altering thereby the cellular microenvironment. This property becomes critical as far as living organisms possess several mechanisms to maintain physiologically-relevant levels of essential metals, whereas the nonessential metals lack of homeostatic regulation (Ballatori, 2002). Taking into consideration the metals coordinated for the enzymes that participate in ethanol metabolism, a short description with supporting evidence for Zn^{2+} and Fe^{2+} interactions with Pb will be presented in the context of the metal effects on ADH and catalase, either alone or in combination with ethanol. Although the evidence regarding the Pb^{2+}/Mg^{2+} interaction has yet to be documented, relevant *in vivo* and *in vitro* data will be discussed considering the key importance of the ALDH2 enzyme in aldehydes removal. Thus, what follows is a description of experimental evidence regarding biochemical assays and ethanol's behavioral effects in invertebrates, vertebrates or cell cultures as a consequence of combined Pb exposure, with a focus on the prosthetic groups of the metalloenzymes that participate in the oxidative ethanol metabolic pathway.

3.1 ADH

As the second most abundant ion in the central nervous system, Zn^{2+} is considered an essential component of numerous proteins involved in defense mechanisms against oxidative stress. In this respect, Zn^{2+} present in 5-aminolevulinic acid dehydratase (ALAD), an important enzyme in heme synthesis, is susceptible to displacement by Pb^{2+}, resulting in anemia secondary to the inability of the enzyme to participate in porphyrin synthesis. In addition, ALA build-up is an important contributor to Pb-induced redox imbalance (Kirberger et al., 2013). Related to ethanol metabolism, Zn^{2+} is part of both, the structural and the catalytic actives sites of the ADH conformation, and its deficiency leads to toxicity upon ethanol abuse (Skalny et al., 2018). In this respect, evidence in the animal model *Caenorhabditis elegans* demonstrated that ADH activity (SODH-1 in the worm) is decreased as a consequence of early Pb exposure as demonstrated both biochemically and behaviorally by the lower formation of acrolein, a toxic metabolite derived from allyl alcohol, a reaction mediated by SODH-1 (Alaimo et al., 2013; Williamson et al., 1991). Thus, this Pb-induced decrease in enzymatic activity results ethanol accumulation in these animals as evidenced by hyperlocomotion in response to ethanol exposure (200 mM) and a reduced acute functional tolerance to the sedative effects of the drug (Albrecht et al., 2022).

3.2 Catalase and CYP2E1

Fe^{2+} is another divalent cation susceptible to interactions with Pb, although substitution does not seem to be the mechanism in this case. Evidence indicates that the shared target is the divalent metal transporter 1 (DMT-1), by a proton-coupled and membrane potential–dependent mechanism, which has been shown to participate in the transport and cellular uptake of the Pb^{2+} ion, probably via a mechanism of molecular mimicry augmented in conditions of Fe deficiency (Kirberger et al., 2013). Intriguingly, it has been demonstrated that Pb^{2+} exposure increases Fe^{2+} content in rat brain and in Pb^{2+}-exposed PC12 cells, an event associated with decreased expression of ferroportin 1 (FP1), a pivotal Fe^{2+} efflux protein whose overexpression can attenuate Fe^{2+} accumulation (Bridges and Zalups, 2005; Zhu et al., 2013; Zhou et al., 2014). As catalases (and CYP450s) contain heme as the prosthetic group, essential for the catalysis and biological activity, it is possible that in the presence of Pb a competition between both cations exists, with consequences ensuing at the biochemical and behavioral level. In this regard, studies published by us in adult Wistar rats demonstrated that perinatal exposure to 220 ppm Pb increases catalase activity levels both in blood and key limbic areas and its expression in the same brain regions. Moreover, the excessive ethanol consumption reported in the Pb-exposed animals further increases catalase activity and expression in these brain areas, an effect that responds to catalase blockers and activators, including a lentiviral shRNA catalase vector administration in tegmental ventral area (Mattalloni et al., 2019b; Mattalloni et al., 2013; Virgolini et al., 2017; Mattalloni et al., 2019a). These results shed relevance in terms of the bicompartmental ethanol metabolism where catalase is the major catalyzer of brain ethanol oxidation to acetaldehyde, with this metabolite considered a positive reinforcement that prompts ethanol consumption. Although the consequence of Pb availability in this case results in enzymatic activation instead of inhibition, a plausible mechanism may be related to the putative "promiscuity" exhibited by the enzyme in terms of the metal cofactor required for its catalytic properties. Further research on catalase activity in terms of metalloenzymes composition is needed to support this hypothesis. Unfortunately, no studies have been performed to evaluate CYP2E1 functionality which would allow for speculation as to whether similar consequences might take place after Pb exposure.

3.3 ALDH

Evidence suggests that competition between Pb^{2+} and Mg^{2+} for binding sites at the molecular and cellular levels, as well as at the systemic level may represent an important aspect of Pb toxicity in the human body. For instance, Pb^{2+} has been found to displace Mg^{2+} in pyrimidine 5'-nucleotidase type 1, inhibiting the activity of the enzyme leading to anemia (Kirberger et al., 2013; Wyparło-Wszelaki et al., 2022). The limited evidence that addressed the interaction between Pb, ethanol and ALDH2 has shown that Pb reduced hepatic ALDH2 enzymatic activity (Flora and Tandon, 1987). Results from our laboratory demonstrated decreased ALDH2 activity in whole brain and reduced expression in mesolimbic areas after ethanol consumption in perinatal Pb-exposed rats (Mattalloni et al., 2017). These results were reproduced in ALDH2 in SH-SY5Y human neuroblastoma cells where ALDH2 enzymatic activity and expression after exposure to Pb, ethanol or their combination were reversed after Alda-1 administration. We can ascribe these differences to Pb's propensity to displace Mg^{2+} from the near catalytic site of the ALDH2 enzyme. Another possibility is related to Pb's affinity for thiol groups, such as those in the amino acid Cys, which is a crucial component of the ALDH2 catalytic site, Cys302 (Deza-Ponzio et al., 2023).

4. Conclusion

Taken together, the data presented herein point out the possibility that non-essential metals competition for prosthetic groups of key metalloenzymes participating in the oxidative ethanol metabolism may cause altered responses to this drug with neurobiological consequences. Above all, it is well-known that excessive alcohol consumption disrupts the concentrations of crucial trace elements, making it critical to understand their importance in early diagnosis and to improve treatment strategies. As many of them are involved in alcohol metabolism by modulating the catalytic activity of enzymes that produce toxic metabolites, they can lead to cell damage due to oxidative stress, interfering in the mechanisms of chronic AUD (Namachivayam and Valsala Gopalakrishnan, 2021). This pro-oxidant environment should be counteracted by endogenous enzymes that protect the body against the adverse effects of free radicals, including glutathione peroxidase (GPx), glutathione reductase (GR), superoxide dismutase (SOD) and catalase (CAT), all containing trace elements such as Se, Mn, Cu, Zn, or Fe which in turn, can be replaced by toxic-non-essential elements (Contreras-Zentella et al., 2022). Thus, chronic alcohol consumption alters essential micro- and macronutrient levels resulting in abnormalities as Fe overload, hypophosphatemia, hypomagnesemia, hypokalemia, hypocalcemia, and hyponatremia, as well as decreased levels of Se, Cr, and Zn (Skalny et al., 2018; Baj et al., 2020). Thus, it can be concluded that overall, metal-bounds to both ethanol-metabolizing and antioxidant enzymes are key components for maintaining homeostasis and critical for the adequate balance between health/disease status in a living organism.

References

Acquas, E., Scheggi, S. and Peana, A.T. 2019. Neurobiological Aspects of Ethanol-derived Salsolinol. Neuroscience of Alcohol: Mechanisms and Treatment. Elsevier Inc. https://doi.org/10.1016/B978-0-12-813125-1.00024-6.

Aebi, H. 1984. Catalase *in vitro*. Methods in Enzymology 105(C): 121–126. https://doi.org/10.1016/S0076-6879(84)05016-3.

Alaimo, J.T., Davis, S.J., Song, S.S., Burnette, C.R., Ph, D., Shelton, K.L. et al. 2013. NIH Public Access 36(11): 1840–1850. https://doi.org/10.1111/j.1530-0277.2012.01799.x.Ethanol.

Albrecht, P.A., Fernandez-Hubeid, L.E., Deza-Ponzio, R., Romero, V.L., Gonzales-Moreno, C., Carranza, A.D.V. et al. 2022. Reduced acute functional tolerance and enhanced preference for ethanol in *Caenorhabditis elegans* exposed to lead during development: Potential role of alcohol dehydrogenase. Neurotoxicology and Teratology 94(October): 107131. https://doi.org/10.1016/j.ntt.2022.107131.

Baj, J., Flieger, W., Teresiński, G., Buszewicz, G., Sitarz, R., Forma, A. et al. 2020. Magnesium, calcium, potassium, sodium, phosphorus, selenium, zinc, and chromium levels in alcohol use disorder: A review. Journal of Clinical Medicine 9(6): 1–24. https://doi.org/10.3390/jcm9061901.

Baker, K., Marcus, C.B., Huffman, K., Kruk, H., Malfroy, B. and Doctrow, S.R. 1998. Synthetic combined superoxide dismutase/catalase mimetics are protective as a delayed treatment in a rat stroke model: a key role for reactive oxygen species in ischemic brain injury. The Journal of Pharmacology and Experimental Therapeutics 284(1): 215–221.

Ballatori, N. 2002. Transport of toxic metals by molecular mimicry. Environmental Health Perspectives 110(SUPPL. 5): 689–694. https://doi.org/10.1289/ehp.02110s5689.

Barth, K.S. and Malcolm, R.J. 2010. Disulfiram: an old therapeutic with new applications. CNS & Neurological Disorders Drug Targets 9(1): 5–12. https://doi.org/10.2174/187152710790966678.

Bennett, A.F., Buckley, P.D. and Blackwell, L.F. 1983. Inhibition of the dehydrogenase activity of sheep liver cytoplasmic aldehyde dehydrogenase by magnesium ions. Biochemistry 22(4): 776–784. https://doi.org/10.1021/bi00273a011.

Binienda, Z., Simmons, C., Hussain, S., Slikker, W. and Ali, S.F. 1998. Effect of acute exposure to 3-nitropropionic acid on activities of endogenous antioxidants in the rat brain. Neuroscience Letters 251(3): 173–176. https://doi.org/10.1016/S0304-3940(98)00539-4.

Black, W.J., Stagos, D., Marchitti, S.A., Nebert, D.W., Keith, F., Bairoch, A. et al. 2012. NIH Public Access 19(11): 893–902. https://doi.org/10.1097/FPC.0b013e3283329023.Human.

Bridges, C.C. and Zalups, R.K. 2005. Molecular and ionic mimicry and the transport of toxic metals. Toxicology and Applied Pharmacology 204(3): 274–308. https://doi.org/10.1016/j.taap.2004.09.007.

Calleja, L.F., Belmont-Díaz, J.A., Medina-Contreras, O., Quezada, H., Yoval-Sánchez, B., Campos-García, J. et al. 2020. Omeprazole as a potent activator of human cytosolic aldehyde dehydrogenase ALDH1A1. Biochimica et Biophysica Acta - General Subjects 1864(1): 129451. https://doi.org/10.1016/j.bbagen.2019.129451.

Cavallaro, A., Lavanco, G., Giammanco, M. and Cannizzaro, E. 2017. Acetaldehyde and salsolinol in ethanol's two-step mechanism of action: An overview. Journal of Biological Research (Italy) 90(2): 109–115. https://doi.org/10.4081/jbr.2017.6751.

Cederbaum, A.I. 2012. Alcohol metabolism. Clinics in Liver Disease 16(4): 667–685. https://doi.org/10.1016/j.cld.2012.08.002.

Chen, A.Y., Adamek, R.N., Dick, B.L., Credille, C.V., Morrison, C.N. and Cohen, S.M. 2019. Targeting metalloenzymes for therapeutic intervention. Chemical Reviews 119(2): 1323–1455. review-article. https://doi.org/10.1021/acs.chemrev.8b00201.

Chen, C., Budas, G.R., Churchill, E.N., Disatnik, M., Thomas, D. and Mochly-rosen, D. 2008. An activator of mutant and wildtype aldehyde dehydrogenase reduces ischemic damage to the heart. Science 321(5895): 1493–1495. https://doi.org/10.1126/science.1158554.

Chen, C.-H., Ferreira, J.C.B., Gross, E.R. and Mochly-Rosen, D. 2014. Targeting aldehyde dehydrogenase 2: new therapeutic opportunities. Physiological Reviews 94(1): 1–34. https://doi.org/10.1152/physrev.00017.2013.

Chen, Y.C., Peng, G.S., Wang, M.F., Tsao, T.P. and Yin, S.J. 2009. Polymorphism of ethanol-metabolism genes and alcoholism: Correlation of allelic variations with the pharmacokinetic and pharmacodynamic consequences. Chemico-Biological Interactions 178(1–3): 2–7. https://doi.org/10.1016/j.cbi.2008.10.029.

Contreras-Zentella, M.L., Villalobos-García, D. and Hernández-Muñoz, R. 2022. Ethanol metabolism in the liver, the induction of oxidant stress, and the antioxidant defense system ethanol metabolism in the liver, the induction of oxidant stress, and the antioxidant defense system. Antioxidants 11(7): 1–26. https://doi.org/10.3390/antiox11071258.

da Silva Cunha, Thayssa Tavares, de Souza, Felipe Rodrigues, de Sena Murteira Pinheiro, Pedro, de Sant'Anna et al. 2020. Investigating the molecular basis for the selective inhibition of aldehyde dehydrogenase 2 by the isoflavonoid daidzin. CNS & Neurological Disorders - Drug Targets 19(6).

Deitrich, R.A., Troxell, P.A. and Worth, W.S. 1976. Inhibition of aldehyde dehydrogenase in brain and liver by cyanamide. Biochem. Pharmacol. 25(24): 2733–2737. Retrieved from http://www.ncbi.nlm.nih.gov/entrez/query.fcgi?cmd=Retrieve&db=PubMed&dopt=Citation&list_uids=1008896.

Deza-Ponzio, R., Albrecht, P.A., Fernandez-Hubeid, L.E., Eichwald, T., Cejas, R.B., Garay, Y.C. et al. 2023. ALDH2 inhibition by lead and ethanol elicits redox imbalance and mitochondrial dysfunction in SH-SY5Y human neuroblastoma cell line: Reversion by Alda-1. Neurotoxicology 97: 12–24. https://doi.org/10.1016/j.neuro.2023.05.001.

Di, L., Balesano, A., Jordan, S. and Shi, S.M. 2021. The role of alcohol dehydrogenase in drug metabolism: beyond ethanol oxidation. AAPS Journal 23(1). https://doi.org/10.1208/s12248-020-00536-y.

Díaz, A., Loewen, P.C., Fita, I. and Carpena, X. 2012. Thirty years of heme catalases structural biology. Archives of Biochemistry and Biophysics 525(2): 102–110. https://doi.org/10.1016/j.abb.2011.12.011.

Dickinson, F.M. and Hart, G.J. 1982. Sheep Liver Cytoplasmic Aldehyde Dehydrogenase 205: 443–448.

Edenberg, H.J. 2007. The genetics of alcohol metabolism: role of alcohol dehydrogenase and aldehyde dehydrogenase variants. Alcohol Research & Health: The Journal of the National Institute on Alcohol Abuse and Alcoholism 30(1): 5–13. https://doi.org/10.3168/jds.2010-3914.

Edenberg, H.J. and McClintick, J.N. 2018. Alcohol dehydrogenases, aldehyde dehydrogenases, and alcohol use disorders: a critical review. Alcoholism: Clinical and Experimental Research 42(12): 2281–2297. https://doi.org/10.1111/acer.13904.

Eom, H. and Song, W.J. 2019. Emergence of metal selectivity and promiscuity in metalloenzymes. Journal of Biological Inorganic Chemistry 24(4): 517–531. https://doi.org/10.1007/s00775-019-01667-0.

Flora, S.J.S. and Tandon, S.K. 1987. Effect of combined exposure to cadmium and ethanol on regional brain biogenic amine levels in the rat. Biochemistry International 15(4): 863–871.

Fujiki, Y. and Bassik, M.C. 2021. A new paradigm in catalase research. Trends in Cell Biology 31(3): 148–151. https://doi.org/10.1016/j.tcb.2020.12.006.

Gebhardt, A.C., Lucas, D., Ménez, J.F. and Seitz, H.K. 1997. Chlormethiazole inhibition of cytochrome p450 2E1 as assessed by chlorzoxazone hydroxylation in humans. Hepatology 26(4): 957–961. https://doi.org/10.1002/hep.510260423.

Gibbons, B.J. and Hurley, T.D. 2004. Structure of three class I human alcohol dehydrogenases complexed with isoenzyme specific formamide inhibitors. Biochemistry 43(39): 12555–12562. https://doi.org/10.1021/bi0489107.

Gonnella, T.P., Keating, J.M., Kjemhus, J.A., Picklo, M.J. and Biggane, J.P. 2013. Fluorescence lifetime analysis and effect of magnesium ions on binding of NADH to human aldehyde dehydrogenase 1. Chemico-Biological Interactions 202(1–3): 85–90. https://doi.org/10.1016/j.cbi.2012.12.008.

Goyal, M.M. and Basak, A. 2010. Human catalase: Looking for complete identity. Protein and Cell 1(10): 888–897. https://doi.org/10.1007/s13238-010-0113-z.
Hambidge, M., Cousins, R.J. and Costello, R.B. 2000. Zinc and health: Current status and future directions: Introduction. Journal of Nutrition 130(5 SUPPL.).
Hansberg, W. 2022. Monofunctional heme-catalases. Antioxidants 11(11). https://doi.org/10.3390/antiox11112173.
Hartman, J.H., Miller, G.P. and Meyer, J.N. 2017. Toxicological implications of mitochondrial localization of CYP2E1. Toxicology Research 6(3): 273–289. https://doi.org/10.1039/c7tx00020k.
Hu, J., Tian, W., Zhou, R., Zhang, Y., Lv, J., Zhu, J. et al. 2019. Design, synthesis, and biological evaluation of new ALDH2 activators. Journal of Saudi Chemical Society 23(3): 255–262. https://doi.org/10.1016/j.jscs.2018.07.001.
Hu, P., Jin, L. and Baillie, T.a. 1997. Studies on the metabolic activation of disulfiram in rat. Evidence for electrophilic S-oxygenated metabolites as inhibitors of aldehyde dehydrogenase and precursors of urinary N-acetylcysteine conjugates. The Journal of Pharmacology and Experimental Therapeutics 281(2): 611–617.
Israel, Y., Quintanilla, M.E., Karahanian, E., Rivera-Meza, M. and Herrera-Marschitz, M. 2015. The "first hit" toward alcohol reinforcement: Role of ethanol metabolites. Alcoholism: Clinical and Experimental Research 39(5): 776–786. https://doi.org/10.1111/acer.12709.
Jackson, B., Brocker, C., Thompson, D.C., Black, W., Vasiliou, K., Nebert, D.W. et al. 2011. Update on the aldehyde dehydrogenase gene (ALDH) superfamily. Human Genomics 5(4): 283–303. https://doi.org/7LM2078310429454 [pii].
Jomova, K., Makova, M., Alomar, S.Y., Alwasel, S.H., Nepovimova, E., Kuca, K. et al. 2022. Essential metals in health and disease. Chemico-Biological Interactions 367(August): 110173. https://doi.org/10.1016/j.cbi.2022.110173.
Keung, W.M., Klyosov, A.A. and Vallee, B.L. 1997. Daidzin inhibits mitochondrial aldehyde dehydrogenase and suppresses ethanol intake of Syrian golden hamsters. Proceedings of the National Academy of Sciences of the United States of America 94(5): 1675–1679. https://doi.org/10.1073/pnas.94.5.1675.
Keung, Wing Ming. 2003. Anti-dipsotropic isoflavones: the potential therapeutic agents for alcohol dependence. Medicinal Research Reviews 23(6): 669–696. https://doi.org/10.1002/med.10049.
Kirberger, M., Wong, H.C., Jiang, J. and Yang, J.J. 2013. Metal toxicity and opportunistic binding of Pb2 + in proteins. Journal of Inorganic Biochemistry. https://doi.org/10.1016/j.jinorgbio.2013.04.002.
Koppaka, V., Thompson, D.C., Chen, Y., Ellermann, M., Nicolaou, K.C., Juvonen, R.O. et al. 2012. Aldehyde dehydrogenase inhibitors: a comprehensive review of the pharmacology, mechanism of action, substrate specificity, and clinical application. Pharmacological Reviews 64(3): 520–539. https://doi.org/10.1124/pr.111.005538.
Liu, L.K. and Tanner, J.J. 2019. Crystal structure of aldehyde dehydrogenase 16 reveals trans-hierarchical structural similarity and a new dimer. Journal of Molecular Biology 431(3): 524–541. https://doi.org/10.1016/j.jmb.2018.11.030.
Liu, Y., Yamanaka, M., Abe-Kanoh, N., Liu, X., Zhu, B., Munemasa, S. et al. 2017. Benzyl isothiocyanate ameliorates acetaldehyde-induced cytotoxicity by enhancing aldehyde dehydrogenase activity in murine hepatoma Hepa1c1c7 cells. Food and Chemical Toxicology 108: 305–313. https://doi.org/10.1016/j.fct.2017.08.016.
Manrique, H.M., Miquel, M. and Aragon, C.M.G. 2006. Acute administration of 3-nitropropionic acid, a reactive oxygen species generator, boosts ethanol-induced locomotor stimulation. New support for the role of brain catalase in the behavioural effects of ethanol. Neuropharmacology 51(7–8): 1137–1145. https://doi.org/10.1016/j.neuropharm.2006.07.022.
Marchitti, S.A., Deitrich, R.A. and Vasiliou, V. 2007. Neurotoxicity and metabolism of the catecholamine-derived 3, 4-dihydroxyphenylacetaldehyde and the role of aldehyde dehydrogenase. Pharmacological Reviews 59(2): 125–150. https://doi.org/10.1124/pr.59.2.1.125.
Margoliash, E. and Novogrodsky, A. 1958. A study of the inhibition of catalase by 3-amino-1: 2: 4-triazole. Biochemical Journal 68(3): 468–475. doi: 10.1042/bj0680468.
Mattalloni, Mara S., De Giovanni, L.N., Molina, J.C., Cancela, L.M. and Virgolini, M.B. 2013. Participation of catalase in voluntary ethanol consumption in perinatally low-level lead-exposed

rats. Alcoholism: Clinical and Experimental Research 37(10): 1632–1642. https://doi.org/10.1111/acer.12150.
Mattalloni, M.S., Deza-Ponzio, R., Albrecht, P.A., Cancela, L.M. and Virgolini, M.B. 2017. Developmental lead exposure induces opposite effects on ethanol intake and locomotion in response to central vs. systemic cyanamide administration. Alcohol 58. https://doi.org/10.1016/j.alcohol.2016.11.002.
Mattalloni, M.S., Albrecht, P.A., Salinas-Luypaert, C., Deza-Ponzio, R., Quintanilla, M.E., Herrera-Marschitz, M. et al. 2019a. Silencing brain catalase expression reduces ethanol intake in developmentally-lead-exposed rats. NeuroToxicology 70. https://doi.org/10.1016/j.neuro.2018.10.010.
Mattalloni, M.S., Deza-Ponzio, R., Albrecht, P.A., Fernandez-Hubeid, L.E., Cancela, L.M. and Virgolini, M.B. 2019b. Brain ethanol-metabolizing enzymes are differentially expressed in lead-exposed animals after voluntary ethanol consumption: Pharmacological approaches. NeuroToxicology 75. https://doi.org/10.1016/j.neuro.2019.09.011.
Morrison, H. 2020. Enzyme Active Sites and Their Reaction Mechanisms. Academic Press.
Namachivayam, A. and Valsala Gopalakrishnan, A. 2021. A review on molecular mechanism of alcoholic liver disease. Life Sciences 274(November 2020). https://doi.org/10.1016/j.lfs.2021.119328.
Oshino, N., Oshino, R. and Chance, B. 1973. The characteristics of the "peroxidatic" reaction of catalase in ethanol oxidation. The Biochemical Journal 131(3): 555–563. Retrieved from http://www.pubmedcentral.nih.gov/articlerender.fcgi?artid=1177502&tool=pmcentrez&rendertype=abstract.
Overstreet, D.H., Keung, W.-M., Rezvani, A.H., Massi, M. and Lee, D.Y.W. 2003. Herbal remedies for alcoholism: promises and possible pitfalls. Alcoholism, Clinical and Experimental Research 27(2): 177–185. https://doi.org/10.1097/01.ALC.0000051022.26489.CF.
Peng, J., Stevenson, F.F., Doctrow, S.R. and Andersen, J.K. 2005. Superoxide dismutase/catalase mimetics are neuroprotective against selective paraquat-mediated dopaminergic neuron death in the substantial Nigra: Implications for Parkinson disease. Journal of Biological Chemistry 280(32): 29194–29198. https://doi.org/10.1074/jbc.M500984200.
Pequerul, R., Vera, J., Giménez-Dejoz, J., Crespo, I., Coines, J., Porté, S. et al. 2020. Structural and kinetic features of aldehyde dehydrogenase 1A (ALDH1A) subfamily members, cancer stem cell markers active in retinoic acid biosynthesis. Archives of Biochemistry and Biophysics 681(January): 108256. https://doi.org/10.1016/j.abb.2020.108256.
Perez-Miller, S.J. and Hurley, T.D. 2003. Coenzyme isomerization is integral to catalysis in aldehyde dehydrogenase. Biochemistry 42(23): 7100–7109. https://doi.org/10.1021/bi034182w.
Perez-Miller, S., Younus, H., Vanam, R., Chen, C.-H., Mochly-Rosen, D. and Hurley, T.D. 2010. Alda-1 is an agonist and chemical chaperone for the common human aldehyde dehydrogenase 2 variant. Nature Structural & Molecular Biology 17(2): 159–164. https://doi.org/10.1038/nsmb.1737.
Plapp, B.V., Savarimuthu, B.R., Ferraro, D.J., Rubach, J.K., Brown, E.N. and Ramaswamy, S. 2017. Horse liver alcohol dehydrogenase: zinc coordination and catalysis. Biochemistry 56(28): 3632–3646. https://doi.org/10.1021/acs.biochem.7b00446.
Rao, P.S.S., Midde, Narasimha M., Miller, Duane D., Chauhan, S., Kumar, A. and Kumar, S. 2015. Diallyl sulfide: potential use in novel therapeutic interventions in alcohol, drugs, and disease mediated cellular toxicity by targeting cytochrome P450 2E1. Current Drug Metabolism 16(6): 486–503.
Rawles, J.W., Rhodes, D.L., Potter, J.J. and Mezey, E. 1987. Characterization of human erythrocyte aldehyde dehydrogenase. Biochemical Pharmacology 36(21): 3715–3722. https://doi.org/10.1016/0006-2952(87)90025-6.
Rezvani, A.H., Overstreet, D.H., Perfumi, M. and Massi, M. 2003. Plant derivatives in the treatment of alcohol dependency. Pharmacology Biochemistry and Behavior 75(3): 593–606. https://doi.org/10.1016/S0091-3057(03)00124-2.
Rivera-Meza, M., Vásquez, D., Quintanilla, M.E., Lagos, D., Rojas, B., Herrera-Marschitz, M. et al. 2019. Activation of mitochondrial aldehyde dehydrogenase (ALDH2) by ALDA-1 reduces both the acquisition and maintenance of ethanol intake in rats: A dual mechanism? Neuropharmacology 146(November 2018): 175–183. https://doi.org/10.1016/j.neuropharm.2018.12.001.
Sanny, C.G. and Rymas, K. 1993. *In vivo* effects of disulfiram and cyanamide on canine liver aldehyde dehydrogenase isoenzymes as detected by high-performance (pressure) liquid chromatography.

Alcoholism, Clinical and Experimental Research 17(5): 982–987. Retrieved from http://www.ncbi.nlm.nih.gov/pubmed/8279685.

Sepasi Tehrani, H. and Moosavi-Movahedi, A.A. 2018. Catalase and its mysteries. Progress in Biophysics and Molecular Biology 140: 5–12. https://doi.org/10.1016/j.pbiomolbio.2018.03.001.

Sheikh, S., Ni, L., Hurley, T.D. and Weiner, H. 1997. The potential roles of the conserved amino acids in human liver mitochondrial aldehyde dehydrogenase. Journal of Biological Chemistry 272(30): 18817–18822. https://doi.org/10.1074/jbc.272.30.18817.

Skalny, A.V., Skalnaya, M.G., Grabeklis, A.R., Skalnaya, A.A. and Tinkov, A.A. 2018. Zinc deficiency as a mediator of toxic effects of alcohol abuse. European Journal of Nutrition 57(7): 2313–2322. https://doi.org/10.1007/s00394-017-1584-y.

Sophos, N.A. and Vasiliou, V. 2003. Aldehyde dehydrogenase gene superfamily: The 2002 update. Chemico-Biological Interactions 143–144: 5–22. https://doi.org/10.1016/S0009-2797(02)00163-1.

Stachowicz, A., Olszanecki, R., Suski, M., Głombik, K., Basta-Kaim, A., Adame, D. et al. 2017. Proteomic analysis of mitochondria-enriched fraction isolated from the frontal cortex and hippocampus of apolipoprotein e knockout mice treated with alda-1, an activator of mitochondrial aldehyde dehydrogenase (ALDH2). International Journal of Molecular Sciences 18(2): 1–16. https://doi.org/10.3390/ijms18020435.

Takahashi, K. and Weiner, H. 1980. Magnesium stimulation of catalytic activity of horse liver aldehyde dehydrogenase. Changes in molecular weight and catalytic sites. Journal of Biological Chemistry 255(17): 8206–8209. https://doi.org/10.1016/s0021-9258(19)70631-0.

Takahashi, Kojiro, Weiner, H. and Filmer, D.L. 1981. Effects of pH on horse liver aldehyde dehydrogenase: alterations in metal ion activation, number of functioning active sites, and hydrolysis of the acyl intermediate. Biochemistry 20(21): 6225–6230. https://doi.org/10.1021/bi00524a049.

Tawa, E.A., Hall, S.D. and Lohoff, F.W. 2016. Overview of the genetics of alcohol use disorder. Alcohol Alcohol 51(5): 507–514.

Teschke, R. 2019. Microsomal ethanol-oxidizing system: success over 50 years and an encouraging future. Alcoholism: Clinical and Experimental Research 43(3): 386–400. https://doi.org/10.1111/acer.13961.

Vallari, R.C. and Pietruszko, R. 1984. Interaction of Mg^{2+} with human liver aldehyde dehydrogenase. II. Mechanism and site of interaction. Journal of Biological Chemistry 259(8): 4927–4933. https://doi.org/10.1016/s0021-9258(17)42935-8.

Vallee, B.L. and Hoch, F.L. 1955. Zinc, a component of yeast alcohol dehydrogenase. Proceedings of the National Academy of Sciences 41(6): 327–338. https://doi.org/10.1073/pnas.41.6.327.

Vaswani, M. 2019. ADH and ALDH polymorphisms in alcoholism and alcohol misuse/dependence. Neuroscience of Alcohol: Mechanisms and Treatment. Elsevier Inc. https://doi.org/10.1016/B978-0-12-813125-1.00004-0.

Vetrano, A.M., Heck, D.E., Mariano, T.M., Mishin, V., Laskin, D.L. and Laskin, J.D. 2005. Characterization of the oxidase activity in mammalian catalase. Journal of Biological Chemistry 280(42): 35372–35381. https://doi.org/10.1074/jbc.M503991200.

Virgolini, M.B., Mattalloni, M.S., Albrecht, P.A., Deza-Ponzio, R. and Cancela, L.M. 2017. Modulation of ethanol-metabolizing enzymes by developmental lead exposure: Effects in voluntary ethanol consumption. Frontiers in Behavioral Neuroscience 11. https://doi.org/10.3389/fnbeh.2017.00095.

Virgolini, Miriam B. and Aschner, M. 2021. Molecular mechanisms of lead neurotoxicity. Adv. Neurotoxicol. 5: 159–213.

Weiner, H. and Takahashi, K. 1983. Effects of magnesium and calcium on mitochondrial and cytosolic liver aldehyde dehydrogenases. Pharmacology, Biochemistry and Behavior 18(SUPPL. 1): 109–112. https://doi.org/10.1016/0091-3057(83)90155-7.

Williamson, V.M., Long, M. and Theodoris, G. 1991. Isolation of *Caenorhabditis elegans* mutants lacking alcohol dehydrogenase activity. Biochemical Genetics 29(7–8): 313–323. https://doi.org/10.1007/BF00554139.

Wyparło-Wszelaki, M., Machoń-Grecka, A., Wąsik, M. and Dobrakowski, M. 2022. Critical aspects of the physiological interactions between lead and magnesium. Journal of Biochemical and Molecular Toxicology 36(2): 22964.

Zalups, R.K. and Koropatnick, J. 2004. Cellular and molecular neurobiology: foreword. Cellular and Molecular Neurobiology (Vol. 24). https://doi.org/10.1023/B:CEMN.0000023697.07673.74.

Zamocky, M., Furtmüller, P.G. and Obinger, C. 2008. Evolution of catalases from bacteria to humans. Antioxidants and Redox Signaling 10(9): 1527–1547. https://doi.org/10.1089/ars.2008.2046.

Zhou, F., Chen, Y., Fan, G., Feng, C., Du, G., Zhu, G. et al. 2014. Lead-induced iron overload and attenuated effects of ferroportin 1 overexpression in PC12 cells. Toxicology *in Vitro* 28(8): 1339–1348. https://doi.org/10.1016/j.tiv.2014.07.005.

Zhu, G., Fan, G., Feng, C., Li, Y., Chen, Y., Zhou, F. et al. 2013. The effect of lead exposure on brain iron homeostasis and the expression of DMT1/FP1 in the brain in developing and aged rats. Toxicology Letters 216(2–3): 108–123. https://doi.org/10.1016/j.toxlet.2012.11.02.

Chapter 7

Toxicology of Vanadium with Emphasis on the Central Nervous System

Ifukibot Levi Usende,[1] *Funmilayo E. Olopade*[2] and *James Olukayode Olopade*[3,*]

1. Introduction

Vanadium (V), a metallic element of the 1st transition series has an atomic number and atomic weight of 23 and 50.9415 respectively and exists in variable oxidation states (–1 to +5) (Olopade and Connor, 2011; Olaolorun et al., 2021). Vanadium is named in honor of Vanadis, the goddess of beauty and fertility (Shaver et al., 1995; Olopade and Connor, 2011). Vanadium is silvery and a metal with corrosion resistant properties and possesses varying industrial uses ranging from automobiles and steel production by construction companies, to productions of ceramic, fertilizer, batteries and pigments/paints (Imtiaz et al., 2015; Olaolorun et al., 2021). Vanadium compounds have been shown to biologically exhibits two opposite effects, being essential as barely detectable amounts (0.05 μM) and very toxic when in excess quantity (> 10 μM) (Scibior et al., 2020a,b; Das et al., 2012; Olaolorun et al., 2014). Vanadium compounds have been used therapeutically, and are well known for their effects against diabetes, infections (fungal, viral, bacterial, or parasitic), cancer, obesity conditions as well as their anti-hyper-cholesterolemic, neuroprotective and cardioprotective roles at very minimal levels (Olopade and Connor, 2011; Scibior et al., 2020a, b), but are capable of inducing severe pathological alterations at high (toxic) doses (Scibior et al., 2020a, b). Vanadium compounds are known to progressively accumulate in different organs (including bone, kidneys, liver, spleen, testes and lungs) (Cortizo et al., 2000; Usende et al., 2017) as well as in the brain (Folarin et al., 2017) and become toxic.

[1] Department of Veterinary Anatomy, University of Abuja, Nigeria.
[2] Department of Anatomy, University of Ibadan, Nigeria.
[3] Department of Veterinary Anatomy, University of Ibadan, Nigeria.
* Corresponding author: jkayodeolopade@yahoo.com

Vanadium compounds' toxicity is dependent on their species (Evangelou, 2002), administration pathway (parenteral > inhalation > ingestion) (Barceloux, 1999), valency (vanadates V+5 is more toxic than vanadyl V+4 compounds) (Domingo, 2002), exposure duration (Folarin et al., 2017; Usende et al., 2018a, b) and dose administered (Paternain et al., 1990) and. The word "Vanadiumism" was coined by Dutton in 1911 to report vanadium toxicity in people working in the industries and inhaled dust particles of V pentoxide. Anemia has been identified as the first symptom of vanadiumism, which was followed by respiratory (an increasing vulnerability to tuberculosis condition), neurological and gastrointestinal problems and accompanying symptoms, in the course of continuous exposure (Dutton, 1911; Nechay, 1984; Olopade and Connor, 2011; Li et al., 2013). Some other toxic consequences of exposure to vanadium metal include tumors in humans and in animal models (Stocks, 1960; Rojas et al., 1999; Hickey et al., 1967; Stern et al., 1993; Parfett and Pilon, 1995), DNA damage and apoptotic effects (Rojas et al., 1996; Sakurai, 1994; Altamirano-Lozano et al., 1999; Usende et al., 2018b), cardiovascular (Soares et al., 2008) and damage of the reproductive system, fetal toxicity (Domingo, 1996; Paternain et al., 1990; Usende et al., 2022a); alteration of biochemical and hematological balance (Adebiyi et al., 2015; Usende et al., 2018a), histopathological lesioning and organotoxicity (Olaolorun et al., 2014; Liu et al., 2012; Usende et al., 2018b), neurobehavioral deficits, neurotoxicity, including neuroinflammation (Usende et al., 2016, 2022b; Olopade and Connor, 2011; Li et al., 2013; Azeez et al., 2016; Folarin et al., 2016, 2017; Ngwa et al., 2009).

Vanadium neurotoxicity is a rising concern and studies on the toxicopathological, neuroinflammatory and behavioral effects of this metal is recently receiving attention (Ngwa et al., 2009; Olopade and Connor, 2011; Li et al., 2013; Azeez et al., 2016; Folarin et al., 2017; Usende et al., 2016, 2018b, 2022b; Olaolorun et al., 2021). Vanadium easily travels across the BBB and accumulates in different brain regions to stir up neuroinflammatory processes and demyelination among others (Folarin et al., 2017; Soazo and Garcia, 2007; Todorich et al., 2011; Olopade and Connor, 2011; Usende et al., 2016, 2022b). In the neuronal cell culture model, vanadium has been shown to be conveyed into the cells through divalent metal ion transporter or transferrin receptor mediated internalization (Olaolorun et al., 2022; Erikson et al., 2004) to induce cell death through oxidative stress mechanistic pathways by depletion of intracellular antioxidants while increasing generation of ROS (Sasi et al., 1994; Olaolorun et al., 2022; Thompson and McNeill, 1993). Ngwa et al. (2009) posited that vanadium exposure may be linked to increased incidence of cases of neurodegenerative diseases particularly Parkinson's disease, because vanadium causes dopaminergic neuronal degeneration and apoptosis in experimental models.

Our focus here is on vanadium's neurotoxic effects and possible explanatory mechanisms, drawing on studies on cell culture and in rodents. We focused on vanadium's entry and accumulation in various brain regions and its induction of oxidative stress, biochemical alterations and neurobehavioral deficits. We then focused on the role and effects of vanadium toxicity on ependymal cells and ependymal layer, neuronal populations sensitive to oxidative stress and glial cells activation and myelin damage.

2. *In vitro* studies of vanadium neurotoxicity

In vitro studies, no doubt, are helping to provide valuable insights into the mechanism of vanadium neurotoxicity. Specifically, reports have shown using different cell culture models, vanadium induced ROS generation (although these reports are controversial), damage of DNA, and apoptosis initiation as key elements in neurotoxicity of this metal (Olaolorun et al., 2021). However, ROS' role in vanadium neurotoxicity and the apoptotic mechanism in cell culture models remains to be fully elucidated as these pathways vary with cell type and the species of vanadium compounds used. Concerning the CNS derived cell culture models used in vanadium neurotoxicity studies, focus has been on the neuronal and oligodendrocytes cell lines, possibly because of the neurobehavioral deficits and the level of demyelination reported in histological sections in *in vivo* experimental models of neurotoxicity of vanadium (Ngwa et al., 2014; Folarin et al., 2016; Azeez et al., 2016; Usende et al., 2022b). Astrocytes have received little attention while no attention is given to microglia cultures.

On neuronal primary cerebellar granule cell progenitors (CGPs) cultures, vanadium has been shown to induce cell death essentially via activation of extrinsic pathway of JNK dependent apoptosis (Luo et al., 2003). According to the authors (Luo et al., 2003), vanadium was able to generate ROS in their CGPs model, however, they concluded that the ROS generated has a partial contribution to cell death (Luo et al., 2003). This was because antioxidants offered little protection while JNK antagonists totally eliminated the CGPs death caused by vanadium (Luo et al., 2003). Contrasting, in other studies using hippocampal neuronal cell of mouse origin (HT22 cells) (Igado et al., 2018) and cell line of dopaminergic neurons (Ngwa et al., 2009), reactive oxygen species have been placed central in cascade of neurotoxicity of vanadium. With HT22 cell line, Igado et al. (2018) showed increased generation of ROS and DNA damage after vanadium exposure to these cells, and posited that the cytotoxicity induced by vanadium was significantly reduced with the novel antioxidant compound, MIMO2, derived from Moringa oleifera. Ngwa et al. (2009) on the other hand, and in a different experiment using dopaminergic neuronal cell line, showed an important role for reactive oxygen species in vanadium induced cytotoxicity and apoptosis. The apoptotic pathway implicated mitochondrial pathway requiring the release of cytochrome c and activation of caspase-9 (Ngwa et al., 2009). The authors (Ngwa et al., 2009) unlike Luo et al. (2003) revealed that antioxidants protected the cells against vanadium induced cytotoxicity.

Concerning astrocytes culture and vanadium toxicity studies, few reports available have shown that these cells are critical neuroprotective cells populations of the CNS, therefore are comparatively more resistant to vanadium toxicity in a time dependent manner *in vitro* (Scibior et al., 2020a, b; Aschner et al., 2010). However, Aschner et al. (2010) alone reported high production of ROS in cultures of primary astrocyte with corresponding elevated erythropoietin and HIF-1α expression after exposure to vanadium. Due to dearth of information available, there is need to explore microglia *in vitro* following vanadium exposure. Also, since cells of the CNS do not exist independently *in vivo* but interact to maintain hemostasis, there is an urgent call to investigate the interactions of these cells in co-culture experimental conditions bearing in mind the cross-talk between them.

3. Vanadium entry and accumulation in different brain regions

Vanadium traverses the BBB (Olopade and Connor, 2011), accumulating in various brain regions (Berman, 1980; Avila-Costa et al., 2005, 2006; Todorich et al., 2011; Azeez et al., 2016; Garcia et al., 2005; Folarin et al., 2017; Usende et al., 2016, 2017, 2022b) bringing about neurologic and neuropathologic consequences (Olopade and Connor, 2011). However, vanadium neurotoxicity depends on several factors including the cumulative dose, rate of dose administration, route of exposure and examination endpoint (Olaolorun et al., 2021). Studies that have demonstrated that vanadium accumulates in different brain regions also report that these accumulations are known to induce important neurochemical alterations such as changes in dopamine, 5-hydroxytryptamine and noradrenaline levels, as well as inhibition of uptake and release of noradrenaline (Ngwa et al., 2014; Olaolorun et al., 2021).

Concerning the effects of dose on vanadium neurotoxicity, Avila-Costa et al. (2005) in their pilot study exposed CD-1 male mice to vanadium pentoxide (V_2O_5) via inhalation at low concentration doses showed no significant changes in exposed mice brains relative to their matched controls. However, at a higher dose, (0.02 M) twice weekly, and after a 1–8 weeks period of exposure, they observed significantly increased vanadium accumulation in the brain only at week one of inhalation treatment and thereafter it became consistent. These findings reported by Avila-Costa et al. (2005) indeed implicate dose and not exposure duration as the important factor in the neurotoxicity of vanadium (Olaolorun et al., 2021). Folarin et al. (2017) explored the intraperitoneal route and exposed male BALB/c mice to 3 mg/kg of body weight of sodium metavanadate ($NaVO_3$) and sacrificed the mice in batches for every 3 months till 18 months, and their investigations with Laser ablation inductively coupled plasma-mass spectrometry (LA-ICP-MS) revealed a progressive high vanadium bio-uptake with increasing time of exposure in various regions of the brain including olfactory bulb, the brain stem and the cerebellum. Their findings (Folarin et al., 2017) corroborated well with the previous reports of Avila-Costa et al. (2004, 2005, 2006) which posit that vanadium crosses the BBB, enters intact the brain parenchyma and it accumulates progressively in various regions of the brain. Folarin et al. (2017) therefore maintained that the progressively increased accumulation of vanadium in the different brain regions of exposed mice is indicative of increased bio-uptake of vanadium into the brain in a timely manner. Interestingly, Folarin et al. (2017) reported that vanadium has a predilection for some brain regions such as the olfactory bulb, the brain stem as well as the cerebellum, thus corroborating with reports of Ngwa et al. (2014), Garcia et al. (2005) and Haider et al. (1998) on various pathologies seen in the olfactory bulb, cerebellum and brain stem following exposure to vanadate.

4. Mechanisms of vanadium induced neurotoxicity

Vanadium, like most other transition metals, has variable oxidation states and exhibits its toxicological effects in biosystems through a number of mechanistic pathways and its effects are dose dependent, although these mechanisms remain to be fully clarified (Scibior et al., 2020a, b). Vanadium is known to participates in redox reactions generating ROS and reduction of intracellular antioxidant levels leading to cell death via oxidative stress (Evangelou, 2002; Olopade and Connor, 2011; Usende et al., 2016, 2018a). As vanadyl ion bounded to transferrin, vanadium enters the cell by passive diffusion or endocytosis; also, as vanadate via anion channels, vanadium can find its way into the cell (Yang et al., 2003). While in the cell, intracellular antioxidants (especially Glutathione, GSH) reduces vanadate to vanadyl, in addition to ROS generation (Shi and Dalal, 1992; Ding et al., 1994). In the presence of hydrogen peroxide (H_2O_2), the vanadyl can be quickly oxidized back to vanadate following a Fenton-like reaction. Thus, though pentavalent and tetravalent vanadium are stable, they can be interconverted in redox cycling (Olaolorun et al., 2021; Beyersmann and Hartwig, 2008). In this process, a hydroxyl radical is produced (Capella et al., 2002) which activates chain reactions that leads to the generation of radicals of superoxide anion that is automatically dismutated to oxygen and H_2O_2 by superoxide anion dismutase (SOD) (Ejeh et al., 2019; Olaolorun et al., 2021). H_2O_2 next reacts with vanadyl yielding more OH^- (Evangelou, 2002). In a similar manner, vanadyl reacts with oxygen generating superoxide anion and vanadate (Stohs and Bagchi, 1995). In summary, these chain reactions ultimately generate reactive oxygen species (ROS) overwhelming the intracellular antioxidants system and causing cell death due to oxidative stress.

The other mechanisms of vanadium induced ROS production is via iron release from ferritin in a Fenton-like Reaction (Todorich et al., 2011; Evangelou, 2002), and the eventual leakage of electrons via the electron transport chain (Complexes II and III) in mitochondria (Olaolorun et al., 2021; Hosseini et al., 2013). Details of the ability of vanadium to release intracellularly stored iron and in the production of ROS in a Fenton-like reaction have been extensively studied by Todorich et al. (2011) and they posited that the oligodendrocyte precursor cells (OPCs) with high ferritin stores are sufferers of lesions of neuroinflammation and hypomyelination caused, than the mature oligodendrocytes and astrocytes (for details refer to Todorich et al., 2011). Interestingly, the reactive oxygen species (ROS) produced resulted in the peroxidation of the lipids of cell membranes seen as demyelinating lesions in nervous system (Azeez et al., 2016; Soazo and Garcia, 2007; Mustapha et al., 2014; Usende et al., 2016, 2022b), protein aggregations, mutagenic DNA processes activation, DNA degradation, and apoptotic induction (Lapenna et al., 2002; Altamirano-Lozano, 1998) following vanadium exposure.

5. Vanadium's oxidative stress induction

Vanadium's induction of oxidative stress is mainly via the generation of ROS and lipid peroxidation (Usende et al., 2016, 2018a; Folarin et al., 2017; Todorich et al., 2011; Jaiswal and Kale, 2020; Azeez et al., 2016) as already discussed above. Several studies in rodent models showed that upon crossing of the BBB, vanadium accumulates in the cerebellum, corpus callosum and hippocampus, and these are major brain regions affected by the oxidative stress induced by vanadium, which leads to dramatic morphophysiological and numerical changes in neuronal populations, microglia, astrocytes and damaging effects on myelin infrastructure (Garcia et al., 2005; Azeez et al., 2016; Soazo and Garcia, 2007; Cuesta et al., 2011; Folarin et al., 2017; Jaiswal and Kale, 2020; Usende et al., 2016, 2022b). Simply put, the oxidative damage affecting the different regions of the brain, including both the lateral hypothalamus, prefrontal cortex, cerebellum and hippocampus following vanadium intoxication is due to the unbalanced redox state of vanadium that led to the generation of ROS and lipid peroxidation hence, its neurotoxicity (Cuesta et al., 2011; Usende et al., 2022b).

6. Vanadium induced biochemical alterations

Studies on brain biochemical changes following vanadium administration implicated oxidative stress and lipid peroxidation as major sequalae (Olopade and Connor, 2011). Recently, Usende et al. (2018a) showed that intraperitoneal exposure to vanadium led to significantly increased markers of oxidative stress including protein carbonyl (PCO), malondialdehyde (MDA), hydrogen peroxide (H_2O_2) as well as total thiol (PT); but significantly decreased brain tissue non-protein thiol (NPT) level. They (Usende et al., 2018a) also reported decreased brain antioxidant markers, including reduced glutathione (GSH), superoxide dismutase (SOD), glutathione peroxidase (GPx) and glutathione-S-transferase (GST) levels, similar to reports of Younce et al. (1991) and Zaporowska (1994). In the brain, like in other biological systems, vanadium as a catalytic metal induces the generation of ROS resulting to cellular damage (Ehrlich et al., 2008; Sasi et al., 1994). This is so because the mammalian brain is known to contain very high amount of lipid substances including polysaturated fatty acids and low levels of radical eliminating enzymes, and therefore, very venerable to free radical attack (Olopade and Connor, 2011). Also, in the brain, lipid membranes are the primary targets of ROS, although carbohydrates, nucleic acids and proteins are not spared (Garcia et al., 2004). Vanadium exposure decreases the brain antioxidative enzyme system (Usende et al., 2018a) via the depletion of glutathione and non-protein sulfhydryl (NP-SH) group (Olopade and Connor, 2011). Also, Haider et al. (1998) proposed that the mechanism of this depletion could be in part due to the inhibition of glutathione reductase enzymes responsible for the reduction of oxidized glutathione to glutathione and the utilization of NP-SH in covalent bonding by vanadium. In summary, the accumulation of ROS and depletion of brain antioxidant enzyme systems following vanadium exposure may be responsible for induction of cellular damage, a major sequalae for development of neurodegenerative diseases (Smith et al., 2000; Practico and Delanty, 2000; Gibson and Huang, 2002), and dysfunction of the olfactory system, an early and classical symptom of PD and AD (Doty, 2001; Haehner et al., 2011; Hawkes, 2006) and neurobehavioral deficits (Garcia et al., 2004; Adebiyi et al., 2019; Azeez et al., 2016; Usende et al., 2016).

7. Vanadium induced neurobehavioral deficits

Several neurobehavioral deficits have been reported both in humans and rodents' experimental models of vanadium intoxication. In humans, chronic vanadium toxicity manifests as tremor with severe central nervous system (CNS) depression (Ehrlich et al., 2008), reduced cognitive abilities as well as psychiatric disorders such as manic-depressive disorders (Olopade and Connor, 2011).

In rodent experimental models, exposure to the heavy metal vanadium by different routes have resulted in severe neurobehavioral deficits. Deficits in locomotory activities including significant decreased line crossings, ambulation, rearing and grooming activities in the open field arena have been documented by Garcia et al. (2004), Azeez et al. (2016), Adebiyi et al. (2019) and Usende et al. (2016) after exposure through the intraperitoneal route of vanadium in rats and mice. Similarly, Soazo and Garcia (2007) together with Olopade et al. (2011), documented deficits in negative geotaxis, surface righting reflex, rearing and center square entering, with loss of grooming and forelimb and hindlimb support in rat pups of suckling age. In their experiments (Olopade et al., 2011; Soazo and Garcia, 2007), they emphasized the importance of lactation route of vanadium neurotoxicity. Also, Usende et al. (2016) and Garcia et al. (2004) documented a significant reduction in line crosses, rearing as well as grooming in open field arena, being indicators for decrease or deficits in locomotory function after vanadium administration. More so, a significant reduced performance in overall rotarod test of muscular strength in rat pups after exposure to intraperitoneal vanadium treatment for 21 days starting from PND1 has been documented by Usende et al. (2016) and Todorich et al. (2011). Furthermore, reports by Adebiyi et al. (2019) and Mustapha et al. (2014) showed decreased locomotory activities and muscular strength in suckling rat pups exposed to vanadium for 2 or 3 weeks and 7 days respectively. Wang et al. (2015) also reported that motor coordination was severely impaired after exposure of rats to vanadium through lactational route. Concerning olfactory function deficits, Colin-Barenque et al. (2014) reported that vanadium inhalation induced olfactory dysfunction following a 28-days inhalation of V_2O_5, which they attributed to the loss of dendritic spines and granule cells death in the olfactory epithelium. Avila-Costa et al. (2006) also showed severe impairment of spatial memory following evaluation using Morris water maze model after vanadium exposure. Studies have shown that temporarily inactivating or causing lesions of the dorsal hippocampus including hippocampal neuronal loss is responsible for impaired acquisition and retrieval of spatial memory using tasks as those evaluated in Morris water maze model (Riedel et al., 1999; Moser and Moser, 1998); Avila-Costa et al. (2006) explored this model to arrive at their conclusion of spatial memory impairment following vanadium intoxication.

8. Vanadium induced neurotoxicity via destruction of ependymal cells and ependymal cell layer

Reports from our laboratory and those of other laboratories have showed that intoxication with vanadium disrupts the CSF-brain barrier and BBB (Garcia et al., 2004; Avila-Costa et al., 2005; Azeez et al., 2016; Todorich et al., 2011; Usende et al., 2016, 2022b) and this is evident by severe histoarchitectural disintegration of the ependymal epithelium. Electron microscopic studies revealed that vanadium neurotoxicity is associated with severe histopathological and morphological changes characterized by mass conglomeration, denudation and cilia loss from areas around the floor of the brain ventricles, and ependymal cell disruption/destruction and sloughing off (Usende et al., 2022b; Avila-Costa et al., 2005). Concerning the ependymal cell layer, evidence of severe detachment of cell layer and basal membrane disintegration as well as dissolution of tight junctions between ependymal cells and layers, evidence of disruption of the BBB; severe edema of the subependymal region, intracytoplasmic membranous vesicles and many vacuolations have been associated with vanadium intoxication (Avila-Costa et al., 2005; Usende et al., 2022b). These intracytoplasmic membranous vesicles reported by Usende et al. (2022b) were large, edematous and containing some protein materials engulfed fully by activated microglia as electron dense granules. Ependymal cell junctions are part of BBB and are severely modified in pathological conditions (Avila-Costa et al., 2005) such as in cases of vanadium intoxication. Of the numerous functions, one vital role of the ependymal cells (which are destroyed following vanadium intoxication as documented by Avila-Costa et al. (2005) as well as Usende et al. (2022b)) is in the transportation of compounds across the blood brain barrier (BBB), either via tight junctions opening, or by increasing vesicular transportation (Avila-Costa et al., 2005). Morphophysiological changes of the capillaries that joined in the formation of the BBB may cause impairment of nutrition of parenchymal tissue, and several points of evidence indicated the protein tyrosine phosphatase inhibition in the ependymal cells. The phosphatase inhibition was associated with increased degree of tyrosine phosphorylation of these junctional proteins, thereby weakening the ependymal cell adhesion (Rubin and Staddon, 1999). Vanadium has been reported to inhibit enzymes taking part in phosphoryl transfer reactions (Zhen et al., 2002), therefore, any changes occurring in these enzymes also decreases the cell adhesion via alterations of cellular morphology (Parsadanian et al., 1998; Avila-Costa et al., 2005).

9. Effects of vanadium on oxidative stress sensitive neuronal populations

9.1 Ventral tegmental area (VTA), Substantia nigra pars compacta (SNC) and olfactory bulb dopaminergic interneurons

Tyrosine hydroxylase (TH) is a crucial dopamine marker (neurotransmitter) containing interneurons (Usende et al., 2022b; Weihe et al., 2006) and recent reports have shown that dopamine biosynthesis down-regulation and decreased distribution of populations of TH-immunoreactive interneurons in the olfactory bulb (Ngwa et al., 2014), VTA and SNC (Usende et al., 2022b) of rodents exposed to vanadium. Immunolabelling of the neuropil and dendrites of TH-immunoreactive interneurons in the VTA and SNC of brains of African Giant Rats (AGR) treated with vanadium appeared significantly reduced and stereological TH^+ cell counts in these regions were significantly low (Usende et al., 2022b). Similarly, intranasal vanadium exposure is known to cause a reduction in the level of tyrosine hydroxylase (TH) of the olfactory bulb (Ngwa et al., 2014) especially in the glomerular layer, which is known for its abundance of dopaminergic interneurons (Davila et al., 2003; Halasz et al., 1981). Dopamine has been documented to play important role in olfaction (Ngwa et al., 2014; Hsia et al., 1999) and behavior (Usende et al., 2022b). Therefore, the down-regulation of dopamine observed by Usende et al. (2022b) and Ngwa et al. (2014) can be linked to the abnormal neurobehavioral phenotypes after exposure to vanadium including locomotor deficits reported by various investigators (Riedel et al., 1999; Moser and Moser, 1998; Soazo and Garcia, 2007; Colin-Barenque et al., 2014; Garcia et al., 2004; Avila-Costa et al., 2006; Olopade et al., 2011; Todorich et al., 2011; Usende et al., 2016; Azeez et al., 2016; Adebiyi et al., 2019).

9.1.1 Hippocampal, dentate gyrus, reticular thalamic nuclei and prefrontal cortex fast spiking GABAminergic interneurons

A recent article by Usende et al. (2022b) implicated the fast spiking GABAminergic parvalbumin interneurons (PV) as target for vanadium toxicity in different brain regions. Specifically, the reticular thalamic nuclei, hippocampus, prefrontal cortex and dentate gyrus are regions of target of these interneurons. Experimentally exposing African Giant Rats (AGR) to 3 mg/kg body weight of vanadium led to reduced distribution of parvalbumin-containing immune-positive interneuron (PV+) populations in these regions of the brain, while immunolabelling of neuropil and dendrites of same regions revealed a decreased number or destroyed dendrites and neuropil (Usende et al., 2022b). Stereological counts of these PV+ cells revealed a significant loss in the reticular thalamic nuclei, hippocampus, dentate gyrus and prefrontal cortex of AGR brains treated with sodium metavanadate (Usende et al., 2022b).

Parvalbumin neurons belong to subpopulation of GABA cells and they control basically principal neuronal output (Cabungcal et al., 2013), therefore necessitating the fast rhythmic neuronal synchrony that facilitates information processing during cognitive tasks (Cabungcal et al., 2013; Whittington et al., 2011; Sohal et al., 2009). Due to the fast spiking attribute of PV interneurons which also requires a high metabolic rate and increase in mitochondrial density, these cells become oxidative stress sensitive (Cabungcal et al., 2013), an established mechanistic pathway for vanadium neurotoxicity. Several experimental documentations reported that vanadium toxicity is linked to increased oxidative stress via the generation of ROS (Todorich et al., 2011; Azeez et al., 2016; Folarin et al., 2018; Usende et al., 2016, 2018a, b). Therefore, the neuronal pathologies associated with PV+ cells reported by Usende et al. (2022b) corroborate other reports by Hu et al. (2010), Grillo et al. (2003) and Schiavone et al. (2009) on PV+ cells after exposure to environmental stressors.

9.1.2 Lateral hypothalamus (LH) melanin concentration hormone (MCH) and orexinergic (OX-A) neurons

Melanin concentration hormone (MCH) and orexinergic (OX-A) neurons located in the lateral hypothalamic region are another group of neurons reported to be susceptible to vanadium exposure. Usende et al. (2022b) reported that in this brain region (lateral hypothalamus) of AGR exposed to 3 mg/kg body weight of vanadium experimentally, the distribution of the populations of MCH+ and OX-A+ neurons were reduced, just as the immunolabelling of neuropil and dendrites of these neurons were significantly scanty. Stereological cell counts of both MCH+ and OX-A neurons showed a significant reduction following vanadium exposure. In addition, the ramification index, intersecting, critical and ending radii of dendritic arbors of OX-A+ neurons in vanadium treated AGR brains were significantly decreased, revealing a marked reduction in the complexity of dendritic arborizations of these OX-A+ neurons following vanadium treatment (Usende et al., 2022b). Neuronal spine loss as well as necrotic-like cell death in the hippocampus have also been reported following inhalation route of vanadium exposure (Avila-Costa et al., 2004, 2006). They (Usende et al., 2022b) attributed this reduction in OX-A and MCH as well as TH and PV neuronal population after vanadium exposure to early cell death, increased peptide release as well as down regulation of peptide expression corroborating earlier reports of Palomba et al. (2015). They also attributed the decreased dendritic arborizations and other morphological damages of OX-A+ neuronal populations of AGR brains treated with sodium metavanadate to the ability of this metal to generate ROS leading to oxidative stress that caused the neuronal damage, and to vanadium's ability to induce neurodegenerative disorders. Interestingly, neurodegenerative anomalies and other conditions including alcoholism, epilepsy, and mental retardation are strongly linked with decrease in neuronal arborization (Usende et al., 2022b; Avila-Costa et al., 1999; Fiala et al., 2002).

10. Effects of vanadium on perineuronal nets (PNNs) and extracellular matrix (ECM)

A recent study showed that exposing AGR model to 3 mg/kg body weight of vanadium is associated with scanty and loss of PPNs and ECM staining intensity around the soma and dendrites of fast spiking PV+ interneurons, implicating vulnerability of these interneurons to vanadium toxicity (Usende et al., 2022b). They showed that one mechanism of vanadium induced destruction of these fast spiking PV interneurons is due to vanadium mediated destructions of this very important and unique protective mechanism in the neurons characterized by its envelopment with specialized extracellular matrix and or aggrecan enriched PNNs (Usende et al., 2022b). Specifically, PNNs is made up of mainly charged chondroitin sulfate proteoglycans hyaluronase tenascin with link proteins (Usende et al., 2022b; Carulli et al., 2010), and functions to promote the maturation of neurons and their network and synaptic stability (Usende et al., 2022b; Sugiyama et al., 2009). Interestingly, PNNs protect neuronal populations against oxidative stress (Suttkus et al., 2012; Morawski et al., 2004) which is a known mechanistic pathway for neuronal damage caused by vanadium (Todorich et al., 2011; Olaolorun et al., 2021; Usende et al., 2016, 2018a, 2022b). The findings of Usende et al. (2022b) on AGR exposed experimentally to vanadium suggest that intact WFA labeled PNNs offer protection to the fast spiking PV+ cells against damaging effects resulting from oxidative stress, implicating PNNs, the target of vanadium intoxication as neuroprotective (Cabungcal et al., 2013; Suttkus et al., 2012; Morawski et al., 2004). They concluded that intact

PNNs and ECM serve to protect neurons via neutralization of ROS generated by vanadium and boost the capacities of cellular antioxidants (Usende et al., 2022b).

11. Vanadium effects on glia cells *in vivo* using rodent models and myelin damage

11.1 Microglia activation

Microglia are resident immunocompetent cells and are the key mediators of the processes of neuroinflammation in the CNS following vanadium toxicity (Hanisch and Kettenmann, 2007; Bwala et al., 2014; Folarin et al., 2017; Azeez et al., 2016; Usende et al., 2016, 2022b); studies have documented that microglia activation is highly neurotoxic (Jha et al., 2016). Vanadium treatment activates (increase in number and change in morphology) microglia populations in the brain (Folarin et al., 2017; Bwala et al., 2014; Azeez et al., 2016; Usende et al., 2016, 2022b). Recently, Usende et al. (2022) exposed AGR to vanadium and using Iba1 immunostaining showed microglial cells activation characterized by their cell body and dendritic hypertrophy, with thicker and shorter branches, presenting a classical bushy appearance in various brain regions (such as the hippocampus, cortex, corpus callosum, lateral hypothalamus and substantia nigra). Other reports such as those of Usende et al. (2016) and Bwala et al. (2014) showed increase in numerical and isoform populations of rounded clear pre-phagocytic amoeboid microglia in the cerebellar brain region of sodium metavanadate exposed rats, suggestive of swift change from resting microglia to activated isoform as a result of neuroinflammation. Furthermore, Folarin et al. (2017) in their BALB/c male mice model documented the activation of microglia as a response to subchronic vanadium neurotoxicity in various regions of the brain (including the genus of the corpus callosum, cerebellum and hippocampus) and correlated the level of activation of these microglia with the duration of exposure and neuronal damage respectively. The report of Folarin et al. (2017) is clearly suggestive of the recruitment and quick activation of microglia resulting from vanadium oxidative stress induction (Usende et al., 2016, 2018a; Block and Calderon-Garciduenas, 2009) consequent to production of free radicals (Usende et al., 2018a, 2022b; Tsuda et al., 2004) as we have previously discussed. Similarly, another study, Azeez et al. (2016) using mice model and CD11b immunostaining approach documented that the brains of mice with vanadium toxicity had activated microglia characterized by cell body and processes hypertrophy with a bushy appearance in gray matter areas, hippocampus, neocortex and the corpus callosum midline commissural portion. The eventual dynamics and occurrence of these microglial cell polarizations following neurotoxicity of vanadium however, remains to be fully investigated. Interestingly, the distributional pattern of activated microglia documented by these authors (Bwala et al., 2014; Azeez et al., 2016; Folarin et al., 2017; Usende et al., 2016, 2022b) correlates with those of astrocyte's activation, and myelin and axons damage (Olaolorun et al., 2021).

11.2 Astrocytes activation

Astrocytes are key regulators of neuronal microenvironment, defending the CNS against possible oxidative and toxic insults (Usende et al., 2016, 2022b; Heller and Rusakov, 2015; Sofroniew and Vinters, 2010), and in coordinating and controlling the CNS (Heller and Rusakov, 2015; Sofroniew and Vinters, 2010). Astrocytes therefore make up the key CNS defense element (Heller and Rusakov, 2015). Astrogliosis is documented as sequalae in neurotoxicity of vanadium by several authors (Usende et al., 2016, 2022b; Mustapha et al., 2014; Todorich et al., 2011; Azeez et al., 2016; Garcia et al., 2005). Recently, following experimental exposure of AGR to 3 mg/kg body weight of vanadium, Usende et al. (2022b), reported vanadium induced astrocytic hypertrophy in all regions of the brain, especially in the cerebral cortex, substantia nigra, hippocampus, corpus callosum and lateral hypothalamus. They also showed similar findings in brain of AGR sampled from natural high vanadium contaminated environment. Similar reports have been shown by several other authors. Rats and mice models exposed either intraperitoneally or through lactational routes to vanadium had severe astrocytic activation in different brain regions (Usende et al., 2016, 2022b; Olopade et al., 2011; Azeez et al., 2016; Todorich et al., 2011; Mustapha et al., 2014; Folarin et al., 2017; Garcia et al., 2005). Administration of sodium metavanadate to adult rats for 5 consecutive days resulted in severe astrogliosis in the hippocampus and cerebellum (Garcia et al., 2005) indicative of acute, quick, astrocytic reactions to vanadium neurotoxic effects (Olaolorun et al., 2021). Also, Todorich et al. (2011) and Usende et al. (2016) showed that vanadium treatment of rat pups from post-natal day (PND)1 to PND14 or PND21 led to astrocytic activation in the genus of the corpus callosum and hippocampus. GFAP upregulation and hypertrophy of astrocytes soma and processes were reported in 21-days-old mice exposed to sodium metavanadate intoxication (Mustapha et al., 2014). Intranasal exposure to vanadium is also known to induce migratory effects and accumulation of astrocytes in the olfactory bulb glomerular layer (Ngwa et al., 2014). Regional astrocytic diversities, inclusive of protoplasmic astrocytes in the gray matter tract and fibrous astrocytes in the white matter tract with noticeable TNF-α expression induction have been documented by Azeez et al. (2016) after 3-months of vanadium intoxication. For example, in the region of the internal capsule, Azeez et al. (2016) explained that astrocyte activation after vanadium intoxication was very marked in regions of microglia activation and myelin/axonal damage. Despite the fact that astrocytes together with microglia are important in the process of homeostasis of iron, necessary for the synthesis of myelin (Lundgaard et al., 2014; Clemente et al., 2013); astrocytes (activated following vanadium intoxication) are known to have inhibitory effects on remyelination and modulate the activity of oligodendrocytes (Alizadeh et al., 2015).

11.3 Oligodendrocytes, myelination and axonal damage

Demyelination and or hypomyelination is a major phenotype of neurotoxicity of vanadium (Usende et al., 2016, 2022b; Garcia et al., 2004; Todorich et al., 2011; Azeez et al., 2016). Todorich et al. (2011) documented selective decrease in the population of pre-myelinating oligodendrocytes in the corpus callosum of rat pups after exposure to vanadium treatment for 2-weeks and concluded that the mechanism of vanadium induced hypomyelination is by destructions of primary oligodendrocytes (oligodendrocytes progenitor cells) resulting in decreased myelinogenesis, in addition to the destruction of myelin due to oxidative stress and ROS generation (Todorich et al., 2011; Folarin et al., 2017; Mustapha et al., 2014; Azeez et al., 2016). Examination of the genus of the corpus callosum and the cerebellar arbor vitae of rat pups at PND15 and 21 exposed to vanadium treatment starting from PND1 also showed a decrease in the number of NG-2 positive oligodendrocyte progenitor cells (Usende et al., 2016). These findings suggest that oligodendrocyte progenitors and immature oligodendrocytes are the major targets of vanadium toxicity, although the mature and myelinating oligodendrocytes are not spared.

Furthermore, myelin is known as the preferential target of lipid peroxidation mediated by vanadium (Ray et al., 2007; Garcia et al., 2005), and Azeez et al. (2016) showed that myelin damage was seen in different regions of brains of mice treated with vanadium. They (Azeez et al., 2016) observed mostly hypomyelination, but demyelinated areas were also seen, especially in the hippocampus and neocortical areas of vanadium treated mice. In another study, Garcia et al. (2004) treated rats with vanadium intraperitoneally for 5-days and reported a robust decrease in myelination fiber density in the corpus callosum and cerebellar regions using anti-myelin basic protein (MBP) immune-labeling. Also, rat pups from dams receiving vanadium treated for 12 days beginning at postnatal day (PND) 10 had significant decreased myelinated fiber density in the corpus callosum and cerebellum at PND 21 (Soazo and Garcia, 2007). The same pattern of hypomyelination of the corpus callosum and cerebellum were also noticed in rat brain after sodium metavanadate toxicity via intraperitoneal route for 7-days (Adebiyi et al., 2019), and 14 and 21-days (Usende et al., 2016). On the basis of regional myelin damage seen in the brains of vanadium treated mice, Azeez et al. (2016) investigated the mechanism of the axonal pathology using SMI-32 immunoreactivity and interestingly showed intense staining of nonphosphorylated neurofilaments in myelin damaged areas, especially fibers of the midline portion of the corpus callosum, and in the deep cortical layer axons. These axons appeared thickened, fragmented and interrupted by several swellings. Based on the axonal pathology documented by Azeez et al. (2016), Usende et al. (2022b) hypothesized that vanadium treatment induces destruction of the myelin sheath. They tested the hypothesis by exposing African giant rats (AGR) intraperitoneally to sodium metavanadate for two weeks and confirmed intense destructions of myelin sheaths consequent to splitting of lamella of the sheath resulting in numerous demyelinated and un-myelinated axons in vanadium treated AGR (Usende et al., 2022b).

Of note, the destroyed myelin (peroxidative damage) caused by vanadium toxicity is as a result of the potential target for myelin membranes, because of its high relative phospholipids content (Garcia et al., 2004; Sasi et al., 1994; Haider et al., 1998; Haider and El-Fakhri, 1991). To be specific, 5 days vanadium treatment to rats led to significant reduced total lipid contents, phospholipids, cerebrosides and cholesterol in different regions of the brain with corresponding depletion of concentration of polyunsaturated fatty acids (arachidonic, linolenic, oleic and linoleic), phosphatidyl ethanolamine, sphingomyelin and phosphatidyl choline (Sasi et al., 1994) due possibly to high levels of lipid peroxides (Haider and El-Fakhri, 1991; Haider et al., 1998). Currently, no data is available on re-myelination events in vanadium induced neurotoxicity, therefore this grey area remains to be investigated.

12. Conclusion

Much research on vanadium has been on its hepatic, renal, gonadal, genotoxic and respiratory effects, whereas comparatively little has been done on its neurotoxic effects. In this review we have covered the points of evidence that supports the idea that vanadium may interfere with various CNS cellular morphophysiology. Vanadium crosses the blood brain barrier and induces oxidative stress, biochemical alterations with consequent lipid peroxidation and neurobehavioral deficits. One mechanism of vanadium-induced neurotoxicity is by destruction of neuronal populations that are very sensitive to oxidative stress with consequent microgliosis, astrogliosis, oligodendrocytes depletion, and myelin and axonal damage. Based on this review, more work is needed to explore potential targets for therapy of vanadium-induced neurotoxicity.

References

Adebiyi, O.E., Obisesan, A.D., Olayemi, F.O. and Olopade, J.O. 2015. Protective effect of ethanolic extract of Grewia carpinifolia leaves on vanadium induced toxicity. Alex. J. Vet. Sci. 47(1): 24–31.

Adebiyi, O.E., Olayemi, F.O., Olopade, J.O. and Tan, N.H. 2019. Beta-sitosterol enhances motor coordination, attenuates memory loss and demyelination in a vanadium-induced model of experimental neurotoxicity. Pathophysiology 26: 21–29. https://doi.org/10.1016/j.pathophys.2018.12.002.

Alizadeh, A., Dyck, S.M. and Karimi-Abdolrezaee, S. 2015. Myelin damage and repair in pathologic CNS: challenges and prospects. Front Mol. Neurosci. 8: 35. https://doi.org/10.3389/fnmol.2015.00035.

Altamirano-Lozano, M. 1998. Genotoxic effects of vanadium compounds. Investig. Clin. 39: 39–47. PMID: 9650459.

Altamirano-Lozano, M., Valverde, M., Alvarez-Barrera, L., Molina, B. and Rojas, E. 1999. Genotoxic studies of vanadium pentoxide (V_2O_5) in male mice. II. Effects in several mouse tissues. Teratog. Carcinog. Mutagen. 19(4): 243–255. https://doi.org/10.1002/(SICI)1520-6866(1999)19:4<243::AID-TCM1>3.0.CO;2-J.

Aschner, M., Levin, E.D., Sunol, C., Olopade, J.O., Helmcke, K.J., Avila, D.S. et al. 2010. Gene–environment interactions: neurodegeneration in non-mammals and mammals. Neurotoxicology 31(5): 582–588. https://doi.org/10.1016/j.neuro.2010.03.008.

Avila-Costa, M.R., Colın-Barenque, L., Fortoul, T.I., Machado-Salas, J.P., Espinosa-Villanueva, J., Rugerio-Vargas, C. et al. 1999. Memory deterioration in an oxidative stress model and its correlation with cytological changes on rat hippocampus CA1. Neurosci. Lett. 270: 107–109. https://doi.org/10.1016/S0304- 3940(99)00458-9.

Avila-Costa, M.R., Flores, E.M., Colin-Barenque, L., Ordonez, J.L., Guti_errez, A.L., Nino-Cabrera, H.G. et al. 2004. Nigrostriatal modifications after vanadium inhalation: an immunocytochemical and cytological approach. Neurochem. Res. 29(7): 1365–1369. https://doi.org/10.1023/B:NERE.0000026398.86113.7d.

Avila-Costa, M.R., Colın-Barenque, L., Zepeda-Rodr´ıguez, A., Antuna, S.B., Saldivar, O.L., Espejel-Maya, G. et al. 2005. Ependymal epithelium disruption after vanadium pentoxide inhalation; Aa mice experimental model. Neurosci. Lett. 381: 21–25. https://doi.org/10.1016/j.neulet.2005.01.072.

Avila-Costa, M.R., Fortoul, T.I., Niño-Cabrera, G., Colín-Barenque, L., Bizarro-Nevares, P., Gutiérrez-Valdez, A.L. et al. 2006. Hippocampal cell alterations induced by the inhalation of vanadium pentoxide ($V_{(2)}O_{(5)}$) promote memory deterioration. Neurotoxicology 27(6): 1007–1012. doi: 10.1016/j.neuro.2006.04.001.

Azeez, I.A., Olopade, F., Laperchia, C., Andrioli, A., Scambi, I., Onwuka, S.K. et al. 2016. Regional myelin and axon damage and neuroinflammation in the mouse brain after long-term postnatal vanadium exposure. J. Neuropathol. Exp. Neurol. 75: 843–854. https://doi.org/10.1093/jnen/nlw058.

Barceloux, D.G. 1999. Vanadium. J. Toxicol. Clin. Toxicol. 37(2): 265–278.

Berman, E. 1980. Toxic Metals and Their Analysis. Wiley-Blackwell, United Kingdom. EDB-81-083144.

Beyersmann, D. and Hartwig, A. 2008. Carcinogenic metal compounds: recent insight into molecular and cellular mechanisms. Arch. Toxicol. 82(8): 493. https://doi.org/10.1007/s00204-008-0313-y.

Block, M.L. and Calderon-Garciduenas, L. 2009. Air pollution: Mechanisms of neuroinflammation and CNS disease. Trends Neurosci. 32: 506–516. https://doi.org/10.1016/j.tins.2009.05.009.

Bwala, D.A., Ladagu, A.D., Olopade, F.E., Siren, A.L., Yahaya, A. and Olopade, J.O. 2014. Neurotoxic profiles of vanadium when administered at the onset of myelination in rats: the protective role of vitamin E. Trop. Vet. 32(1&2): 47–56.

Cabungcal, J., Steullet, P., Morishita, H., Kraftsik, R., Cuenod, M., Hensch, T.K. et al. 2013. Perineuronal nets protect fast-spiking interneurons against oxidative stress. Proc. Natl. Acad. Sci. U. S. A. 110(22): 9130–1935. https://doi.org/10.1073/ pnas.1300454110.

Capella, L.S., Gefe, M.R., Silva, E.F., Affonso-Mitidieri, O., Lopes, A.G., Rumjanek, V.M. et al. 2002. Mechanisms of vanadate-induced cellular toxicity: role of cellular glutathione and NADPH. Arch. Biochem. Biophys. 406(1): 65–72. https://doi.org/10.1016/S0003-9861(02)00408-3.

Carulli, D., Pizzorusso, T., Kwok, J.C.F., Putignano, E., Poli, A., Forostyak, S. et al. 2010. Animals lacking link protein have attenuated perineuronal nets and persistent plasticity. Brain 133(8): 2331–2347. https://doi.org/10.1093/brain/awq145.

Clemente, D., Ortega, M.C., Melero-Jerez, C. and de Castro, F. 2013. The effect of glia-glia interactions on oligodendrocyte precursor cell biology during development and in demyelinating diseases. Front Cell Neurosci. 7: 268. https://doi.org/10.3389/fncel.2013.00268.

Colin-Barenque, L., Pedraza-Chaverri, J., Medina-Campos, O., Jimenez-Martínez, R., Bizarro-Nevares, P., González-Villalva, A. et al. 2014. Functional and morphological olfactory bulb modifications in mice after vanadium inhalation. Toxicol. Pathol. 43(2): 282–291. https://doi.org/10.1177/01926233145486.

Cortizo, A.M., Bruzzone, L., Molinuevo, S. and Etcheverry, S.B. 2000. A possible role of oxidative stress in the vanadium-induced cytotoxicity in the MC3T3E1 osteoblast and UMR106 osteosarcoma cell lines. Toxicology 147(2): 89–99. https://doi.org/10.1016/S0300-483X(00)00181-5.

Cuesta, S., Frances, D. and Garcıa, G.B. 2011. ROS formation and antioxidant status in brain areas of rats exposed to sodium metavanadate. Neurotoxicol. Teratol. 33(2): 297–302. https://doi.org/10.1016/j.ntt.2010.10.010.

Das, S., Chatterjee, M., Janarthan, M., Ramachandran, H. and Chatterjee, M. 2012. Vanadium in cancer prevention. pp. 163–185. *In*: Michibata, H. (ed.). Vanadium: Biochemical and Molecular Biological Approaches, Chapter 8. Springer, Dordrecht.

Davila, N.G., Blakemore, L.J. and Trombley, P.Q. 2003. Dopamine modulates synaptic transmission between rat olfactory bulb neurons in culture. J. Neurophysiol. 90: 395–404. https://doi.org/10.1152/jn.01058.2002.

Ding, M., Gannett, P.M., Rojanasakul, Y., Liu, K. and Shi, X. 1994. One-electron reduction of vanadate by ascorbate and related free radical generation at physiological pH. J. Inorg. Biochem. 55(2): 101–112. https://doi.org/10.1016/0162-0134(94)85032-1.

Domingo, J.L. 1996. Vanadium: a review of the reproductive and developmental toxicity. Reprod. Toxicol. 10(3): 175–182. https://doi.org/10.1016/0890-6238(96)00019-6.

Domingo, J.L. 2002. Vanadium and tungsten derivatives as antidiabetic agents. Biol. Trace Elem. Res. 88(2): 97–112. https://doi.org/10.1385/BTER:88:2:097.

Doty, R.L. and Hastings, L. 2001. Neurotoxic exposure and olfactory impairment. Clin. Occupat. Environ. Med. 1: 547–575.

Dutton, W.F. 1911. Vanadiumism. J. Am. Med. Assoc. 56(22): 1648–1652.

Ehrlich, V.A., Nersesyan, A.K., Hoelzl, C., Ferk, F., Bichler, J., Valic, E. et al. 2008. Inhalative exposure to vanadium pentoxide causes DNA damage in workers: results of a multiple end point study. Environ Health Perspect. 116(12): 1689–1698. https:// doi. org/ 10. 1289/ ehp.11438.

Ejeh, S.A., Abalaka, S.E., Usende, I.L., Alimi, Y.A. and Oyelowo, F.O. 2019. Acute toxicity, oxidative stress response and clinicopathological changes in Wistar rats exposed to aqueous extract of Uvaria chamae leaves. Scientific African 3(1): 1–9. https://doi.org/10.1016/j.sciaf.2019.e00068.
Erikson, K.M., Syversen, T., Steinnes, E. and Aschner, M. 2004. Globus pallidus: a target brain region for divalent metal accumulation associated with dietary iron deficiency. J. Nutr. Biochem. 15(6): 335–341. https://doi.org/10.1016/j.jnutbio.2003.12.006.
Evangelou, A.M. 2002. Vanadium in cancer treatment. Crit. Rev. Oncol. Hematol. 42(3): 249–265. https://doi.org/10.1016/S1040-8428(01)00221-9.
Fiala, J.C., Spacek, J. and Harris, K.M. 2002. Dendritic spine pathology: cause or consequence of neurological disorders? Brain Res. Rev. 39(1): 29–54. https://doi.org/10.1016/S0165-0173(02)00158-3.
Folarin, O., Olopade, F., Onwuka, S. and Olopade, J. 2016. Memory deficit recovery after chronic vanadium exposure in mice. Oxidative Med. Cell. Longev. https://doi.org/10.1155/2016/4860582.
Folarin, O.R., Snyder, A.M., Peters, D.G., Olopade, F., Connor, J.R. and Olopade, J.O. 2017. Brain metal distribution and neuro-inflammatory profiles after chronic vanadium administration and withdrawal in mice. Front. Neuroanat. 11: 58. https://doi.org/10.3389/fnana.2017.00058.
Folarin, O.R., Adaramoye, O.A., Akanni, O.O. and Olopade, J.O. 2018. Changes in the brain antioxidant profile after chronic vanadium administration in mice. Metab. Brain Dis. 33(2): 377–385. https://doi.org/10.1007/s11011-017-0070-9.
Garcia, G.B., Quiroga, A.D., Sturtz, N., Martinez, A.I. and Biancardi, M.E. 2004. Morphological alterations of central nervous system (CNS) myelin in vanadium (V)-exposed adult rats. Drug Chem. Toxicol. 27: 281–293. https://doi.org/10.1081/DCT-120037747.
Garcia, G.B., Biancardi, M.E. and Quiroga, A.D. 2005. Vanadium (V)-induced neurotoxicity in the rat central nervous system: a histo-immunohistochemical study. Drug Chem. Toxicol. 28: 329–344. https://doi.org/10.1081/DCT-200064496.
Gibson, G.E. and Huang, H.M. 2002. Oxidative processes in the brain and non-neuronal tissues as biomarkers of Alzheimer's disease. Frontiers in Bioscience 7(4): 1007–1015; https://doi.org/10.2741/A827.
Grillo, C.A., Piroli, G.G., Rosell, D.R., Hoskin, E.K., Mcewen, B.S. and Reagan, L.P. 2003. Region specific increases in oxidative stress and superoxide dismutase in the hippocampus of diabetic rats subjected to stress. Neuroscience 121(1): 133–140. https://doi.org/10.1016/S0306-4522(03)00343-9.
Haehner, A., Hummel, T. and Reichman, H. 2011. Olfactory loss in Parkinson's disease. Parkinson's Dis. https://doi.org/10.4061/2011/450939.
Haider, S.S. and El-Fakhri, M. 1991. Action of alpha-tocopherol on vanadium-stimulated lipid peroxidation in rat brain. Neurotoxicology 12(1): 79–85. PMID: 2014070.
Haider, S.S., Abdel-Gayoum, A.A. and El-Fakhri Ghwarsha, K.M. 1998. Effect of selenium on vanadium toxicity in different regions of rat brain. Hum Exp Toxicol. 17(1): 23–8. doi: 10.1177/096032719801700104.
Halasz, N., Johansson, O., Hokfelt, T., Ljungdahl, A. and Goldstein, M. 1981. Immunohistochemical identification of two types of dopamine neuron in the rat olfactory bulb as seen by serial sectioning. J. Neurocytol. 10: 251–259. https://doi.org/10.1007/BF01257970.
Hanisch, U.K. and Kettenmann, H. 2007. Microglia: Active sensor and versatile effector cells in the normal and pathologic brain. Nat. Neurosci. 10: 1387–1394. https://doi.org/ 10.1038/nn1997.
Hawkes, C. 2006. Olfaction in neurodegenerative disorder. Mov Disord. 18: 364–372. https://doi.org/10.1159/000093759.
Heller, J.P. and Rusakov, D.A. 2015. Morphological plasticity of astroglia: understanding synaptic microenvironment. Glia 63: 2133–2151. https://doi.org/10.1002/ glia.22821.
Hickey, R.J., Schoff, E.P. and Clelland, R.C. 1967. Relationship between air pollution and certain chronic disease death rates: multivariate statistical studies. Arch. Environ. Health 15(6): 728–738. https://doi.org/10.1080/00039896.1967.10664990.
Hosseini, M.J., Shaki, F., Ghazi-Khansari, M. and Pourahmad, J. 2013. Toxicity of vanadium on isolated rat liver mitochondria: a new mechanistic approach. Metallomics 5(2): 152–166. https://doi.org/10.1039/c2mt20198d.
Hsia, A.Y., Vincent, J.D. and Lledo, P.M. 1999. Dopamine depresses synaptic inputs into the olfactory bulb. J. Neurophysiol. 82: 1082–1085. https://doi.org/10.1152/jn.1999.82.2.1082.

Hu, W., Zhang, M., Cz′eh, B., Flügge, G. and Zhang, W. 2010. Stress impairs GABAergic network function in the hippocampus by activating nongenomic glucocorticoid receptors and affecting the integrity of the parvalbumin-expressing neuronal network. Neuropsychopharmacol 35(8): 1693–1707. https://doi.org/10.1038/ npp.2010.31.

Igado, O.O., Glaser, J., Ramos-Tirado, M., Banko_glu, E.E., Atiba, F.A., Holzgrabe, U. et al. 2018. Isolation of a novel compound (MIMO2) from the methanolic extract of Moringa oleifera leaves: protective effects against vanadium-induced cytotoxicity. Drug Chem. Toxicol. 41(3): 249–258. https://doi.org/10.1080/01480545.2017.1366504.

Imtiaz, M., Rizwan, M.S., Xiong, S., Li, H., Ashraf, M., Shahzad, S.M. et al. 2015. Vanadium, recent advancements and research prospects: a review. Environ. Int. 80: 79–88. https://doi.org/10.1016/j.envint.2015.03.018.

Jaiswal, M.R. and Kale, P.P. 2020. Mini review–vanadium-induced neurotoxicity and possible targets. Neurol. Sci. 41(4): 763–768. https://doi.org/10.1007/s10072-019-04188-5.

Jha, M.K., Lee, W.H. and Suk, K. 2016. Functional polarization of neuroglia: implications in neuroinflammation and neurological disorders. Biochem. Pharmacol. 103: 1–16. https://doi.org/10.1016/j.bcp.2015.11.003.

Lapenna, D., Ciofani, G., Bruno, C., Pierdomenico, S.D., Giuliani, L., Giamberardino, M.A. et al. 2002. Vanadyl as a catalyst of human lipoprotein oxidation. Biochem. Pharmacol. 63(3): 375–380. https://doi.org/10.1016/S0006-2952(01)00849-8.

Li, H., Zhou, D., Zhang, Q., Feng, C., Zheng, W., He, K. et al. 2013. Vanadium exposure-induced neurobehavioral alterations among Chinese workers. Neurotoxicology 36: 49–54. https://doi.org/10.1016/j.neuro.2013.02.008.

Liu, J., Cui, H., Liu, X., Peng, X., Deng, J., Zuo, Z. et al. 2012. Dietary high vanadium causes oxidative damage-induced renal and hepatic toxicity in broilers. Biol. Trace Elem. Res. 145(2): 189–200. https://doi.org/10.1007/s12011-011-9185-8.

Lundgaard, I., Osório, M.J., Kress, B.T., Sanggaard, S. and Nedergaard, M. 2014. White matter astrocytes in health and disease. Neuroscience 276: 161–173. https://doi.org/10.1016/j.neuroscience.2013.10.050.

Luo, J., Sun, Y., Lin, H., Qian, Y., Li, Z., Leonard, S.S. et al. 2003. Activation of JNK by vanadate induces a Fas-associated death domain (FADD)-dependent death of cerebellar granule progenitors *in vitro*. J. Biol. Chem. 278(7): 4542–4551. https://doi.org/10.1074/jbc.M208295200.

Morawski, M., Brückner, M.K., Riederer, P., Brückner, G. and Arendt, T. 2004. Perineuronal nets potentially protect against oxidative stress. Exp. Neurol. 188(2): 309–315. https://doi.org/10.1016/j.expneurol.2004.04.017.

Moser, M.B. and Moser, E.I. 1998. Distributed encoding and retrieval of spatial memory in the hippocampus. J. Neurosci. 18: 7535–7542. https://doi.org/10.1523/JNEUROSCI.18-18-07535.1998.

Mustapha, O., Oke, B., Offen, N., Sirén, A. and Olopade, J. 2014. Neurobehavioral and cytotoxic effects of vanadium during oligodendrocyte maturation: a protective role for erythropoietin. Environ Toxicol. Pharmacol. 38: 98–111. doi: 10.1016/j.etap.2014.05.001.

Nechay, B.R. 1984. Mechanisms of action of vanadium. Annu. Rev. Pharmacol. Toxicol. 24(1): 501–524. https://doi.org/10.1146/annurev.pa.24.040184.002441.

Ngwa, H.A., Kanthasamy, A., Anantharam, V., Song, C., Witte, T., Houk, R. et al. 2009. Vanadium induces dopaminergic neurotoxicity via protein kinase Cdelta dependent oxidative signaling mechanisms: relevance to etiopathogenesis of Parkinson's disease. Toxicol. Appl. Pharmacol. 240(2): 273–285. https://doi.org/10.1016/j.taap.2009.07.025.

Ngwa, H.A., Kanthasamy, A., Jin, H., Anantharam, V. and Kanthasamy, A.G. 2014. Vanadium exposure induces olfactory dysfunction in an animal model of metal neurotoxicity. Neurotoxicology 43: 73–81. https://doi.org/10.1016/j.neuro.2013.12.004.

Olaolorun, F.A., Obasa, A.A., Balogun, H.A., Aina, O.O. and Olopade, J.O. 2014. Lactational vitamin E protects against the histotoxic effects of systemically administered vanadium in neonatal rats. Niger. J. Physiol. Sci. 29(2): 125–129. PMID: 26196578.

Olaoloruna, F.A., Olopade, F.E., Usende, I.L., Lijoka, A.D., Ladagu, A.D. and Olopade, J.O. 2021. Neurotoxicity of vanadium. Advances in Neurotoxicology, https://doi.org/10.1016/bs.ant.2021.01.002.

Olopade, J.O. and Connor, J.R. 2011. Vanadium and neurotoxicity: A review. Curr. Top Toxicol. 7: 33–38.

Olopade, J.O., Fatola, I.O. and Olopade, F.E. 2011. Vertical administration of vanadium through lactation induces behavioural and neuromorphological changes: protective role of Vitamin E. Nig. J. Physiol. Sci. 26: 55–60. PMID: 22314988.

Palomba, M., Seke-Etet, P.F., Laperchia, C., Tiberio, L., Xu, Y.Z., Colavito, V. et al. 2015. Alterations of orexinergic and melanin concentrating hormone neurons in experimental sleeping sickness. Neuroscience 290: 185–195. https://doi.org/10.1016/j.neuroscience.2014.12.066.

Parfett, C.L.J. and Pilon, R. 1995. Oxidative stress-regulated gene expression and promotion of morphological transformation induced in C3H/10T1/2 cells by ammonium metavanadate. Food Chem. Toxicol. 33(4): 301–308. https://doi.org/10.1016/0278-6915(94)00141-A.

Parsadanian, H.K., Marchenko, S.N., Parsadanian, K.H. and Barilyak, I.R. 1998. Vanadium as a factor that disturbs phosphorus metabolism in nervous tissue. Neurotoxicology 19: 561–564. PMID: 9745912.

Paternain, J.L., Domingo, J.L., Gomez, M., Ortega, A. and Corbella, J. 1990. Developmental toxicity of vanadium in mice after oral administration. J. Appl. Toxicol. 10(3): 181–186. https://doi.org/10.1002/jat.2550100307.

Practico, D. and Delanty, N. 2000. Oxidative injury in disease of the central nervous system: Focus on Alzheimer's disease. American Journal Medicine 109: 577–585. https://doi.org/10.1016/S0002-9343(00)00547-7.

Ray, R.S., Ghosh, B., Rana, A. and Chatterjee, M. 2007. Suppression of cell proliferation, induction of apoptosis and cell cycle arrest: chemopreventive activity of vanadium *in vivo* and *in vitro*. Int. J. Cancer 120(1): 13–23. https://doi.org/10.1002/ijc.22277.

Riedel, G., Micheau, J., Lam, A.G., Roloff, E., Martin, S.J., Bridge, H. et al. 1999. Reversible neural inactivation reveals hippocampal participation in several memory processes. Nat. Neurosci. 2: 898–905. https://doi.org/10.1038/13202.

Rojas, E., Valverde, M., Herrera, L.A., Altamirano-Lozano, M. and Ostrosky-Wegman, P. 1996. Genotoxicity of vanadium pentoxide evaluate by the single cell gel electrophoresis assay in human lymphocytes. Mutat. Res. 359(2): 77–84. https://doi.org/10.1016/S0165-1161(96)90254-X.

Rojas, E., Herrera, L.A., Poirier, L.A. and Ostrosky-Wegman, P. 1999. Are metals dietary carcinogens? Mutat. Res. 443(1–2): 157–181. https://doi.org/10.1016/S1383-5742(99)00018-6.

Rubin, L.L. and Staddon, J.M. 1999. The cell biology of the blood–brain barrier. Annu. Rev. Neurosci. 22: 11–28. https://doi.org/10.1146/annurev.neuro.22.1.11.

Sabbioni, E., Pozzi, G., Devos, S., Pintar, A., Casella, L. and Fischbach, M. 1993. The intensity of vanadium (V)-induced cytotoxicity and morphological transformation in BALB/3T3 cells is dependent on glutathione-mediated bioreduction to vanadium (IV). Carcinogenesis 14(12): 2565–2568. https://doi.org/10.1093/carcin/14.12.2565.

Sakurai, H. 1994. Vanadium distribution in rats and DNA cleavage by vanadyl complex: implication for vanadium toxicity and biological effects. Environ. Health Perspect. 102(Suppl. 3): 35–36. https://doi.org/10.1289/ehp.94102s335.

Sasi, M.M., Haider, S.S., El-Fakhri, M. and Ghwarsha, K.M. 1994. Microchromatographic analysis of lipids, protein, and occurrence of lipid peroxidation in various brain areas of vanadium exposed rats: a possible mechanism of vanadium neurotoxicity. Neurotoxicology 15(2): 413–420. PMID: 8361679.

Schiavone, S., Sorce, S., Dubois-Dauphin, M., Jaquet, V., Colaianna, M., Zotti, M. et al. 2009. Involvement of NOX2 in the development of behavioral and pathologic alterations in isolated rats. Biol. Psychiatry 66(4): 384–392. https://doi.org/10.1016/j.biopsych.2009.04.033.

Scibior, A., Pietrzyk, L., Plewa, Z. and Skiba, A. 2020a. Vanadium: risks and possible benefits in the light of a comprehensive overview of its pharmacotoxicological mechanisms and multi-applications with a summary of further research trends. J. Trace Elem. Med. Biol. 61: 126508. https://doi.org/10.1016/j.jtemb.2020.126508.

Scibior, A., Szychowski, K.A., Zwolak, I., Dachowska, K. and Gminski, J. 2020b. *In vitro* effect of vanadyl sulfate on cultured primary astrocytes: cell viability and oxidative stress markers. J. Appl. Toxicol. 40(6): 737–747. https://doi.org/10.1002/jat.3939.

Shaver, A., Ng, J.B., Hall, D.A. and Posner, B.I. 1995. The chemistry of peroxovanadium compounds relevant to insulin mimesis. Mol Cell Biochem. 153: 5–15. https://doi.org/10.1007/BF01075913.

Shi, X. and Dalal, N.S. 1992. Hydroxyl radical generation in the NADH/microsomal reduction of vanadate. Free Radic Res. Commun. 17: 369–376. https://doi.org/10.3109/10715769209083141.

Smith, M.A., Rottkamp, C.A., Nunomura, A., Raina, A.K. and Perry, G. 2000. Oxidative stress in Alzheimer's disease. Biochimia Biophysica Acta 1502: 139–144. https://doi.org/10.1016/S0925-4439(00)00040-5.

Soares, S.S., Henao, F., Aureliano, M. and Gutierrez-Merino, C. 2008. Vanadate induces necrotic death in neonatal rat cardiomyocytes through mitochondrial membrane depolarization. Chem. Res. Toxicol. 21(3): 607–618. https://doi.org/10.1021/tx700204r.

Soazo, M. and Garcia, G.B. 2007. Vanadium exposure through lactation produces behavioral alterations and CNS myelin deficit in neonatal rats. Neurotoxicol. Teratol. 29: 503–510. https://doi.org/10.1016/j.ntt.2007.03.001.

Sofroniew, M.V. and Vinters, H.V. 2010. Astrocytes: Biology and pathology. Acta Neuropathol. 119: 7–35. https://doi.org/10.1007/s00401-009-0619-8.

Sohal, V.S., Zhang, F., Yizhar, O. and Deisseroth, K. 2009. Parvalbumin neurons and gamma rhythms enhance cortical circuit performance. Nature 459: 698–702. https://doi.org/10.1038/nature07991.

Stern, A., Yin, X., Tsang, S.S., Davison, A. and Moon, J. 1993. Vanadium as a modulator of cellular regulatory cascades and oncogene expression. Biochem. Cell Biol. 71(3–4): 103–112. https://doi.org/10.1139/o93-018.

Stocks, P. 1960. On the relations between atmospheric pollution in urban and rural localities and mortality from cancer, bronchitis and pneumonia, with particular reference to 3: 4 benzopyrene, beryllium, molybdenum, vanadium and arsenic. Br. J. Cancer 14(3): 397–418. https://doi.org/10.1038/bjc.1960.45.

Stohs, S.J. and Bagchi, D. 1995. Oxidative mechanisms in the toxicity of metal ions. Free Radic. Biol. Med. 18(2): 321–336. https://doi.org/10.1016/0891-5849(94)00159-H.

Sugiyama, S., Prochiantz, A. and Hensch, T.K. 2009. From brain formation to plasticity: Insights on Otx2 homeoprotein. Dev. Growth Differ. 51(3): 369–377. https://doi.org/10.1111/j.1440-169X.2009.01093.x.

Suttkus, A., Rohn, S., J¨ager, C., Arendt, T. and Morawski, M. 2012. Neuroprotection against iron-induced cell death by perineuronal nets - an *in vivo* analysis of oxidative stress. Am. J. Neurodegener. Dis. 1(2): 122–129. PMID: 23383386.

Thompson, K.H. and McNeill, J.H. 1993. Effect of vanadyl sulfate feeding on susceptibility to peroxidative change in diabetic rats. Res. Commun. Chem. Pathol. Pharmacol. 80(2): 187–200. PMID: 8100638.

Todorich, B., Olopade, J.O., Surguladze, N., Zhang, X., Neely, E. and Connor, J.R. 2011. The mechanism of vanadium-mediated developmental hypomyelination is related to destruction of oligodendrocyte progenitors through a relationship with ferritin and iron. Neurotox. Res. 19: 361–373. https://doi.org/10.1007/s12640-010-9167-1.

Tsuda, M., Mizokoshi, A., Shigemoto-Mogami, Y., Koizumi, S. and Inoue, K. 2004. Activation of p38 mitogen-activated protein kinase in spinal hyperactive microglia contributes to pain hypersensitivity following peripheral nerve injury. Glia 45(1): 89–95. https://doi.org/10.1002/glia.10308.

Usende, I.L., Leitner, D.F., Neely, E., Connor, J.R. and Olopade, J.O. 2016. The deterioration seen in myelin related morpho-physiology in vanadium exposed rats is partially protected by concurrent iron deficiency. Niger. J. Physiol. Sci. 31(1): 11–22.

Usende, I.L., Emikpe, B.O. and Olopade, J.O. 2017. Heavy metal pollutants in selected organs of African giant rats from three agro-ecological zones of Nigeria: evidence for their role as an environmental specimen bank. Environ. Sci. Pollut. Res. 24: 22570–22578. https://doi.org/10.1007/s11356-017-9904-6.

Usende, I.L., Olopade, J.O., Emikpe, B.O., Oyagbemi, A.A. and Adedapo, A.A. 2018a. Oxidative stress changes observed in selected organs of African giant rats (*Cricetomys gambianus*) exposed to sodium metavanadate. Inter J. Vet Sci. Med. 6: 80–89. https:// doi. org/ 10.1016/j. ijvsm. 2018. 03. 004.

Usende, I.L., Alimba, C.G., Emikpe, B.O., Bakare, A.A. and Olopade, J.O. 2018b. Intraperitoneal sodium metavanadate exposure induced severe clinicopathological alterations, hepato-renal toxicity and cytogenotoxicity in African giant rats (Cricetomys gambianus, Waterhouse, 1840). Environ Sci. Poll Res. 25: 26383–26393. https://doi.org/10.1007/s11356-018-2588-8.

Usende, I.L., Oyelowo, F.O., Adikpe, A.O., Emikpe, B.O., Nafady, A.H.M. and Olopade, J.O. 2022a. Reproductive hormones imbalance, germ cell apoptosis, abnormal sperm morphophenotypes and ultrastructural changes in testis of African giant rats (*Cricetomys gambianus*) exposed to sodium metavanadate intoxication. Environ. Sci. Pollut. Res. https://doi.org/10.1007/s11356-021-18246-z.

Usende, I.L., Olopade, J.O., Azeez, I.A., Andrioli, A., Bankole, M.O., Olopade, F.E. et al. 2022b. Neuroecotoxicology: Effects of environmental heavy metal exposure on the brain of African giant rats and the contribution of vanadium to the neuropathology. IBRO Neuroscience Reports 13: 215–234. https://doi.org/10.1016/j.ibneur.2022.08.008.

Wang, D.C., Lin, Y.Y. and Lin, H.T. 2015. Recovery of motor coordination after exercise is correlated to enhancement of brain-derived neurotrophic factor in lactational vanadium-exposed rats. Neurosci Lett. 600: 232–237. https://doi.org/10.1016/j.neulet.2015.06.036.

Weihe, E., Depboylu, C., Schütz, B., Sch¨afer, M. and Eiden, L.E. 2006. Three types of tyrosine hydroxylase-positive CNS neurons distinguished by dopa decarboxylase and VMAT2 Co-expression. Cell Mol. Neurobiol. 26: 657–676. https://doi.org/10.1007/ s10571-006-9053-9.

Whittington, M.A., Cunningham, M.O., LeBeau, F.E., Racca, C. and Traub, R.D. 2011. Multiple origins of the cortical γ rhythm. Dev. Neurobiol. 71(1): 92–106. https:// doi.org/10.1002/dneu.20814.

Yang, X., Wang, K., Lu, J. and Crans, D.C. 2003. Membrane transport of vanadium compounds and the interaction with the erythrocyte membrane coordination. Chemistry Rev. 237(1-2): 103–111. https://doi.org/10.1016/S0010-8545(02)00247-3.

Younes, M., Kayser, E. and Strubelt, O. 1991. Effect of antioxidant on vanadate-induced toxicity towards isolated perfused rat livers. Toxicology 70: 141–149. https://doi.org/10.1016/0300-483X(91)90041-X.

Zaporowska, H. 1994. Effects of vanadium on L-ascorbic acid concentration in rat tissues. Gen Pharmacol. 25: 467–470. https://doi.org/10.1016/0306-3623(94)90199-6.

Zhen, X., Torres, C., Cai, G. and Friedman, E. 2002. Inhibition of protein tyrosine/mitogen-activated protein kinase phosphatase activity is associated with d2 dopamine receptor supersensitivity in a rat model of Parkinson's disease. Mol. Pharmacol. 62: 1356–1363. doi: 10.1124/mol.62.6.1356.

CHAPTER 8

Toxicity of Aluminum

Analytical, Environmental and Biochemical Aspects

Charles Elias Assmann,[1,†] *Karine Paula Reichert,*[1,†] *Valderi Luiz Dressler,*[2] *Maria Rosa Chitolina Schetinger*[1] and *Vera Maria Morsch*[1,*]

1. Introduction

Aluminum (Al) is a chemical element occupying a position in group 13 of the periodic table. Aluminum is classified as a metal with atomic number 13 and an atomic mass of 26.98. At room temperature, aluminum is solid, and it has a melting point of 932.15 K. Aluminum is a white metal with low density, and its powder is gray in color; it is ductile and malleable. When exposed to air, aluminum objects are oxidized on the surface, however, the oxide layer protects from further oxidation, which gives this element good resistance to corrosion processes (O'Neil et al., 2001).

Aluminum is the most abundant metal in the Earth's crust (8%) and the third most abundant of all chemical elements, following oxygen (46%) and silicon (26%) (Huheey et al., 1993). Aluminum occurs naturally in air, soil, and water. Despite this, biological systems have developed in the absence of this metal during evolution and to this day it does not seem to exert any physiological functions in animals and plants.

[1] Departamento de Bioquímica e Biologia Molecular, Programa de Pós-graduação em Ciências Biológicas: Bioquímica Toxicológica, CCNE, Universidade Federal de Santa Maria, RS, Brasil, 97105-900.

[2] Departamento de Química, Programa de Pós-graduação em Química, CCNE, Universidade Federal de Santa Maria, RS, Brasil, 97105-900.

† Both authors contributed equally.

* Corresponding author: veramorsch@gmail.com

Aluminum is a light, resistant, and industrially used element, and is currently one of the most-used metals in the world. It is a good heat conductor and a great metallic conductor of electric currents. The name aluminum derives from the Latin "alumen" and was given by Davy in 1807. There are historical records of its use by Greek and Roman societies. It was initially isolated in its impure form in the 1820s by two researchers, Hans Christian Orsted (1825) and Friedrich Wohler (1827) (Atwood and Yearwood, 2000; Soni et al., 2001).

In the environment, aluminum appears predominantly in the oxidation state +3. The destination and transport of the element depend on environmental factors, such as salinity, pH of the medium and in a very important way, on the presence of species that have the capacity to form complexes with it (Yokel and McNamara, 2001).

Aluminum is a very reactive element, widely distributed in nature mainly as a component of various silicates, very stable oxides, and hydroxides. Its main mineral is bauxite ($Al_2O_3.XH_2O$). Cryolite is a complex fluoride of aluminum, also found in nature (Na_3AlF_6) but in low amounts. Many of its compounds, found in nature, have value as precious stones. Among these are rubies, sapphires, topazes, and chrysoberyl. The most used non-ferrous metal by man is Aluminum (Staley and Haupin, 1992).

In the environment, aluminum is released both by natural and by anthropogenic processes (Lantzy and MacKenzie, 1979). The main source of this element extraction is bauxite mining and most of its worldwide extraction is intended to obtain this element. To be economically usable, bauxite must have at least 30% aluminum oxide. The process of obtaining aluminum from bauxite occurs in three steps: Extraction of bauxite, obtaining alumina (aluminum oxide - Al_2O_3), and finally, obtaining aluminum.

Aluminum is widely used in household items, packaging, construction, electrical industry, shipbuilding, the airline industry, printing ink, electric insulators, pottery, ceramics, explosives, cement, as adjuvant in vaccines, and in water treatment and purification (Lewis, 2011; Sayed and Yokel, 2005). The density of aluminum is about one-third of steel. For this reason, for many years the automotive industry has been trying to increase the use of this metal in the manufacture of vehicles, replacing steel. Also, aluminum is present in toothpaste, deodorant/antiperspirant roll-ons, cosmetics (blush, eye shadow, lipstick, base, compact powder, mascara, eyeliner, eye and mouth liner, moisturizer for body and face), nail polish, liquid and bar soaps, medicines (e.g., in antacids), and kitchen utensils. In our food, aluminum is present in food additives such as anticoagulants, hardeners, fermenters, emulsifiers, colorants and acidulants. Some are soluble and can cross the intestinal wall. Currently, the packaging industry is a major consumer of aluminum (Soni et al., 2001). The metal's characteristics provide protection against moisture, oxygen, and UV rays.

Although aluminum has a high concentration in the environment, it is not essential to living beings (Williams, 1999; Exley, 2003). Biological systems do not require aluminum to perform their functions (Cirovic et al., 2021). From an evolutionary point of view, this is strange, since nature usually chooses more abundant elements to follow its life cycle.

For a long time, aluminum was considered inoffensive to living beings. However, the aluminum trivalent cationic form (Al^{3+}) is now well-known for being toxic for biological systems. Several studies point to the impact of excess of this metal on the body and the occurrence of neurodegenerative diseases, especially Alzheimer's, autoimmune diseases, breast cancer, and lung lesions. The presence of aluminum can also lead to chronic changes in intestinal problems such as irritable bowel syndrome or worsening of hemorrhoids, abdominal swelling and poor digestion, skin problems, joint and muscle pain, hair loss, weight loss, and tiredness, among others (Berthon, 2002; Geyikoglu et al., 2013; Igbokwe et al., 2019; Exley and Clarkson, 2020).

2. Aluminum determination in biological samples

As mentioned above, aluminum was discovered in the 19th century. In 1880 an accumulation of the element was verified in organs of dogs treated with baking powder containing the element. This is probably the first experiment with aluminum and its "determination" in biological samples. In 1913, a series of experiments on the chemical impact of aluminum in cooking were described in The Lancet journal (The Lancet, 1913). After that, several studies were done on the effect of aluminum on health, including mainly aspects in patients with renal failure (in 1972), where the first account of an outbreak of encephalopathy in a dialysis unit was observed, and in 1973 presence of aluminum was associated with the brain of patients with Alzheimer's disease (Broe and D'Haese, 1990). However, as mentioned above, aluminum can have several other health implications. Therefore, there is an increasing demand to study the effects of aluminum in humans, which requires precise and accurate analytical results.

2.1 Contamination sources

Aluminum is the most abundant metal of the Earth's crust (comprising around 8%). It is found only in its combined form of bauxite, the primary ore, which is composed of impure oxides found in extensively weathered rocks in the tropics. Aluminum also occurs in mixed oxides with other elements, and as complex aluminosilicates such as micas and feldspars. Metallic aluminum is used in a huge variety of products including cans, foils, kitchen utensils, window frames, beer kegs, and airplane parts, among others. It has low density, has a high thermal conductivity, excellent corrosion resistance, and can be easily cast, machined, and formed. It is also non-magnetic, non-sparking, and the second most malleable metal and the sixth most ductile. Aluminum, as aluminum sulphate, is widely used as coagulant in the process of water clarifying, as well as for wastewater treatment. Aluminum is naturally present in foods such as vegetables, where its concentration can reach up to 3% in older plants. Salts of the element are added to foods, such as frozen strawberries, maraschino cherries, and pickles, to improve their appearance (Martin, 1990). In addition, aluminum salts are used in cosmetics and as antiperspirants to control sweat, and in the early 2000s, with the prohibition of the use of mercury, aluminum began to be used as an adjuvant in several vaccines to elicit a more robust immune response and increase vaccine efficacy. In several countries, infants and young children receive high quantities of aluminum from multiple inoculations (Miller, 2016). On the other hand, aluminum is present in air associated with particulate matter, and the concentration varies in the range from 0.005 to 0.18 μg m^{-3}, depending on location, weather conditions, type, and level of human activity in the area. Most of the aluminum in the air is small and associated with suspended particles of soil (dust) (Alasfar and Isaifan, 2021). Furthermore, with the advent of materials in nanometric dimensions, aluminum also has diverse applications as a material in nanoparticulate form where the element has unique characteristics, different from those of materials of conventional size (Alghriany et al., 2022). Despite the large use and natural presence of aluminum and its low toxicity, some worries have been raised about aluminum's role in breast cancer, breast cyst, Alzheimer's disease, and autism. The mechanism of aluminum action is poorly understood, but it can cause gene instability, alter gene expression, or enhance oxidative stress, but carcinogenicity of aluminum has not been proved yet. In a similar form, aluminum can accumulate in the body, and is recognized as a potential neurotoxin, as it is reported to adversely affect several important biological reactions in the brain including neurotransmitter biosynthesis, synaptic transmission, oxidative stress, inflammatory responses, and neuronal development.

2.2 Sample collection and preparation

Sample collection must be performed to ensure that the sample portion is representative. In addition, the sample preservation method and storage must ensure that no *post-mortem* changes occur so the specimen is preserved as close to the *in vivo* condition as possible and without deterioration that might invalidate any subsequent analysis. Also, the sample must be preserved in a way that allows a variety of chemical analyses so that the sample can be used for several purposes within an experiment. These principles have been applied for the determination of elements in biological samples. In addition, the materials for sample storage, cutting, grinding, and other procedures necessary for its preparation must ensure that the sample will not be contaminated with the element to be determined. In the case of aluminum, this aspect is very relevant, since it is an element present in considerable concentrations in many materials routinely used in chemical analysis laboratories, for which care must be taken from the collection to the analysis of the sample. Therefore, it is important to ensure the purity of all materials that will come in contact with the sample. Polytetrafluoroethylene (PTFE and its variants), high density polyethylene (HDP) suitable for use in chemical analysis, quartz, zirconia, and titanium are among the high purity materials indicated for sample handling. Likewise, reagents such as water, acids (such as nitric, hydrochloric, perchloric, and sulfuric acid), and hydrogen peroxide used for biological sample decomposition and preparing solutions must be of high purity. In addition, care must be taken in relation to additives for sample preservation, such as anticoagulants for blood. Water can be distilled and further purified with ion exchange resins, reverse osmosis, among others. Some acids can be purified by distillation at low temperatures (sub-boiling). When applicable, these reagents must be kept in a clean place and stored in appropriate bottles (such as PTFE or quartz) (San Martín et al., 2022).

After collection, it is best to analyze the samples as soon as possible. However, this is not always possible, and samples must be preserved until analysis. Keeping samples at low temperatures (frozen) is perhaps the most common method when the aim is to determine most chemical elements. In this case, care must be taken to not lose part of the sample at the time of thawing, mainly for tissues and organs. For blood and urine, the sample may be subjected to a process soon after collection that may involve filtration, centrifugation, and other processes before freezing.

When the aim is to determine the total aluminum concentration in a biological sample, most of the analytical techniques require sample decomposition. In general, this is achieved by treating the sample with concentrated acids and hydrogen peroxide at high temperature and pressure. Treatment can be done in open or closed vessels. Closed vessels are indicated more since they avoid external contamination and the reaction conditions are more drastic, leading to a better sample decomposition. By using this condition, organic compounds are decomposed in water and carbon dioxide and the element becomes a free ion (Al^{3+}) in the solution. However, several other procedures for sample preparation can be used for specific analyses. Mild extraction conditions using diluted acids or alkaline solutions, or organic solvents, are used for analyte extraction when speciation analysis is the aim of the work. All these procedures can be assisted by conventional thermal heating, microwave heating, or ultrasonic or electromagnetic radiation (infrared and ultraviolet radiation). In some specific cases, solid samples can be analyzed without decomposition or in the form of slurry.

2.3 Techniques for aluminum determination

Aluminum is determined in different specimens of the human body, including, serum, blood, tissues, hair, etc., for different purposes. Although aluminum is abundant in the environment, its concentration is usually low in human specimens. This brings a great challenge to its quantitative determination, given the need to determine low concentrations of the element, ease of sample contamination during handling, both from the environment itself and from materials and reagents used during the analytical process, and the availability of a technique sensitive enough to determine low concentrations of the element. This difficulty was already reported in 1985, where an interlaboratory study on aluminum in serum found considerable disparity between reported results that was not attributable to the nature of the samples' distribution, nor to the standard aluminum solutions used in different laboratories. For the analyses, all laboratories used electrothermal atomic absorption spectrometry (ET AAS) but sample preparation procedures and temperature programs differed. While some laboratories performed better than others, no laboratory showed consistent acceptable performance (Taylor et al., 1985). However, with improved knowledge and analytical techniques, more and more reliable results are currently being achieved and at low concentrations of aluminum, revealing with greater precision the real effects of this element on humans.

Currently, there are a variety of well-established analytical techniques for single-element and multi-element determination, including aluminum. The techniques provide accurate and precise results in a wide range of element concentrations. However, as previously mentioned, the concentration of aluminum in humans is relatively low, requiring techniques that provide low detection limits (LOD). In this way, some techniques are no longer used due to their insufficient

LOD, interferences, or other unfavorable characteristics. Such is the case of UV-Vis spectrometry, fluorescence spectroscopy, flame atomic absorption spectrometry, X-ray fluorescence spectrometry, and neutron activation analysis. On the other hand, ET AAS, inductively coupled plasma emission spectrometry (ICP OES), and inductively coupled plasma mass spectrometry (ICP-MS) are the techniques most used for element determination in the analysis of biological materials (San Martín et al., 2022). Electrothermal atomic absorption spectrometry is a monoelemental and well-established technique for aluminum determination. Although it takes about five minutes per determination, it requires a small amount of sample (typically less than 20 μL or less than 10 mg), allows the analysis of liquid and solid samples, allows many samples to be analyzed without decomposition prior to determination, provides for LOD (on the order of 1 μg L^{-1} or 0.04 μmol L^{-1}), good precision, and accuracy. It is an important technique when the determination of only one element is required. Inductively coupled plasma emission spectrometry and ICP-MS are multi-element techniques that allow the determination of almost all chemical elements. In both techniques, the most common is the analysis of solutions, preferably samples with low carbon content. Therefore, a decomposition step is required for most biological samples. Exceptions may be blood, serum, and urine, but usually they must be properly diluted with water, diluted alkali solution (such as tetramethylammonium hydroxide) or diluted acids (such as nitric acid). Both techniques allow for a very fast analysis, taking 1–2 min, even for multielement determination. The LOD for aluminum in ICP OES is similar to that of ET AAS, while ICP-MS provides LOD about 100 times lower. Some difficulties related to interferences, mainly spectral, are observed, but ICP OES is perhaps more robust than ICP-MS. In ICP-MS, since aluminum is monoisotopic (m/z 27), there is no alternative isotope that can be used to circumvent spectral interference. However, current instruments are equipped with technology to eliminate most spectral interference. To this end, most ICP-MS equipment is equipped with collision/reaction cells, which consist of quadrupoles or hexapoles (electrodes subjected to direct and/or alternating electrical current), pressurized with reactive (NH_3, O_2, CH_4, among others) or non-reactive (He, H_2, etc.) gases. The combination of the applied electric potential and the gas allows for preferential reactions or collisions to occur with interfering ions (like ArO^+, $ArCl^+$, $ArNa^+$, $ArAr^+$, ArC^+, CN^+, CO^+, etc.) than with the analyte in the cell, thus eliminating the interfering ion.

The ICP OES and ICP-MS techniques are quite versatile in terms of the use of different sample introduction systems into the plasma (ICP). Pneumatic nebulizers are the standard devices used for solution introduction into the plasma but ultrasonic nebulizers, pneumatic micronebulizers, electrothermal vaporizers, and laser ablation systems can be used. In particular, laser ablation systems allow direct analysis of solid biological samples with high spatial resolution (on the order of 1 μm). Thus, it is possible to evaluate the distribution of elements, including aluminum, in each sample, both on the surface of a sample and in depth. In this analysis, it is possible to obtain two- or three-dimensional images of the distribution of the element of the analyzed material.

3. Aluminum and biological systems

3.1 Exposure/Intake

Aluminum is considered ubiquitous in the environment. High levels of aluminum in biological systems have been linked to the development of several diseases such as Alzheimer's disease (AD), autism, multiple sclerosis, epilepsy, breast cancer, depression, anxiety among others (Becaria et al., 2002; Exley, 2013; McLachlan et al., 2019).

In view of the deleterious effects of aluminum on human health, important healthcare committees established limits on the daily intake of aluminum. In 2008, the European Food Safety Authority (EFSA) established a tolerable weekly intake (TWI) of 1 milligram (mg) per kilogram (kg) of body weight per week. In 2012, the JECFA (Joint Expert Committee on Food Additives) of the Food and Agriculture Organization (FAO) and the World Health Organization (WHO) derived a TWI of 2 mg per kg body weight per week. Both were based on studies on developmental disorders in young rats (WHO, 2011).

The primary oral source of aluminum for the typical human is foods contributing to ~ 5–10 mg of aluminum daily, whereas drinking water provides ~ 0.1 mg of aluminum (Yokel et al., 2001).

The ubiquitous property of this metal makes it present in air, water, foodstuffs, cosmetics and personal care products, and drugs (Exley, 2013).

3.1.1 Air

Aluminum is released into the environment by both natural processes and anthropogenic sources (ATSDR, 2008). Aluminum particulates in the atmosphere are mainly derived from soil (weathering), mining, agriculture, and industrial processes. The major anthropogenic sources of aluminum-containing particulate matter include coal combustion, aluminum production, iron and steel foundries, brass and bronze refineries, motor vehicle emissions, and other industrial activities such as smelting, filing, sawing, welding of aluminum metals, and cigarette smoke (ATSDR, 2008; Yokel, 2012).

3.1.2 Water

Aluminum occurs in natural surface waters through normal weathering of rocks and minerals (Becaria et al., 2002). Thus, the mobilization of aluminum to the aquatic environment cannot be avoided, being that in extremely acidic waters dissolved aluminum can exceed a concentration of 90 mg/L (ATSDR, 2008).

In natural water, aluminum is accumulated in different species including green algae (*Pseudokirchneriella subcapitata*), cladocerans (*Ceriodaphnia dubia*), and a species of fish (*Pimephales promelas*) (DeForest et al., 2018). Moreover, aluminum also can be distributed in natural foods, for instance, by the irrigation systems (Mousavi et al., 2013) as well as culture maintained in hydroponic conditions (Gallon et al., 2004). Furthermore, potable drinking water is treated with aluminum salt coagulants (Letterman and Driscoll, 1988). There are estimates that oral aluminum bioavailability from drinking water contributes to 1–2% of the typical human's daily aluminum intake (Yokel and Florence, 2006).

The use of aluminum compounds as coagulating agents in the treatment of water for drinking could increase its aluminum content (Gallon et al., 2004). In pure water, aluminum has a minimum solubility in the pH range of 5.5–6.0 and concentrations of dissolved aluminum increase at higher or lower pH values (WHO, 2010).

3.1.3 Foods

Aluminum occurs naturally in foods as contaminants, cooking utensils, or food additives. Foods naturally high in aluminum include potatoes, spinach, and tea (Stahl et al., 2011). The concentration of aluminum in different fruits and vegetables can be different according to regions and soil conditions including acidity of soils, water irrigation, and plant variety (Mohammad et al., 2014). However, the major intake of aluminum in foods is aluminum-based additives, mainly acidic sodium aluminum phosphate (acidic SALP). Experimental studies on oral aluminum bioavailability from acidic SALP in a biscuit suggested that ~ 0.1% of the aluminum was orally absorbed and provides considerably greater potential to contribute to an aluminum body burden than does drinking water (Yokel and Florence, 2006).

The foodstuff sources include processed foods, food packaging, and cooking utensils. The consumption of foods containing aluminum food additives are the highest sources of aluminum in the diet. Studies indicate that cooking utensils have the highest concentrations of aluminum when compared to other metals, representing 1.64646 ppm in solution from aluminum cooking utensils. In addition, the aluminum element was more stable or inert in Teflon cooking utensils compared to stainless steel and titanium coated stainless steel cooking utensils when the cooking utensil was heated (Sianturi et al., 2020). Thus, the use of aluminum utensils for cooking meals can be associated with chronic exposure to aluminum and, eventually, with aluminum toxicity.

3.1.4 Cosmetics

Topically applied cosmetics and personal care products that are used daily include antiperspirants, sunscreens, body creams, toothpaste, shampoos, and make-up products. Topical exposure to aluminum can range from 2 to 5 g of aluminum per day when applied directly to skin for long periods (Exley, 2013). Furthermore, damage in the skin such as desquamation, fissures, and cuts can facilitate the rate of dermal aluminum absorption (Sanajou et al., 2021).

3.1.5 Drugs

Aluminum is widely used as an adjuvant in vaccines. The main forms of aluminum that have been used to improve the immunogenicity of vaccines include aluminum oxyhydroxide, hydroxyphosphate, phosphate, and potassium sulfate against several diseases (Masson et al., 2018). These vaccines present high efficiency according to official health authorities. However, studies demonstrated that the inclusion of aluminum salts in vaccine administration can interfere with cellular and metabolic processes potentially leading to severe neurologic diseases, stimulating the immune system through the activation of antigen-presenting cells, and consequently, the induction of chemokine secretion (Hunter, 2002). The major problem is for young children, who receive several vaccine doses over a short period of time and are considered to be at the highest risk for Al-dependent, vaccine-related complications including possibly neurodevelopmental delay, autism spectrum disorder, and Alzheimer's disease (AD) (Morris et al., 2017; Mold et al., 2018).

Aluminum hydroxide also is used widely in antacids and buffered aspirin, contributing to the intake of aluminum. However, absorption of aluminum hydroxide is usually less than 0.01% of the intake dose (Niu, 2018).

3.1.6 Dialysis

The main source of aluminum exposure among patients on hemodialysis includes dialysis fluid. Studies have shown abnormal blood aluminum levels in dialysis patients (Chuang et al., 2022). Recent data indicate that a serum aluminum level of ≥ 6 ng/mL was independently associated with all-cause death in patients on chronic hemodialysis. Moreover, the results suggest a dose-dependent effect of serum aluminum levels on mortality (Tsai et al., 2022). Other studies showed that dialysis fluid quality was not in compliance with international standards; dialysis fluid in 63% of the samples contained high aluminum concentrations and usually were associated with elevated concentrations in dialysis fluid in some hospitals as well (Humudat and Al-Naseri, 2020).

3.2 Absorption

Aluminum compounds are usually poorly absorbed after oral ingestion (maximum 1%), while the oral absorption from drinking water is approximately 0.3% (SCHEER, 2017) and absorption from food is 0.1% in the gastrointestinal tract (EFSA, 2008). Absorption rates depend on chemical species of aluminum, being the Al^{3+} bioavailable and most toxic form of aluminum for living organisms. The speciation of aluminum depends on the aluminum compound ingested/exposed, as well as the pH of biological systems and the presence or absence of organic acids. The important factors that favor oral aluminum absorption by the gastrointestinal tract are outlined in Table 1.

Due to the strong selectivity of the cation exchanger, aluminum gradually replaces the divalent essential cations, affecting the homeostasis of various ions. For example, Mg^{2+} is one of the most affected cations, since the two cations are similar in size, which is a dominant factor over the charge identity in terms of metal ion competition (Rezabal et al., 2006). Thus, the Al^{3+} and Mg^{2+} ions compete for membrane transporters and metal-binding sites on enzymes (Chen and Ma, 2013).

In addition, aluminum has high affinity for oxygen-donor ligands, i.e., carboxylates, organic and inorganic phosphates, nucleotides, and polynucleotides such as DNA and RNA, deprotonated hydroxyl groups and phosphate groups of nucleoside di and triphosphates, such as ADP and ATP (Wang, 2018). Therefore, ATP can bind aluminum in cells, forming an Al-ATP complex and Al^{3+} competes with Mg^{2+} for binding to ATP. Al^{3+} also impairs biological processes involving rapid Ca^{2+} exchange (Exley, 1999; Burnstock, 2014).

Table 1. Factors that contribute to the absorption and accumulation of aluminum in the human body.

Factors	Effects	References
Citrate	Formation of Al-citrate complex was more bioavailable than other aluminum chemical species. Citrate presence increases the oral absorption of aluminum.	(Wang, 2018)
Silicon	Silicon (Si)-containing compounds could reduce aluminum absorption and facilitate aluminum excretion.	(Dorneles et al., 2016)
Fluoride	Fluoride forms numerous complexes with aluminum and can increase aluminum absorption; it can solubilize the aluminum as F-Al compounds in the gastric phase.	(Yin et al., 2021)
Iron	Iron competes with aluminum; thus, aluminum absorption generally increases with iron deficiency.	(Ittel et al., 1996)
Calcium	A calcium-deficient diet increased aluminum absorption and accumulation.	(Yokel and Mcnamara, 1990)
ATP	ATP released from mucosal epithelial cells can bind with aluminum. The formation of Al-ATP complex can also affect different regions of the human body.	(Exley 1999; Burnstock, 2014)
Ethanol	Both aluminum and ethanol are pro-oxidant and neurotoxins; ethanol enhanced the aluminum's effects.	(Ghosh et al., 2021)
Food additives	Several food additives are aluminum salts (sodium aluminum phosphate, sodium aluminum sulfate, sodium aluminosilicate, calcium aluminum silicate, aluminum silicate, aluminum ammonium sulphate) thus contributing to the total daily intake of aluminum.	(EFSA, 2008; Glynn and Lignell, 2019)
Age	Studies indicated that aluminum concentration in the body increases with aging.	(Nie, 2018)

Moreover Al^{3+} binds to various proteins, inducing the oligomerization and conformational changes of the proteins. The changes can potentially inhibit different enzymes (such as Na^+/K^+ATPase activity, transferrin, plasma Ca^{2+}-ATPase, etc.), Additionally, Al^{3+} can bind to phosphorylated amino acids in proteins and can promote the self-aggregation and accumulation of highly phosphorylated cytoskeleton proteins, for instance, highly phosphorylated neurofilaments and microtubule-associated proteins (Ittel et al., 1996; Silva and Gonçalves, 2003; Kawahara and Kato-Negishi, 2011). Like this, the Al^{3+} can compete with metals in some metalloenzymes, causing their inactivation and compromising crucial processes and pathways.

3.3 Deposit

When absorbed, aluminum can accumulate in all tissues of the human body, with total aluminum concentration ranging from 30 to 50 mg/kg of body weight in healthy human subjects. However, aluminum distribution is not homogeneous, i.e., normal serum aluminum concentrations ranged from 1 to 3 μg/L, which are bound and transported between all major tissues/organs by the iron transport protein transferrin as a transferrin-Al complex (Exley et al., 2007a). On the other hand, in the human body, approximately half of the aluminum is deposited in the skeleton (range of 5–10 mg/kg) and approximately one-fourth accumulates in the lungs (from accumulation of inhaled insoluble aluminum compounds). This is followed by the liver, kidney and brain with 1 μg/g of dry weight (Exley, 2013).

However, Exley and Mold (2019) recently confirmed by aluminum-specific fluorescence microscopy that the metal is distributed unequally in brain tissues. When an aluminum concentration of 1.00 μg/g dry weight (1 ppm) is determined for 0.500 g fresh weight of tissue, there are regions within this sample of tissue where the concentration of aluminum is significantly higher than 1 ppm, mainly in non-neuronal cells such as microglia (Exley and Mold, 2018, 2019).

In addition, aluminum can also accumulate in breast tissue where it can be genotoxic and potentially stimulate the development of breast cancer (Exley et al., 2007b). The major constituent of antiperspirants is aluminum salts and daily application of aluminum-based antiperspirants with high durability on the skin can accumulate in breast tissue (Exley et al., 2007b). Studies point to a range of values of 1.68 μg/g for normal breast specimens and 1.88–2.1 μg/g for breast cancer specimens (Rodrigues-Peres et al., 2013). Furthermore, the breast accumulated may even be distributed to the milk of lactating mothers. In fact, some studies have demonstrated an alarming Al concentration of 0.191 μg/mL of milk (Javad et al., 2018).

Furthermore, studies have shown that aluminum can cross the placenta and reach the fetus. Of particular significance, aluminum in maternal placental tissue can be associated with increased risk for fetal neural tube defects (Liu et al., 2021) and affects neurodevelopment of the fetus as aluminum impairs the neural fate of progenitor neural cells (Reichert et al., 2019). Recent literature data shows that the aluminum concentration in placenta tissue was about 2.8 mg/g of dry weight (Liu et al., 2021). Other organs and tissues that can accumulate aluminum include the spleen, muscle, heart, and skin (Yokel, 2012).

3.4 Excretion

Unabsorbed aluminum is eliminated through the alimentary tract in the feces. Under normal conditions, aluminum is primarily excreted through the renal route. Aluminum is also excreted in the milk, bile, feces, sweat, hair, nails, sebum, and semen (Exley, 2013). Urinary excretion of aluminum is enhanced by chemical chelators such as deferoxamine and organic acids. There are significant differences in the excretion half-life of aluminum, ranging from hours, days, and months to years, suggesting that there is more than one compartment of aluminum storage from which aluminum is eliminated (Schafer and Jahreis, 2006). Figure 1 summarizes the main routes of exposure, absorption, accumulation, and excretion of aluminum in biological systems.

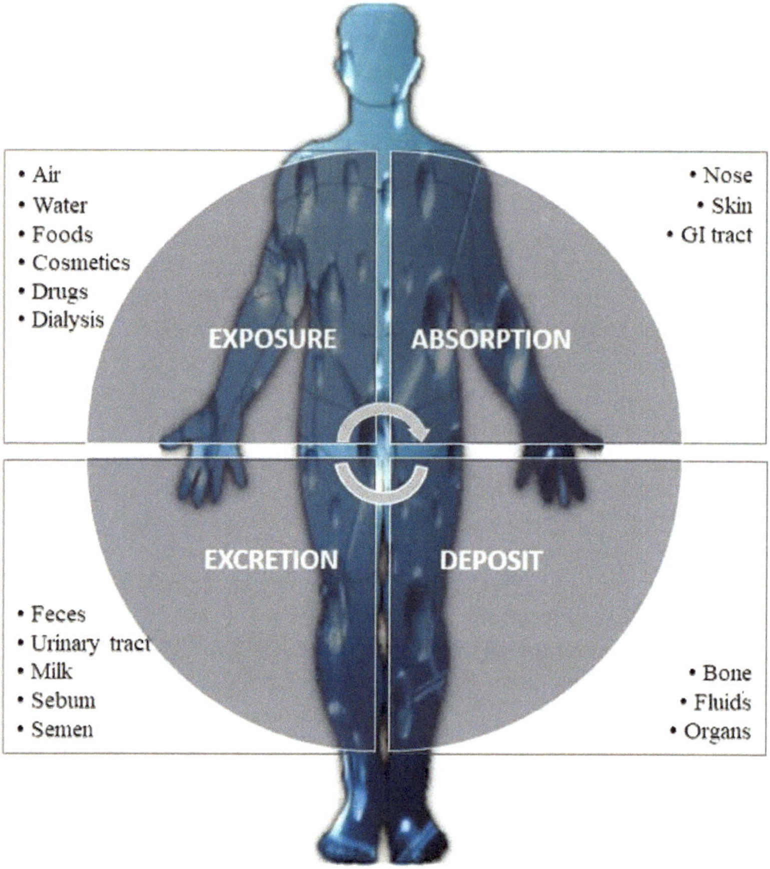

Figure 1. Main routes of exposure, absorption, accumulation, and excretion of aluminum in biological systems. GI: Gastrointestinal tract.

4. Toxic effects

4.1 Peripheral toxicity

4.1.1 Aluminum toxicity in intestinal microbiota

Gut microbiota is associated with an important role in essential physiological functions and overall health of different species. Recent evidence has pointed out that aluminum intake affects the regulation of the permeability, the microflora and the immune function of the intestine (Yu et al., 2019; Weng et al., 2020). For instance, Nie et al. (2022) showed that aluminum chloride ($AlCl_3$) decreased Gram-positive bacteria and increased proinflammatory Gram-negative bacteria in intestinal flora of the zebrafish (Nie et al., 2022). Similarly, another study revealed that the incubation of the *Bacteroides fragilis*—abundant Gram-negative in the human gastrointestinal (GI) tract microbiome—with aluminum sulfate acts like a potent inducer of pro-inflammatory glycolipid lipopolysaccharide (BF-LPS) synthesis (Alexandrov et al., 2020). In addition, aluminum exposure can influence the gut microbiota composition in mice (Zhai et al., 2017) and reduce intestinal barrier function and immune function (Vignal et al., 2016). Aluminum exposure also induced oxidative stress, intestinal changes and immune responses in red swamp crayfish (*Procambarus clarkii*) (Zhang et al., 2023).

On the other hand, reports suggest that the dietary supplementation with probiotics regulates gut microbiota structure and function in Nile tilapia exposed to aluminum (Yu et al., 2019). Besides, in a mice model of $AlCl_3$ and d-galactose-induced AD, the treatment with oral resveratrol-selenium-peptide nanocomposites blunted the cognitive disorder caused by aluminum and galactose by regulating gut microbiota disruption, particularly by normalizing end-points of oxidative stress by decreasing reactive oxygen species (ROS) and increasing activity of antioxidant enzymes and alleviating gut microbiota disorder, particularly with respect to oxidative stress and inflammatory-related bacteria such as Alistipes, Helicoba *Alistipes, Helicobacter, Rikenella, Desulfovibrio,* and *Faecalibaculum* (Li et al., 2021).

Thus, the neurotoxins exposure (i.e., aluminum) can stimulate the profusion of toxins derived from anaerobic Gram-negative bacteria and other constituents of the gastrointestinal-tract microbiome and may significantly contribute to the initiation, development and/or propagation of inflammatory conditions linked with several diseases.

4.1.2 Nephrotoxicity and hepatotoxicity

The evidence for the hepatotoxic and nephrotoxic actions of aluminum in animals and humans is still elusive. Some studies have indicated that the intoxication of rats with $AlCl_3$ resulted in its accumulation in the hepatic and renal tissues (Othman et al., 2020). In addition, there have been acute effects of $AlCl_3$ as pro-oxidant, inducing the lipid peroxidation, reduced glutathione, and antioxidant enzymes of the kidneys of Swiss albino mice in addition to pathological changes in cortical renal tissue including shrunken glomeruli, intraglomerular congestion, mesangial hyperplasia, obliteration of the filtration slits, degeneration of mitochondria in proximal convoluted tubules, and cell necrosis (Benzaid et al., 2021; Al Kahtani, 2010).

Other symptoms of aluminum hepato- and nephrotoxicity are elevated levels of liver and kidney function markers, such as increase in the ALT, AST, bilirubin, creatinine and urea serum levels (Al-Kahtani and Morsy, 2019; Attiah Al-Hazmi and Rawi, 2021; Al-Kahtani, 2010; Abdel Moneim et al., 2013; Ghorbel et al., 2015). Biochemical alterations in the liver were also associated with upregulation of caspase-3 and decreased expression of Bcl-2, contributing to apoptosis (Al-Kahtani et al., 2020).

Yousef et al. (2019) showed that co-exposure to aluminum oxide nanoparticles and zinc oxide nanoparticles resulted in more pronounced hepatorenal toxicities and systemic inflammation associated with epigenetic changes in the gene expression of mtTFA and PGC-1α. The authors hypothesized the activation of these genes could cause mitochondrial dysfunction, inducing the generation of ROS instigating the oxidative stress (Wang et al., 2018; Ibrahim Yousef et al., 2019). In addition, literature has indicated a central role of ROS in attacks of almost all cellular components, including membrane phospholipids, inducing damage to cell membranes, which eventually leads to apoptosis, causing liver dysfunction and mitochondrial energy metabolism disorder in rats exposed to aluminum (Xu et al., 2017; Abu-Elfotuh et al., 2022). Acute aluminum poisoning can cause pathological changes in the structure of the kidney tissue of chicks, resulting in damage to function as well as triggering inflammation and oxidative stress in this organ (Chen et al., 2011).

The morphological lesions in the hepatic and renal tissues of $AlCl_3$-intoxicated rats include increased levels of proinflammatory cytokines such as TNF-α, IL-6, Nod-like receptor protein 3 (NLRP3), cysteinyl aspartate specific proteinase 1 (caspase-1), autism spectrum conditions, interleukin-18 (IL-18), and interleukin-1β (IL-1β) (Al Dera, 2016; Abu-Elfotuh et al., 2022; Xiao et al., 2022). In addition, evidence showed a disruptive effect of aluminum on electrolyte homeostasis due to inhibited activities of Na^+/K^+ and Ca^{2+}/Mg^{2+} ATPases and increasing serum levels of potassium, calcium, phosphate, and chloride, with concomitant decreases in serum levels of magnesium and bicarbonate responsible to nephrotoxic effects on renal functions (Obafemi, 2022; Sautu et al., 2023).

4.2 Neurotoxicity

Aluminum is a neurotoxic element to the human brain (Bondy, 2010; Exley, 2013, 2014; Kumar and Gill, 2009). The cationic form, Al^{3+}, is reactive to biological tissues and deposits in the brain throughout life, making this structure highly susceptible to the effects of this hazardous element (Exley, 2014). In this sense, studies have suggested the involvement of aluminum in the development of neurodegenerative conditions such as Alzheimer's disease (Exley, 2017; Huat et al., 2019; Wang et al., 2016). Evidence has supported this hypothesis by indicating the presence of high aluminum quantities in the brain tissue of Alzheimer's patients (Exley and Clarkson, 2020; Mirza et al., 2017). Aluminum-evoked changes in the brain may include alterations typically found in Alzheimer's disease, for instance, memory and learning deficits, neurodegeneration, and brain aging (Bondy, 2014, 2016; Exley, 2014), which are attributed to this neurotoxin (Bondy, 2010; Exley, 2014; Kumar and Gill, 2009). Moreover, aluminum may also be involved with the aggregation and toxicity of β-amyloid peptides, a hallmark of Alzheimer's disease (Kawahara and Kato-Negishi, 2011; Mold et al., 2020; Walton and Wang, 2009; Zhao et al., 2014).

4.2.1 Oxidative stress and inflammation

One of the mechanisms by which aluminum may be exerting its toxic effects in the brain is by promoting oxidative stress (Dey and Singh, 2022; Exley, 2013; Igbokwe et al., 2019; Kumar and Gill, 2009, 2014; Maya et al., 2016). The hippocampus and prefrontal cortex brain structures of rats exposed to low aluminum doses showed increased oxidative stress markers such as levels of reactive oxygen species (ROS). Subchronic aluminum exposure to low doses also augmented lipid peroxidation in the hippocampus (Martinez et al., 2017). In plasma samples of rats exposed to a low aluminum dose, levels of ROS and lipid peroxidation were higher when compared to the control group. Additionally, the increase in oxidative stress seen in animals exposed to the low-dose protocol was comparable to that seen in animals treated with a higher dose (Martinez et al., 2018).

In another study performed with adult rats, aluminum was shown to promote oxidative stress in the prefrontal cortex and hippocampus and disturb memory and learning. Overall, aluminum exposure increased markers of lipid peroxidation and decreased catalase activity in the prefrontal cortex and hippocampus when compared to control animals (Fernandes et al., 2020). Isolated mitochondria from brain regions of rats submitted to a chronic aluminum treatment showed increased ROS levels. Additionally, the production of ROS was higher in a time-dependent manner in the corpus striatum and hippocampus regions when compared to the control group. Besides increasing ROS levels, aluminum may also reduce mitochondrial biogenesis (Sharma et al., 2013).

Another pathway by which aluminum may induce its neurotoxic effects is by prompting inflammatory states (Igbokwe et al., 2019; Maya et al., 2016). Human neural cells cultured in the presence of aluminum, even at a nanomolar concentration, have shown increased expression of pro-inflammatory and pro-apoptotic genes (Lukiw et al., 2005). Also, recent evidence has shown the involvement of aluminum in promoting inflammation in the central nervous system (CNS). For example, rats exposed to $AlCl_3$ for 90 days presented elevated neuroinflammation in the hippocampus. Notably, the expression of some pro-inflammatory genes such as IL-1β, IL-6, and TNF-α was seen augmented following $AlCl_3$ treatment. These changes in the pattern of gene expression were accompanied by memory and learning deficits (Cao et al., 2016). In fact, IL-1β and IL-1 pathways are suggested to be associated with $AlCl_3$-triggered damage in the hippocampus of rats. Besides increasing pro-inflammatory cytokines, $AlCl_3$ also promoted microglial activation (Zhang et al., 2018). In another work, the authors indicated that $AlCl_3$ triggered the IL-1β/JNK signaling pathway which was also associated with the promotion of apoptosis and necroptosis of hippocampal neural cells. $AlCl_3$ also caused depression in rats (Zhang et al., 2020). Another study showed that aluminum triggered cognitive deficits and caused the activation of glial cells. Moreover, pyroptosis and activation of the DDX3X-NLRP3 inflammasome signaling pathway may be relevant to the neuroinflammatory processes elicited by aluminum in the brain (Hao et al., 2021).

4.2.2 Cholinergic and purinergic systems

Cholinergic signaling is fundamental for brain neurotransmission and is involved in many functions such as memory and learning (Picciotto et al., 2012). Furthermore, the purinergic system is a signaling pathway which is central in adenosine and adenine nucleotide metabolism and increasing evidence has shown its participation in several diseases (Huang et al., 2021; Reichert et al., 2021). Next, some investigations are presented which have studied aluminum's effects, simultaneously or not, on the purinergic and cholinergic signaling pathways.

In the 1990s, Schetinger and collaborators provided some initial evidence on the possible alterations prompted by aluminum in the central nervous system (CNS) related to the purinergic system. By using synaptosomes obtained from the cerebral cortex of adult rats exposed to different doses of $AlCl_3$, the authors assessed the activity of the ATP diphosphohydrolase, as it was known at the time, using ATP and ADP as substrates. Based on the kinetics results, the authors pointed out the inhibitory outcome of aluminum on the activity of ATP diphosphohydrolase, which participates in the degradation of ATP and can act as a neurotransmitter in the CNS (Schetinger et al., 1995).

Other studies have explored the mechanisms underlying aluminum toxicity in the CNS and its relation to Alzheimer's disease. Long-term exposure to low levels of aluminum has been shown to alter acetylcholinesterase (AChE) activity in the mouse brain. In a protocol following 12 weeks of exposure, mice that received aluminum plus citrate solution presented an increase in AChE activity in brain structures, namely the cortex, hippocampus, hypothalamus, and striatum (Kaizer et al., 2005). The activity of AChE has also been shown to be altered after long-term exposure to aluminum (50 mg/kg) in rats. Results varied among the brain structures that were investigated, showing an increase in AChE activity in the hypothalamus and striatum for groups that received aluminum and aluminum plus citrate. However, groups that received only aluminum showed decreased AChE activity in the cerebellum, cortex, and hippocampus. Interestingly, groups treated with aluminum and aluminum plus citrate demonstrated enhanced erythrocyte AChE activity (Kaizer et al., 2008).

Kaizer and collaborators also assessed the behavior of purinergic enzymes, namely nucleoside triphosphate diphosphohydrolase (NTPDase) (ATP and ADP as substrates) and 5'-nucleotidase (AMP as substrate), in synaptosomal fractions (cerebral cortex and hippocampus) and platelets of rats treated with aluminum ($AlCl_3$; 50 mg/kg/day). Overall, the authors observed an increase in the activities of NTPDase and 5'-nucleotidase in the investigated structures for groups treated with aluminum, indicating that this element may alter purinergic signaling in the brain and peripheral tissues (Kaizer et al., 2007). Another study assessed the effects of aluminum on the activities of NTPDase and AChE in samples of lymphocytes of rats from protocols conducted *in vitro* and *in vivo*. *In vitro*, aluminum decreased the hydrolysis of ATP and ADP when performing NTPDase activity while AChE activity increased. However, *in vivo*, groups that received treatment with aluminum had increased NTPDase activity, observed by the hydrolysis of ATP and ADP, and AChE activity from peripheral lymphocytes of rats was augmented as well. In this sense, the authors suggested an effect of aluminum on the immune system (Kaizer et al., 2010).

More recently, aluminum was shown to interfere *in vitro* with the purinergic signaling of neural progenitor cells (NPCs). The hydrolysis of nucleotides (ATP, ADP, and AMP) and ATP release from neurospheres was reduced. Furthermore, differentiated neurospheres treated with the cationic ion Al^{3+} presented decreased expression of P2Y1 and A2A receptors (Reichert et al., 2020). In addition, aluminum was recently shown to alter *in vitro* purinergic system components in BV-2 microglial cells. Nucleotide (ATP, ADP, and AMP) hydrolysis was reduced while adenosine breakdown was augmented when the authors investigated enzyme activities. Here, purinoceptor expression and density were also found to be significantly changed. Aluminum raised A2A and P2X7 receptor expression and reduced A1 receptor expression, suggesting that this metal interferes with the purinergic signaling of brain microglial cells (Assmann et al., 2021).

Aluminum has been shown to trigger cognitive dysfunctions by altering cholinergic neurotransmission in the brain. In rats, intraperitoneal injection with $AlCl_3$ for 7 days was able to change behavioral parameters, specifically $AlCl_3$ treatment impaired recognition memory. When analyzing cholinergic signaling, the group that received $AlCl_3$ presented decreased levels of acetylcholine and AChE activity was increased in cortex and hippocampus structures when compared to the control group (Liaquat et al., 2019).

A study adopting another administration route of aluminum, via drinking water, also pointed out the adverse effects of this element in the brain related to cholinergic neurotransmission. Mice that received aluminum treatment showed reduced acetylcholine levels in the hippocampus and cortex and overall reduced acetylcholine synthesis. Moreover, some memory impairments were also observed in animals receiving aluminum, which also affected sociability (Farhat et al., 2017). Aluminum treatment, also via drinking water, has been implicated in memory deficits in male rats. To reflect human dietary intake, male rats were given orally low doses of aluminum via water; after treatment periods, some cognitive tasks were performed and the AChE activity in the prefrontal cortex and the hippocampus was analyzed. The authors observed impaired cognitive functions in the low-dose aluminum-treated animals and decreased AChE activity in the hippocampus when compared to the control group (Martinez et al., 2017).

Taken together, these studies highlight the impact of aluminum as an element that can alter cholinergic and purinergic signaling pathways, which could also be related to effects on cognitive and immune functions. Nevertheless, further investigations are required to assess the adverse effects of this environmentally abundant element. Figure 2 depicts some of the harmful outcomes of aluminum exposure in biological systems.

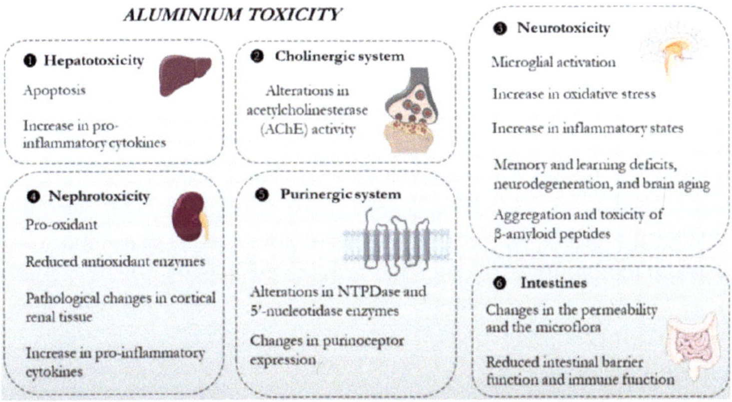

Figure 2. Aluminum presents several negative effects on health in different organisms. Toxic effects might occur in various organs, including the liver (1), brain (3), kidneys (4), and intestines (6) by different mechanisms. Alterations in signaling pathways can also be found due to aluminum exposure, such as in cholinergic (2) and purinergic (5) systems. This figure was created using Servier Medical Art templates, which are licensed under a Creative Commons Attribution 3.0 Unported License (http://smart.servier.com).

5. Conclusion

The present review discussed important findings regarding aluminum determination and toxicity. The determination of aluminum depends on several factors such as its physical and chemical forms, sample preparation and technique of determination. The toxic responses in animals' models vary with the route of administration (i.p., gavage, administered in the water, food, etc.), dose, duration of the treatment and the age of the animals. Also, in humans there are many health risk issues about aluminum exposure, which were discussed in this chapter. In fact, some of them deserve particular attention, mainly because nowadays the exposure to aluminum can occur on a daily basis (for instance, from drinking water, foods, cosmetics, and medicines among others). Applying the concept of One Health (Amuasi et al., 2020) which comprises the health of humans, animals, and the environment it is important to raise the discussion about aluminum toxicity. Not only humans and animals suffer from aluminum exposure, but also plants which are not tolerant to aluminum can have their development impaired by this metal.

It is interesting to discuss the toxicity of aluminum from an evolutionary point of view particularly because aluminum is the third most abundant element in nature, but it has no biological function known until now. In fact, it is considered a toxic element being able to disrupt cellular processes impairing different signaling pathways for example. Some tissues are most sensitive to aluminum and here we focused on the central nervous system, liver, and kidney toxicity.

Aluminum is considered a neurotoxin and has been suggested to be involved in the development of neurodegenerative conditions such as Alzheimer's disease. In the brain and in liver and kidney aluminum can promote inflammation and oxidative stress impairing their functionality. In addition, the enzymes of the purinergic and cholinergic systems are affected showing dysregulation in the hydrolysis of the neurotransmitters ATP and acetylcholine, as well as the production of the neuromodulator adenosine. In this context, both purinergic and cholinergic systems could be related with the effect of aluminum as a neurotoxin.

More investigations are needed to better understand the toxicology of this metal and to propose new strategies to prevent exposure and tissue damages including oxidative stress, apoptosis, necrosis, mitochondrial dysfunction, inflammation, and cell permeability. Perhaps, working with the concept of One Health can help to diminish the gaps around the understanding of the effects of aluminum exposure to humans, animals and the environment (including plants, soil, air and water). In conclusion, the primary objective of this review was to contribute to a deeper comprehension of aluminum toxicity and provide insights on effective strategies for managing it within a broader context.

References

Abu-Elfotuh, K., Hussein, F.H., Abbas, A.N., Al-Rekabi, M.D., Barghash, S.S., Zaghlool, S.S. et al. 2022. Melatonin and zinc supplements with physical and mental activities subside neurodegeneration and hepatorenal injury induced by aluminum chloride in rats: Inclusion of GSK-3β-Wnt/β-catenin signaling pathway. Neurotoxicology 2022 Jul; 91: 69–83. doi: 10.1016/j.neuro.2022.05.002.

Agency for Toxic Substances and Disease Registry (ATSDR). 2008. Toxicological Profile for Aluminum.

Al Dera, H.S. 2016. Protective effect of resveratrol against aluminum chloride induced nephrotoxicity in rats. Saudi Med J. 2016 Apr; 37(4): 369–78. doi: 10.15537/smj.2016.4.13611.

Al Kahtani, M.A. 2010. Renal damage mediated by oxidative stress in mice treated with aluminum chloride: protective effects of Taurine. J. Biol. Sci. 10: 584–595. DOI: 10.3923/jbs.2010.584.595.

Alasfar, R.H. and Isaifan, R.J. 2021. Aluminum environmental pollution: the silent killer. Environ Sci. Pollut. Res. Int. 2021 Sep; 28(33): 44587–44597. doi: 10.1007/s11356-021-14700-0.

Alexandrov, P.N., Hill, J.M., Zhao, Y., Bond, T., Taylor, C.M., Percy, M.E. et al. 2020. Aluminum-induced generation of lipopolysaccharide (LPS) from the human gastrointestinal (GI)-tract microbiome-resident *Bacteroides fragilis*. J. Inorg. Biochem. 2020 Feb; 203: 110886. doi: 10.1016/j.jinorgbio.2019.110886.

Alghriany, A.A.I., Omar, H.E.M., Mahmoud, A.M. and Atia, M.M. 2022. Assessment of the toxicity of aluminum oxide and its nanoparticles in the bone marrow and liver of male mice: ameliorative efficacy of curcumin nanoparticles. ACS Omega 2022 Apr 14; 7(16): 13841–13852. doi: 10.1021/acsomega.2c00195.

Al-Hazmi, M.A., Rawi, S.M. and Hamza, R.Z. 2021. Biochemical, histological, and neuro-physiological effects of long-term aluminum chloride exposure in rats. Metab Brain Dis. 2021 Mar; 36(3): 429–436. doi: 10.1007/s11011-020-00664-6.

Al-Kahtani, M. and Morsy, K. 2019. Ameliorative effect of selenium nanoparticles against aluminum chloride-induced hepatorenal toxicity in rats. Environ Sci. Pollut. Res. Int. 2019 Nov; 26(31): 32189–32197. doi: 10.1007/s11356-019-06417-y.

Al-Kahtani, M., Abdel-Daim, M.M., Sayed, A.A., El-Kott, A. and Morsy, K. 2020. Curcumin phytosome modulates aluminum-induced hepatotoxicity via regulation of antioxidant, Bcl-2, and caspase-3 in rats. Environ Sci. Pollut. Res. Int. 2020 Jun; 27(17): 21977–21985. doi: 10.1007/s11356-020-08636-0.

Amuasi, J.H., Lucas, T., Horton, R. and Winkler, A.S. 2020. Reconnecting for our future: The Lancet One Health Commission. Lancet. 2020 May 9; 395(10235): 1469–1471. doi: 10.1016/S0140-6736(20)31027-8.

Assmann, C.E., Mostardeiro, V.B., Weis, G.C.C., Reichert, K.P., de Oliveira Alves, A., Miron, V.V. et al. 2021. Aluminum-induced alterations in purinergic system parameters of BV-2 brain microglial cells. J. Immunol. Res. 2021 Jan 12; 2021: 2695490. doi: 10.1155/2021/2695490.

Atwood, D.A. and Yearwood, B.C. 2000. The future of aluminum chemistry. J. Organomet. Chem. 600(1-2): 186–197. doi: 10.1016/S0022-328X(00)00147-9.

Becaria, A., Campbell, A. and Bondy, S.C. 2002. Aluminum as a toxicant. Toxicol. Ind. Health. 2002 Aug; 18(7): 309–20. doi: 10.1191/0748233702th157oa.

Benzaid, C., Tichati, L., Trea, F., Rouabhia, M. and Ouali, K. 2021. *Rhamnus alaternus* aqueous extract enhances the capacity of system redox defence and protects hepatocytes against aluminum chloride toxicity in rats. Clin. Phytosci. 7: 67. https://doi.org/10.1186/s40816-021-00302-3.

Berthon, G. 2002. Aluminium speciation in relation to aluminium bioavailability, metabolism and toxicity. Coord. Chem. Reviews 228(2): 319–341. doi: 10.1016/S0010-8545(02)00021-8.

Bichu, S., Tilve, P., Kakde, P., Jain, P., Khurana, S., Ukirade, V. et al. 2019. Relationship between the use of aluminium utensils for cooking meals and chronic aluminum toxicity in patients on maintenance hemodialysis: a case control study. J. Assoc. Physicians India 2019 Apr; 67(4): 52–56.

Bondy, S.C. 2010. The neurotoxicity of environmental aluminum is still an issue. Neurotoxicology 2010 Sep; 31(5): 575–81. doi: 10.1016/j.neuro.2010.05.009.

Bondy, S.C. 2014. Prolonged exposure to low levels of aluminum leads to changes associated with brain aging and neurodegeneration. Toxicology 2014 Jan 6; 315: 1–7. doi: 10.1016/j.tox.2013.10.008.

Bondy, S.C. 2016. Low levels of aluminum can lead to behavioral and morphological changes associated with Alzheimer's disease and age-related neurodegeneration. Neurotoxicology 2016 Jan; 52: 222–9. doi: 10.1016/j.neuro.2015.12.002.

Burnstock, G. 2014. Purinergic signalling in the gastrointestinal tract and related organs in health and disease. Purinergic Signal. 2014 Mar; 10(1): 3–50. doi: 10.1007/s11302-013-9397-9.

Cao, Z., Yang, X., Zhang, H., Wang, H., Huang, W., Xu, F. et al. 2016. Aluminum chloride induces neuroinflammation, loss of neuronal dendritic spine and cognition impairment in developing rat. Chemosphere 2016 May; 151: 289–95. doi: 10.1016/j.chemosphere.2016.02.092.

Chen, B., Chen, L., Yang, Z., Fu, Q., Li, X. and Cao, C. 2023. Acute aluminum sulfate triggers inflammation and oxidative stress, inducing tissue damage in the kidney of the chick. Biol Trace Elem Res. 2023 Mar; 201(3): 1442–1450. doi: 10.1007/s12011-022-03260-0.

Chen, Z.C. and Ma, J.F. 2013. Magnesium transporters and their role in Al tolerance in plants. Plant Soil 368: 51–56. doi: 10.1007/s11104-012-1433-y.

Chuang, P.H., Tsai, K.F., Wang, I.K., Huang, Y.C., Huang, L.M., Liu, S.H. et al. 2022. Blood aluminum levels in patients with hemodialysis and peritoneal dialysis. Int. J. Environ. Res. Public Health 2022 Mar 24; 19(7): 3885. doi: 10.3390/ijerph19073885.

Ćirović, A., Ćirović, A., Nikolić, D., Ivanovski, A. and Ivanovski, P. 2021. The adjuvant aluminum fate—Metabolic tale based on the basics of chemistry and biochemistry. J. Trace Elem. Med. Biol. 2021 Dec; 68: 126822. doi: 10.1016/j.jtemb.2021.126822.

De Broe, M.E. and D'Haese, P.C. 1990. Historical survey of aluminum-related diseases. *In*: De Broe, M.E. and Coburn, J.W. (eds.). Aluminum and Renal Failure. Developments in Nephrology, vol 26. Springer, Dordrecht. doi: 10.1007/978-94-009-1868-9_1.

DeForest, D.K., Brix, K.V., Tear, L.M. and Adams, W.J. 2018. Multiple linear regression models for predicting chronic aluminum toxicity to freshwater aquatic organisms and developing water quality guidelines. Environ. Toxicol. Chem. 2018 Jan; 37(1): 80–90. doi: 10.1002/etc.3922.

Dey, M. and Singh, R.K. 2022. Neurotoxic effects of aluminium exposure as a potential risk factor for Alzheimer's disease. . Rep. 2022 Jun; 74(3): 439–450. doi: 10.1007/s43440-022-00353-4.

Dorneles, A.O.S., Pereira, A.S., Rossato, L.V., Possebom, G., Sasso, V.M., Bernardy, K. et al. 2016. Silicon reduces aluminum content in tissues and ameliorates its toxic effects on potato plant growth. Rural Sci. 46(3): 506–512. doi: 10.1590/0103-8478cr20150585.

European Food Safety Authority (EFSA). 2008. Safety of aluminium from dietary intake - Scientific Opinion of the Panel on Food Additives, Flavourings, Processing Aids and Food Contact Materials (AFC). The EFSA Journal 754: 1–34. doi: 10.2903/j.efsa.2008.754.

Exley, C. 1999. A molecular mechanism of aluminium-induced Alzheimer's disease? J. Inorg. Biochem. 1999 Aug 30; 76(2): 133–40. doi: 10.1016/s0162-0134(99)00125-7.

Exley, C. 2003. A biogeochemical cycle for aluminium? J. Inorg. Biochem. 2003 Sep 15; 97(1): 1–7. doi: 10.1016/s0162-0134(03)00274-5.

Exley, C., Beardmore, J. and Rugg, G. 2007a. Computational approach to the blood-aluminum problem? Int. J. Quantum Chem. 107: 275–278. doi: 10.1002/qua.21190.

Exley, C., Charles, L.M., Barr, L., Martin, C., Polwart, A. and Darbre, P.D. 2007b. Aluminium in human breast tissue. J. Inorg. Biochem. 2007b Sep; 101(9): 1344–6. doi: 10.1016/j.jinorgbio.2007.06.005.

Exley, C. 2013. Human exposure to aluminium. Environ. Sci. Process Impacts 2013 Oct; 15(10): 1807–1816. doi: 10.1039/c3em00374d.

Exley, C. 2014. What is the risk of aluminium as a neurotoxin? Expert Rev. Neurother. 2014 Jun; 14(6): 589–91. doi: 10.1586/14737175.2014.915745.

Exley, C. 2017. Aluminum should now be considered a primary etiological factor in Alzheimer's Disease. J. Alzheimers Dis. Rep. 2017 Jun 8; 1(1): 23–25. doi: 10.3233/ADR-170010.

Exley, C. and Mold, M.J. 2019. Aluminium in human brain tissue: how much is too much? J. Biol. Inorg. Chem. 2019 Dec; 24(8): 1279–1282. doi: 10.1007/s00775-019-01710-0.

Exley, C. and Clarkson, E. 2020. Aluminium in human brain tissue from donors without neurodegenerative disease: A comparison with Alzheimer's disease, multiple sclerosis and autism. Sci. Rep. 2020 May 8; 10(1): 7770. doi: 10.1038/s41598-020-64734-6.

Farhat, S.M., Mahboob, A., Iqbal, G. and Ahmed, T. 2017. Aluminum-induced cholinergic deficits in different brain parts and its implications on sociability and cognitive functions in mouse. Biol. Trace Elem. Res. 2017 May; 177(1): 115–121. doi: 10.1007/s12011-016-0856-3.

Fernandes, R.M., Corrêa, M.G., Aragão, W.A.B., Nascimento, P.C., Cartágenes, S.C., Rodrigues, C.A. et al. 2020. Preclinical evidences of aluminum-induced neurotoxicity in hippocampus and pre-frontal cortex of rats exposed to low doses. Ecotoxicol. Environ. Saf. 2020 Dec 15; 206: 111139. doi: 10.1016/j.ecoenv.2020.111139.

Gallon, C., Munger, C., Prémont, S. and Campbell, P.G.C. 2004. Hydroponic study of aluminum accumulation by aquatic plants: effects of fluoride and pH. Water, Air, & Soil Pollution 153: 135–155. https://doi.org/10.1023/B:WATE.0000019943.67578.ed.

Geyikoglu, F., Türkez, H., Bakir, T.O. and Cicek, M. 2013. The genotoxic, hepatotoxic, nephrotoxic, haematotoxic and histopathological effects in rats after aluminium chronic intoxication. Toxicol. Ind. Health 2013 Oct; 29(9): 780–91. doi: 10.1177/0748233712440140.
Ghorbel, I., Maktouf, S., Kallel, C., Ellouze Chaabouni, S., Boudawara, T. and Zeghal, N. 2015. Disruption of erythrocyte antioxidant defense system, hematological parameters, induction of pro-inflammatory cytokines and DNA damage in liver of co-exposed rats to aluminium and acrylamide. Chem. Biol. Interact. 2015 Jul 5; 236: 31–40. doi: 10.1016/j.cbi.2015.04.020.
Ghosh, B., Sharma, R.K. and Yadav, S. 2021. Aluminium induced neurodegeneration in rat cerebellum in the presence of ethanol coexposure. J. Pharm. Bioallied Sci. 2021 Nov; 13(Suppl 2): S1228–S1233. doi: 10.4103/jpbs.jpbs_377_21.
Glynn, A. and Lignell, S. 2019. Increased urinary excretion of aluminium after ingestion of the food additive sodium aluminium phosphate (SALP)—a study on healthy volunteers. Food Addit Contam Part A Chem. Anal. Control Expo Risk Assess 2019 Aug; 36(8): 1236–1243. doi: 10.1080/19440049.2019.1626998.
Hao, W., Hao, C., Wu, C., Xu, Y., Wu, S., Lu, X. et al. 2021. Aluminum impairs cognitive function by activating DDX3X-NLRP3-mediated pyroptosis signaling pathway. Food Chem. Toxicol. 2021 Nov; 157: 112591. doi: 10.1016/j.fct.2021.112591.
Hao, W., Hao, C., Wu, C., Xu, Y. and Jin, C. 2022. Aluminum induced intestinal dysfunction via mechanical, immune, chemical and biological barriers. Chemosphere 2022 Feb; 288(Pt 2): 132556. doi: 10.1016/j.chemosphere.2021.132556.
Huang, Z., Xie, N., Illes, P., Di Virgilio, F., Ulrich, H., Semyanov, A. et al. 2021. From purines to purinergic signalling: molecular functions and human diseases. Signal Transduct Target Ther. 2021 Apr 28; 6(1): 162. doi: 10.1038/s41392-021-00553-z.
Huat, T.J., Camats-Perna, J., Newcombe, E.A., Valmas, N., Kitazawa, M. and Medeiros, R. 2019. Metal toxicity links to Alzheimer's disease and neuroinflammation. J. Mol. Biol. 2019 Apr 19; 431(9): 1843–1868. doi: 10.1016/j.jmb.2019.01.018.
Huheey, J.E., Keiter, E.A. and Keiter, R.L. 1993. Inorganic Chemistry: Principles of Structure and Reactivity, Fourth ed. Harper Collins, New York.
Humudat, Y.R. and Al-Naseri, S.K. 2019. Heavy metals in dialysis fluid and blood samples from hemodialysis patients in dialysis centers in Baghdad, Iraq. J. Health Pollut. 2019 Aug 19; 10(27): 200901. doi: 10.5696/2156-9614-10.27.200901.
Hunter, R.L. 2002. Overview of vaccine adjuvants: present and future. Vaccine. 2002 May 31; 20 Suppl 3: S7–12. doi: 10.1016/s0264-410x(02)00164-0.
Igbokwe, I.O., Igwenagu, E. and Igbokwe, N.A. 2019. Aluminium toxicosis: a review of toxic actions and effects. Interdiscip Toxicol. 2019 Oct; 12(2): 45–70. doi: 10.2478/intox-2019-0007.
Ittel, T.H., Kinzel, S., Ortmanns, A. and Sieberth, H.G. 1996. Effect of iron status on the intestinal absorption of aluminum: a reappraisal. Kidney Int. 1996 Dec; 50(6): 1879–88. doi: 10.1038/ki.1996.509.
Javad, M.T., Vahidinia, A., Samiee, F., Elaridi, J., Leili, M., Faradmal, J. et al. 2018. Analysis of aluminum, minerals and trace elements in the milk samples from lactating mothers in Hamadan, Iran. J. Trace Elem. Med. Bio. 50: 8–15. doi: 10.1016/j.jtemb.2018.05.016.
Kaizer, R.R., Corrêa, M.C., Spanevello, R.M., Morsch, V.M., Mazzanti, C.M., Gonçalves, J.F. et al. 2005. Acetylcholinesterase activation and enhanced lipid peroxidation after long-term exposure to low levels of aluminum on different mouse brain regions. J. Inorg. Biochem. 2005 Sep; 99(9): 1865–70. doi: 10.1016/j.jinorgbio.2005.06.015.
Kaizer, R.R., Maldonado, P.A., Spanevello, R.M., Corrêa, M.C., Gonçalves, J.F., Becker, L.V. et al. 2007. The effect of aluminium on NTPDase and 5'-nucleotidase activities from rat synaptosomes and platelets. Int. J. Dev. Neurosci. 2007 Oct; 25(6): 381–6. doi: 10.1016/j.ijdevneu.2007.06.002.
Kaizer, R.R., Corrêa, M.C., Gris, L.R., da Rosa, C.S., Bohrer, D., Morsch, V.M. et al. 2008. Effect of long-term exposure to aluminum on the acetylcholinesterase activity in the central nervous system and erythrocytes. Neurochem. Res. 2008 Nov; 33(11): 2294–301. doi: 10.1007/s11064-008-9725-6.
Kaizer, R.R., Gutierres, J.M., Schmatz, R., Spanevello, R.M., Morsch, V.M., Schetinger, M.R. et al. 2010. *In vitro* and *in vivo* interactions of aluminum on NTPDase and AChE activities in lymphocytes of rats. Cell Immunol. 265(2): 133–8. doi: 10.1016/j.cellimm.2010.08.001.

Kawahara, M. and Kato-Negishi, M. 2011. Link between aluminum and the pathogenesis of Alzheimer's Disease: The integration of the aluminum and amyloid cascade hypotheses. Int. J. Alzheimers Dis. 2011 Mar 8; 2011: 276393. doi: 10.4061/2011/276393.

Kumar, V. and Gill, K.D. 2009. Aluminium neurotoxicity: neurobehavioural and oxidative aspects. Arch Toxicol. 2009 Nov; 83(11): 965–78. doi: 10.1007/s00204-009-0455-6.

Kumar, V. and Gill, K.D. 2014. Oxidative stress and mitochondrial dysfunction in aluminium neurotoxicity and its amelioration: a review. Neurotoxicology 2014 Mar; 41: 154–66. doi: 10.1016/j.neuro.2014.02.004.

Lantzy, R.J. and Mackenzie, F.T. 1979. Atmospheric trace metals: global cycles and assessment of man's impact. Geochim Cosmochim Acta 43(4): 511–525. doi: 10.1016/0016-7037(79)90162-5.

Letterman, R.D. and Driscoll, C.T. 1988. Survey of residual aluminum in filtered water. J. Am. Water Work Assoc. 80(4): 154–158.

Lewis, R.J. 2001. Hawley's Condensed Chemical Dictionary, 14th ed, pp. 39–46, Wiley-Interscience, New Jersey, USA.

Li, C., Wang, N., Zheng, G. and Yang, L. 2021. Oral administration of resveratrol-selenium-peptide nanocomposites alleviates alzheimer's disease-like pathogenesis by inhibiting Aβ aggregation and regulating gut microbiota. ACS Appl. Mater Interfaces 2021 Oct 6; 13(39): 46406–46420. doi: 10.1021/acsami.1c14818.

Liaquat, L., Sadir, S., Batool, Z., Tabassum, S., Shahzad, S., Afzal, A. et al. 2019. Acute aluminum chloride toxicity revisited: Study on DNA damage and histopathological, biochemical and neurochemical alterations in rat brain. Life Sci. 2019 Jan 15; 217: 202–211. doi: 10.1016/j.lfs.2018.12.009.

Liu, M., Wang, D., Wang, C., Yin, S., Pi, X., Li, Z. et al. High concentrations of aluminum in maternal serum and placental tissue are associated with increased risk for fetal neural tube defects. Chemosphere 2021 Dec; 284: 131387. doi: 10.1016/j.chemosphere.2021.131387.

Lukiw, W.J., Percy, M.E. and Kruck, T.P. 2005. Nanomolar aluminum induces pro-inflammatory and pro-apoptotic gene expression in human brain cells in primary culture. J. Inorg. Biochem. 2005 Sep; 99(9): 1895–8. doi: 10.1016/j.jinorgbio.2005.04.021.

Martin, R.B. 1990. Chemistry of aluminum. *In*: De Broe, M.E. and Coburn, J.W. (eds.). Aluminum and renal failure. Developments in Nephrology, vol 26. Springer, Dordrecht. doi: 10.1007/978-94-009-1868-9_2.

Martinez, C.S., Alterman, C.D., Peçanha, F.M., Vassallo, D.V., Mello-Carpes, P.B., Miguel, M. et al. 2017. Aluminum exposure at human dietary levels for 60 days reaches a threshold sufficient to promote memory impairment in rats. Neurotox Res. 2017 Jan; 31(1): 20–30. doi: 10.1007/s12640-016-9656-y.

Martinez, C.S., Vera, G., Ocio, J.A.U., Peçanha, F.M., Vassallo, D.V., Miguel, M. et al. 2018. Aluminum exposure for 60 days at an equivalent human dietary level promotes peripheral dysfunction in rats. J. Inorg. Biochem. 2018 Apr; 181: 169–176. doi: 10.1016/j.jinorgbio.2017.08.011.

Masson, J.D., Crépeaux, G., Authier, F.J., Exley, C. and Gherardi, R.K. 2018. Critical analysis of reference studies on the toxicokinetics of aluminum-based adjuvants. J. Inorg. Biochem. 2018 Apr; 181: 87–95. doi: 10.1016/j.jinorgbio.2017.12.015.

Maya, S., Prakash, T., Madhu, K.D. and Goli, D. 2016. Multifaceted effects of aluminium in neurodegenerative diseases: A review. Biomed. Pharmacother. 2016 Oct; 83: 746–754. doi: 10.1016/j.biopha.2016.07.035.

McLachlan, D.R.C., Bergeron, C., Alexandrov, P.N., Walsh, W.J., Pogue, A.I., Percy, M.E. et al. 2019. Aluminum in neurological and neurodegenerative disease. Mol. Neurobiol. 2019 Feb; 56(2): 1531–1538. doi: 10.1007/s12035-018-1441-x.

Miller, N.Z. 2016. Aluminum in childhood vaccines is unsafe. Journal of American Physicians and Surgeons 21(4): 109–117.

Mirza, A., King, A., Troakes, C. and Exley, C. 2017. Aluminium in brain tissue in familial Alzheimer's disease. J. Trace Elem. Med. Biol. 2017 Mar; 40: 30–36. doi: 10.1016/j.jtemb.2016.12.001.

Mohammad, F.S., Al Zubaidy, I.A.H. and Bassioni, G. 2014. A comparison of aluminum leaching processes in tap and drinking water. Int. J. Electrochem. Sci. 9: 3118–3129.

Mold, M., Umar, D., King, A. and Exley, C. 2018. Aluminium in brain tissue in autism. J. Trace Elem. Med. Biol. 2018 Mar; 46: 76–82. doi: 10.1016/j.jtemb.2017.11.012.

Mold, M., Linhart, C., Gómez-Ramírez, J., Villegas-Lanau, A. and Exley, C. 2020. Aluminum and amyloid-β in familial Alzheimer's disease. J. Alzheimers Dis. 73(4): 1627–1635. doi: 10.3233/JAD-191140.

Morris, G., Puri, B.K. and Frye, R.E. 2017. The putative role of environmental aluminium in the development of chronic neuropathology in adults and children. How strong is the evidence and what could be the mechanisms involved? Metab. Brain Dis. 2017 Oct; 32(5): 1335–1355. doi: 10.1007/s11011-017-0077-2.

Mousavi, S.R., Balali-Mood, M., Riahi-Zanjani, B., Yousefzadeh, H. and Sadeghi, M. 2013. Concentrations of mercury, lead, chromium, cadmium, arsenic and aluminum in irrigation water wells and wastewaters used for agriculture in Mashhad, northeastern Iran. Int. J. Occup. Environ. Med. 2013 Apr; 4(2): 80–6.

Nie, J. 2018. Exposure to aluminum in daily life and Alzheimer's Disease. Adv. Exp. Med. Biol. 1091: 99–111. doi: 10.1007/978-981-13-1370-7_6.

Nie, Y., Yang, J., Zhou, L., Yang, Z., Liang, J., Liu, Y. et al. 2022. Marine fungal metabolite butyrolactone I prevents cognitive deficits by relieving inflammation and intestinal microbiota imbalance on aluminum trichloride-injured zebrafish. J. Neuroinflammation 2022 Feb 7; 19(1): 39. doi: 10.1186/s12974-022-02403-3.

Niu, Q. 2018. Overview of the relationship between aluminum exposure and health of human being. Adv. Exp. Med. Biol. 1091: 1–31. doi: 10.1007/978-981-13-1370-7_1.

Obafemi, T.O. 2022. Gallic and hesperidin ameliorate electrolyte imbalances in $AlCl_3$-induced nephrotoxicity in wistar rats. Biochem. Res. Int. 2022 Oct 10; 2022: 6151684. doi: 10.1155/2022/6151684.

O'Neil, M.J., Smith, A., Heckelman, P.E., Koch, C.B. and Roman, K.J. 2001. Aluminum and aluminum compounds. The Merck Index. An Encyclopedia of Chemicals, Drugs, and Biologicals. Whitehouse Station, NJ: Merck & Co., Inc., 59–65.

Othman, M.S., Fareid, M.A., Abdel Hameed, R.S. and Abdel Moneim, A.E. 2020. The protective effects of melatonin on aluminum-induced hepatotoxicity and nephrotoxicity in rats. Oxid. Med. Cell Longev. 2020 Oct 19; 2020: 7375136. doi: 10.1155/2020/7375136.

Picciotto, M.R., Higley, M.J. and Mineur, Y.S. 2012. Acetylcholine as a neuromodulator: cholinergic signaling shapes nervous system function and behavior. Neuron. 2012 Oct 4; 76(1): 116–29. doi: 10.1016/j.neuron.2012.08.036.

Reichert, K.P., Schetinger, M.R.C., Pillat, M.M., Bottari, N.B., Palma, T.V., Gutierres, J.M. et al. 2019. Aluminum affects neural phenotype determination of embryonic neural progenitor cells. Arch Toxicol. 2019 Sep; 93(9): 2515–2524. doi: 10.1007/s00204-019-02522-6.

Reichert, K.P., Pillat, M.M., Schetinger, M.R.C., Bottari, N.B., Palma, T.V., Assmann, C.E. et al. 2020. Aluminum-induced alterations of purinergic signalling in embryonic neural progenitor cells. Chemosphere 2020 Jul; 251: 126642. doi: 10.1016/j.chemosphere.2020.126642.

Reichert, K.P., Castro, M.F.V., Assmann, C.E., Bottari, N.B., Miron, V.V., Cardoso, A. et al. 2021. Diabetes and hypertension: Pivotal involvement of purinergic signaling. Biomed Pharmacother. 2021 May; 137: 111273. doi: 10.1016/j.biopha.2021.111273.

Rezabal, E., Mercero, J.M., Lopez, X. and Ugalde, J.M. 2006. A study of the coordination shell of aluminum(III) and magnesium(II) in model protein environments: thermodynamics of the complex formation and metal exchange reactions. J. Inorg. Biochem. 2006 Mar; 100(3): 374–84. doi: 10.1016/j.jinorgbio.2005.12.007.

Rodrigues-Peres, R.M., Cadore, S., Febraio, S., Heinrich, J.K., Serra, K.P., Derchain, S.F. et al. 2013. Aluminum concentrations in central and peripheral areas of malignant breast lesions do not differ from those in normal breast tissues. BMC Cancer. 2013 Mar 8; 13: 104. doi: 10.1186/1471-2407-13-104.

Saiyed, S.M. and Yokel, R.A. 2005. Aluminium content of some foods and food products in the USA, with aluminium food additives. Food Addit Contam. 2005 Mar; 22(3): 234–44. doi: 10.1080/02652030500073584.

San Martín, S., Bauçà, J. and Martinez-Morillo, E. 2022. Determination of aluminum concentrations in biological specimens: application in the clinical laboratory. Advances in Laboratory Medicine/Avances en Medicina de Laboratorio. 3(2): 153–159. doi: 10.1515/almed-2022-0056.

Sanajou, S., Şahin, G. and Baydar, T. 2021. Aluminium in cosmetics and personal care products. J. Appl. Toxicol. 2021 Nov; 41(11): 1704–1718. doi: 10.1002/jat.4228.

Schafer, U. and Jahreis, G. 2006. Exposure, bioavailability, distribution and excretion of aluminum and its toxicological relevance to humans. Trace Elem. Electrolytes 23(3): 162–172.
Schetinger, M.R., Wyse, A.T., Da Silva, L.B., Barcellos, C.K., Dias, R.D. and Sarkis, J.J. 1995. Effects of aluminum chloride on the kinetics of rat cortex synaptosomal ATP diphosphohydrolase (EC 3.6.1.5). Biol. Trace Elem. Res. 1995 Dec; 50(3): 209–19. doi: 10.1007/BF02785411.
Scientific Committee on Health, Environmental and Emerging Risks (SCHEER). 2017. Request for a scientific opinion on the tolerable intake of aluminium with regards to adapting the migration limits for aluminium in toys.
Sharma, D.R., Sunkaria, A., Wani, W.Y., Sharma, R.K., Kandimalla, R.J., Bal, A. et al. 2013. Aluminium induced oxidative stress results in decreased mitochondrial biogenesis via modulation of PGC-1α expression. Toxicol. Appl. Pharmacol. 2013 Dec 1; 273(2): 365–80. doi: 10.1016/j.taap.2013.09.012.
Sianturi, M., Gultom, F.L., Faradiba, F., Heumasse, P.V. and Febriza, F. 2020. Study of elements released from various cooking utensil after heating on cooking utensil of aluminum, stainless steel, titanium-coated stainless steel and teflon and their potential health hazards. Int. J. Progress Sci. Technol. 23(1): 459–467. doi: 10.52155/ijpsat.v23.1.2306.
Silva, V.S. and Gonçalves, P.P. 2003. The inhibitory effect of aluminium on the (Na+/K+)ATPase activity of rat brain cortex synaptosomes. J. Inorg. Biochem. 2003 Sep 15; 97(1): 143–50. doi: 10.1016/s0162-0134(03)00257-5.
Soni, M.G., White, S.M., Flamm, W.G. and Burdock, G.A. 2001. Safety evaluation of dietary aluminum. Regul. Toxicol. Pharmacol. 2001 Feb; 33(1): 66–79. doi: 10.1006/rtph.2000.1441.
Stahl, T., Taschan, H. and Brunn, H. 2011. Aluminum content of selected foods and food products. Environ Sci. Eur. 23: 1–11. doi: 10.1186/2190-4715-23-37.
Staley, J.T. and Haupin, W. 1992. Aluminum and aluminum alloys. pp. 248–249. *In*: Kroschwitz, J.I. and Howe-Grant, M. (eds.). Kirk-Othmer Encyclopedia of Chemical Technology. Vol. 2: Alkanolamines to Antibiotics (glycopeptides). New York: John Wiley & Sons, Inc.
Taylor, A., Starkey, B.J. and Walker, A.W. 1985. Determination of aluminium in serum: findings of an external quality assessment scheme. Ann. Clin. Biochem. 1985 Jul; 22(Pt 4): 351–8. doi: 10.1177/000456328502200403.
The Lancet. 1913. Some Kitchen Experiments with Aluminium. The Lancet 181(4662): P54–55. DOI: 10.1016/S0140-6736(01)47783-X.
Tsai, M.-H., Fang, Y.-W., Liou, H.-H., Leu, J.-G. and Lin, B.-S. 2018. Association of serum aluminum levels with mortality in patients on chronic hemodialysis. Sci. Rep. 8: 16729. doi: 10.1038/s41598-018-34799-5.
Vignal, C., Desreumaux, P. and Body-Malapel, M. 2016. Gut: An underestimated target organ for Aluminum. Morphologie. 2016 Jun; 100(329): 75–84. doi: 10.1016/j.morpho.2016.01.003.
Walton, J.R. and Wang, M.X. 2009. APP expression, distribution and accumulation are altered by aluminum in a rodent model for Alzheimer's disease. J. Inorg. Biochem. 2009 Nov; 103(11): 1548–54. doi: 10.1016/j.jinorgbio.2009.07.027.
Wang, L. 2018. Entry and deposit of aluminum in the brain. Adv. Exp. Med. Biol. 1091: 39–51. doi: 10.1007/978-981-13-1370-7_3.
Wang, X., Gong, J., Gui, Z., Hu, T. and Xu, X. 2018. Halloysite nanotubes-induced Al accumulation and oxidative damage in liver of mice after 30-day repeated oral administration. Environ. Toxicol. 2018 Jun; 33(6): 623–630. doi: 10.1002/tox.22543.
Wang, Z., Wei, X., Yang, J., Suo, J., Chen, J., Liu, X. et al. 2016. Chronic exposure to aluminum and risk of Alzheimer's disease: A meta-analysis. Neurosci. Lett. 2016 Jan 1; 610: 200–6. doi: 10.1016/j.neulet.2015.11.014.
Weng, M.H., Chen, S.Y., Li, Z.Y. and Yen, G.C. 2020. Camellia oil alleviates the progression of Alzheimer's disease in aluminum chloride-treated rats. Free Radic Biol. Med. 2020 May 20; 152: 411–421. doi: 10.1016/j.freeradbiomed.2020.04.004.
Williams, R.J. 1999. What is wrong with aluminium? The J.D. Birchall Memorial Lecture. J. Inorg. Biochem. 1999 Aug 30; 76(2): 81–8. doi: 10.1016/s0162-0134(99)00118-x.
World Health Organization (WHO). 2010. Aluminum in drinking-water. Guidel Drink Qual.
World Health Organization (WHO). 2011. Evaluation of Certain Food Additives and Contaminants Seventy-fourth report of the Joint FAO/WHO Expert Committee on Food Additives Food and Agriculture Organization of the United Nations.

Xiao, B., Cui, Y., Li, B., Zhang, J., Zhang, X., Song, M. et al. 2022. ROS antagonizes the protection of Parkin-mediated mitophagy against aluminum-induced liver inflammatory injury in mice. Food Chem. Toxicol. 2022 Jul; 165: 113126. doi: 10.1016/j.fct.2022.113126.

Xu, F., Liu, Y., Zhao, H., Yu, K., Song, M., Zhu, Y. et al. 2017. Aluminum chloride caused liver dysfunction and mitochondrial energy metabolism disorder in rat. J. Inorg. Biochem. 2017 Sep; 174: 55–62. doi: 10.1016/j.jinorgbio.2017.04.016.

Yin, N., Li, Y., Yang, Y., Fan, C., Li, Y., Du, X. et al. 2021. Human health risk assessment in aluminium smelting site: Soil fluoride bioaccessibility and relevant mechanism in simulated gastrointestinal tract. J. Hazard Mater. 2021 Aug 15; 416: 125899. doi: 10.1016/j.jhazmat.2021.125899.

Yokel, R.A. and McNamara, P.J. 1990. The influence of dietary calcium reduction on aluminum absorption and kinetics in the rabbit. Biol. Trace Elem. Res. 1990 Winter; 23: 109–17. doi: 10.1007/BF02917182.

Yokel, R.A. and McNamara, P.J. 2001. Aluminium toxicokinetics: an updated minireview. Pharmacol. Toxicol. 2001 Apr; 88(4): 159–67. doi: 10.1034/j.1600-0773.2001.d01-98.x.

Yokel, R.A., Rhineheimer, S.S., Brauer, R.D., Sharma, P., Elmore, D. and McNamara, P.J. 2001. Aluminum bioavailability from drinking water is very low and is not appreciably influenced by stomach contents or water hardness. Toxicology 2001 Mar 21; 161(1-2): 93–101. doi: 10.1016/s0300-483x(01)00335-3.

Yokel, R.A. and Florence, R.L. 2006. Aluminum bioavailability from the approved food additive leavening agent acidic sodium aluminum phosphate, incorporated into a baked good, is lower than from water. Toxicology. 2006 Oct 3; 227(1-2): 86–93. doi: 10.1016/j.tox.2006.07.014.

Yokel, R.A. 2012. Aluminum. *In*: Encyclopedia of Human Nutrition.

Yousef, M.I., Mutar, T.F. and Kamel, M.A.E. 2019. Hepato-renal toxicity of oral sub-chronic exposure to aluminum oxide and/or zinc oxide nanoparticles in rats. Toxicol. Rep. 2019 Apr 19; 6: 336–346. doi: 10.1016/j.toxrep.2019.04.003.

Yu, L., Qiao, N., Li, T., Yu, R., Zhai, Q., Tian, F. et al. 2019. Dietary supplementation with probiotics regulates gut microbiota structure and function in Nile tilapia exposed to aluminum. PeerJ. 2019 Jun 3; 7: e6963. doi: 10.7717/peerj.6963.

Zhai, Q., Li, T., Yu, L., Xiao, Y., Feng, S., Wu, J. et al. 2017. Effects of subchronic oral toxic metal exposure on the intestinal microbiota of mice. Sci. Bull (Beijing) 2017 Jun 30; 62(12): 831–840. doi: 10.1016/j.scib.2017.01.031.

Zhang, H., Wang, P., Yu, H., Yu, K., Cao, Z., Xu, F. et al. 2018. Aluminum trichloride-induced hippocampal inflammatory lesions are associated with IL-1β-activated IL-1 signaling pathway in developing rats. Chemosphere 2018 Jul; 203: 170–178. doi: 10.1016/j.chemosphere.2018.03.162.

Zhang, H., Wei, M., Lu, X., Sun, Q., Wang, C., Zhang, J. et al. 2020. Aluminum trichloride caused hippocampal neural cells death and subsequent depression-like behavior in rats via the activation of IL-1β/JNK signaling pathway. Sci.Total Environ. 2020 May 1; 715: 136942. doi: 10.1016/j.scitotenv.2020.136942.

Zhang, X., Shen, M., Wang, C., Gao, M., Wang, L., Jin, Z. et al. 2023. Impact of aluminum exposure on oxidative stress, intestinal changes and immune responses in red swamp crayfish (*Procambarus clarkii*). Sci. Total Environ. 855: 158902. doi: 10.1016/j.scitotenv.2022.158902.

Zhao, Y., Hill, J.M., Bhattacharjee, S., Percy, M.E., Pogue, A.I. and Lukiw, W.J. 2014. Aluminum-induced amyloidogenesis and impairment in the clearance of amyloid peptides from the central nervous system in Alzheimer's disease. Front. Neurol. 2014 Sep 5; 5: 167. doi: 10.3389/fneur.2014.00167.

CHAPTER 9

The Differential Role of Toxic and Essential Metals in Formation and Toxicity of Advanced Glycation End-Products

Anatoly V. Skalny,[1] *Michael Aschner,*[2] *Abel Santamaria,*[3] *Rongzhu Lu,*[4] *Joao B. T. Rocha,*[5] *Svetlana V. Zalavina,*[6] *Vladimir I. Korchin,*[7] *Tao Ke*[2] and *Alexey A. Tinkov*[1,8,*]

1. Introduction

Advanced glycation end-products (AGEs) are a highly heterogeneous group of molecules formed during non-enzymatic interaction of carbonyl groups of reducing carbohydrates with free amine groups of proteins, lipids, and nucleic acids (Kuzan, 2021). AGEs exert significant toxicity through a number of mechanisms including oxidative stress, endoplasmic reticulum stress, apoptosis, and inflammation to name

[1] IM Sechenov First Moscow State Medical University (Sechenov University), Moscow, 119435, Russia.
[2] Department of Molecular Pharmacology, Albert Einstein College of Medicine, Bronx, NY, 10461, USA.
[3] Laboratorio de Aminoácidos Excitadores/Laboratorio de Neurofarmacología Molecular y Nanotecnología, Instituto Nacional de Neurología y Neurocirugía, Mexico City 14269, Mexico.
[4] Department of Preventive Medicine and Public Health Laboratory Sciences, School of Medicine, Jiangsu University, Zhenjiang, 212013, Jiangsu, China.
[5] Departamento de Bioquímica e Biologia Molecular, Centro de Ciências Naturais e Exatas, Universidade Federal de Santa Maria, Santa Maria 97105-900, RS, Brazil.
[6] Department of Histology, Embryology and Cytology named after Prof. M. Ya. Subbotin, Novosibirsk State Medical University, Novosibirsk 630091, Russia.
[7] Department of physiology and sport medicine, Khanty-Mansiysk State Medical Academy, Khanty-Mansiysk 628011, Russia.
[8] Laboratory of Ecobiomonitoring and Quality Control, Yaroslavl State University, Yaroslavl 150003, Russia.
* Corresponding author: tinkov.a.a@gmail.com

a few (Nowotny et al., 2015; Adamopoulos et al., 2014). Induction of inflammatory response upon AGEs accumulation was shown to be dependent on activation of the receptor for advanced glycation end-products (RAGE) that is tightly associated with NF-κB activation (Tóbon-Velasco et al., 2014). AGEs are mainly formed in hyperglycemia characteristic for diabetes mellitus due to insulin resistance, playing a significant role in progression of diabetes and development of its complications (Vlassara and Striker, 2013). In addition, AGE toxicity has been shown to be involved in pathogenesis of other diseases including cardiovascular diseases (Hegab et al., 2012), neurodegeneration (Pinkas and Aschner, 2016), eye (Kandarakis et al., 2014) and kidney diseases (Rabbani and Thornalley, 2018) to name a few. Targeting AGE/RAGE signaling is therefore considered as the potential therapeutic strategy (Abdelkader et al., 2022).

AGEs are formed endogenously through Maillard reaction, protein modification by reactive carbonyls, and other mechanisms (Figure 1). During Maillard reaction interaction between carbonyl group and free amino group results in the formation of a Schiff base that is rearranged into a more stable Amadori product that subsequently yields AGEs. Non-enzymatic interaction of reactive carbonyls with protein amino-groups also results in AGEs formation (Aaseth et al., 2021) which is catalyzed by redox-active metals like Cu (Sajithlal et al., 1998) and Fe (Takagi et al., 1995). It has been also demonstrated that AGE (carboxymethyllysine) formation from Amadori products may be also promoted by transition metal ions (Voziyan et al., 2003).

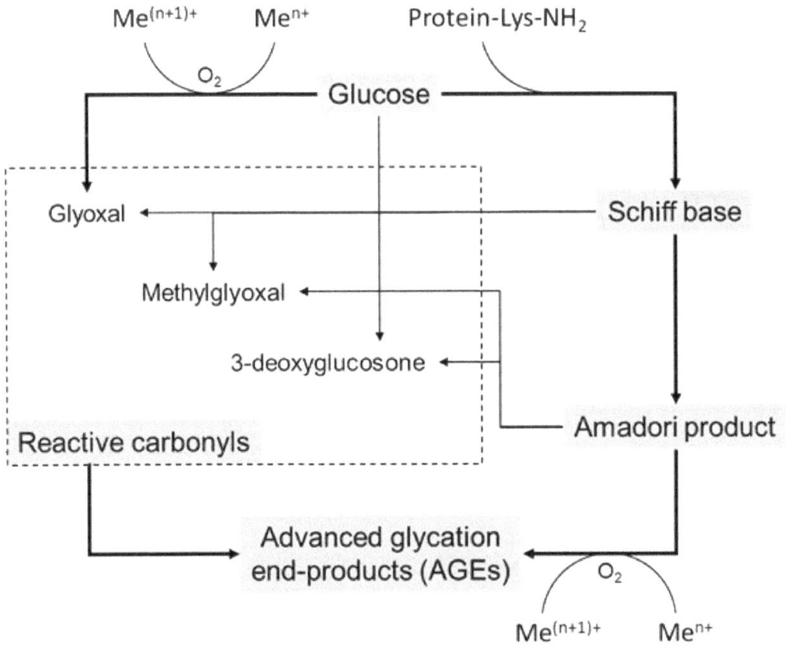

Figure 1. A simplified scheme of AGEs formation involving Maillard reaction and reactive carbonyl formation with the involvement of transition metal ($Me^{n+}/Me^{(n+1)+}$) ions.

The increasing body of evidence demonstrate that other metals than typical redox-active ones Cu and Fe are associated with AGEs accumulation and toxicity. Specifically, *in vitro* studies demonstrated that toxic Cd (Kim et al., 2018) and Hg (de Magalhães Silva et al., 2020), as well as essential Cu (Sajithlal et al., 1998) and Fe (Xiao et al., 2007) may promote protein glycation and AGE formation upon high-glucose levels. In contrast, selenium (Se) (Yu et al., 2015) and zinc (Zn) (Sengani et al., 2023) were shown to reduce AGE formation during protein glycation in the model systems. In addition to *in vitro* model systems, epidemiological findings demonstrate that circulating metal levels may be directly associated with AGE accumulation, as indicated for Cd (Al-Malki et al., 2017), Pb (Huang et al., 2020), Al (Wei et al., 2017) and other toxic and essential metals. The earlier demonstrated role of toxic and essential metals in insulin resistance and subsequent diabetes mellitus (Chen et al., 2009) characterized by hyperglycemia, that acts as a key factor for AGE formation (Yamagishi and Matsui, 2018), also supports the hypothesis of the interference between AGE accumulation and toxicity and metal exposure. Nonetheless, the role of metal exposure in modulation of AGE/RAGE pathway signaling and its contribution to metal-associated diseases is insufficiently studied. Specifically, existing gaps include the particular role of metal overload and deficiency in AGE production in biological systems, as well as contribution of AGEs toxicity and RAGE signaling in metal toxicity or essentiality.

Therefore, the objective of this study is to review the existing data on the association between toxic and essential metal levels with AGE formation, protein glycation, AGEs toxicity, and RAGE signal modulation along with mechanisms sustaining high glucose conditions that promote AGE accumulation.

2. Essential metals and AGE/RAGE signaling

Essential metals and metalloid Se are involved in a variety of physiological functions due to their role as structural components of enzymes and other proteins involved in multiple metabolic pathways including redox homeostasis. Nonetheless, excessive exposure to essential metals leading to its overload may also result in metal toxicity due to increased metal-dependent reactive oxygen species (ROS) production through Fenton reaction (characteristic for Fe, Cu, and Mn), overproduction of metal(loid) containing molecules as observed for selenoproteins, excessive activation of metal- and redox-dependent pathways. Similarly, the influence of essential metals on AGE/RAGE pathway appears to be non-linear varying from antiglycative effects to stimulation of protein glycation and RAGE production with subsequent RAGE activation. The impact on AGE production seems to be specific for particular metals.

2.1 Zinc (Zn)

Zn is involved in regulation of carbohydrate metabolism, being also used as an agent for treatment of glucometabolic disorders through a number of mechanisms (Skalny et al., 2021a). The accumulating body of evidence demonstrate that modulation of AGE formation and toxicity may be considered as the potential mechanism of Zn biological activity. Epidemiological data demonstrate that serum Zn levels in T2DM patients inversely correlates with circulating AGEs concentrations (Grădinaru et al., 2021).

Inverse association between Zn levels and AGEs production was confirmed in *in vitro* glycation system models. It has been demonstrated that zinc oxide nanoparticles (ZnONPs) prevented methylglyoxal (MGO)-induced AGEs formation in human serum albumin (HSA) along with reduction of reactive lysine residues being the primary location for non-enzymatic glycation (Sengani et al., 2023). Zn promoted antiglycative activity of carnosine by reducing AGEs formation from exposure of bovine serum albumin (BSA) to glucose in a dose-dependent manner (125–500 μM) at both co- and post-treatment (after 24 hours), also reducing ROS generation, advanced oxidation protein products formation, as well as cross-linking and dityrosine formation (Moulahoum et al., 2020). Finally, Zn was shown to be a more potent antiglycating agent as compared to other trace elements (Fe, Se, Mn) in its nearly physiological concentrations (Tarwadi and Agte, 2011).

In addition to reduction of AGEs formation, Zn was shown to reduce cellular toxicity of glycated proteins. Specifically, Zn prevented glucose-induced albumin glycation and β aggregation thus reducing its toxicity and increasing viability of glycated albumin-treated HepG2 cells (Tupe and Agte, 2010). ZnONPs from *Morus indica* also possessed antiglycative effects by reducing AGEs formation from MGO-exposed BSA, as well as prevented MGO toxicity in the exposed RBCs (Anandan et al., 2019).

In addition to direct effects of Zn in preventing AGEs formation, Zn was shown to modulate activity of enzymes involved in detoxication of MGO and glyoxal involved in AGEs formation. Specifically, Zn bound to glutamate 172 (Glu172) residue of the active center is directly involved in catalytic activity of glyoxalase I (GloI) (Ridderström et al., 1998). Zn was also shown to be essential for human erythrocyte GloI activity by restoring the activity of the enzyme inhibited by a strong reversible inhibitor S-p-bromobenzylglutathione (Aronsson et al., 1981). In addition to its role in GloI functioning, Zn is also considered essential for human glyoxalase II (GloII) active site that contains a binuclear Zn-binding site (Cameron et al., 1999). Although GloII also contains a Fe-Zn center, it possesses catalytic activity only as mononuclear Zn enzyme (Limphong et al., 2009).

Zn is known to possess protective effects against AGEs toxicity through reducing its accumulation and subsequent AGE/RAGE proinflammatory signaling (Kheirouri et al., 2018), as clearly demonstrated in diabetic models characterized by excessive AGEs levels. In particular, treatment with ZnONPs significantly improved glycemic control markers and reduced circulating AGEs levels in streptozotocin (STZ)-diabetic rats in parallel with improved kidney histology and functioning due to an increase in renal Nuclear factor erythroid 2-related factor 2 (Nrf2) DNA-binding activity, improved autophagy, and down-regulation of NOD-like receptor family pyrin domain containing 3 (NLRP3) inflammasome activation (Abd El-Khalik et al., 2022). In another study using STZ-diabetic nephropathy model Zn supplementation significantly improved kidney weight, creatinine clearance, and lipid and fatty acid content, as well as reduced lipid peroxidation, AGEs accumulation and RAGE expression with subsequent nuclear factor κB (NF-κB) proinflammatory signaling

(Barman et al., 2018). Similarly, Zn supplementation reduced AGEs accumulation and RAGE mRNA expression along with polyol pathway metabolites and enzymes in diabetic rat lens (Barman and Srinivasan, 2019a) and small intestine (Barman and Srinivasan, 2019b). In addition, Zn treatment was shown to protect osteoblastic cells from AGE-induced apoptosis by inhibiting ROS overproduction, cytochrome c leakage, and caspase 3 and 9 activation, that may be at least partially mediated by up-regulation of mitogen-activated protein kinases (MAPK)/extracellular signal-regulated kinase (ERK) and phosphoinositide 3-kinase (PI3K)/Akt signaling (Xiong et al., 2015).

Consistent with protective effect of Zn against AGEs toxicity, Zn deficiency was characterized by increased RAGE levels colocalized with overexpressed S100A8 protein in esophagus characterized by NF-κB p65 and cyclooxygenase-2 (COX-2) protein overexpression, whereas Zn supplementation significantly reduced expression of these proteins thus possessing anti-inflammatory activity (Taccioli et al., 2009).

It is also notable that AGEs may significantly affect Zn metabolism in the cell. Specifically, a study by Shan et al. (2019) demonstrated that AGE exposure (100 μg/mL for 24 h) significantly reduced intracellular Zn levels in human fibroblast-like cells (Shan et al., 2019). Correspondingly, exposure of BAEC cells to AGEs resulted in a significant decrease in cellular Zn content and cell viability, whereas Zn treatment increased cell viability along with restoring cellular Zn levels in AGE-exposed cells without a clear effect on AGE production. Zn treatment in AGE-exposed cells also down-regulated NF-κB and RAGE protein expression, whereas endothelial nitric oxide synthase (eNOS) mRNA and protein expression was increased (Zhuang et al., 2012). Such alterations in Zn metabolism may be at least partially mediated by alteration of Zn trafficking. In particular, albumin glycation was shown to result in reduced Zn interaction with the protein molecule and its binding with subsequent alterations of Zn transport and its tissue distribution (Iqbal et al., 2018). It is proposed that reduction of Zn-binding properties of albumin molecule may be mediated by AGE-induced decrease in histidine residues (Münch et al., 1999; Luevano-Contreras et al., 2013) that are involved in Zn binding (Lu et al., 2008). According to Pearson's theory of hard and soft (Lewis) acids and bases, imidazole group of histidine is considered as a borderline base that is expected to bind with primarily borderline acids like Zn (Wang et al., 2020), and reduced abundance of histidine (His) residues is expected to reduce the metal affinity of the molecule.

The role of Zn as a regulator of carbohydrate homeostasis and the hypoglycemic factor could be an additional link between Zn and AGE/RAGE signaling. Zn is known to play a significant role in insulin production, secretion, and signaling (Maret et al., 2017). This is the only essential metal that has been successfully used for improvement of insulin sensitivity and glycemic control in diabetic patients (Wang et al., 2019). In addition, both type 1 and 2 diabetes are associated with altered Zn metabolism (Jansen et al., 2012) being indicative of the role of Zn deficiency in altered insulin signaling and diabetogenesis (Fukunaka and Fujitani, 2018). The Insulinomimetic effect of Zn was shown to involve up-regulation of the insulin receptor substrate (IRS)-1/PI3K/Akt/glucose transporter (GluT) 4 pathway as well as other mechanisms that are discussed in detail in a number of excellent reviews (Maret, 2017; Sun et al., 2018).

Taken together, the existing studies demonstrate that physiological levels of Zn possess an inhibitory effect on protein glycation, AGEs formation and toxicity, as well as stimulation of RAGE signaling, that may be at least partially involved in antioxidant and anti-inflammatory effects of Zn(II). Such an effect could be also mediated by insulinomimetic effects of the metal reducing circulating glucose levels and decreasing the risk of protein glycation. In contrast, protein glycation and AGEs overaccumulation were shown to have adverse effects of cellular Zn trafficking due to reduction of protein Zn-binding affinity.

2.2 Selenium (Se)

Given the role of Se in the regulation of redox homeostasis, as well as involvement of ROS in AGEs formation, Se may be considered protective against AGEs formation and toxicity. Epidemiological evidence demonstrates that low plasma Se is associated with increased carboxymethyllysine levels in older anemic community-dwelling adults (Roy et al., 2012). However, clinical trials on the effect of Se supplementation on oxidative stress and inflammation markers in patients with diabetic nephropathy did not reveal any statistically significant impact of 200 μg/day Se supplements as Se yeast on circulating AGEs levels (Bahmani et al., 2016).

Results from *in vitro* glycation systems demonstrated that Se may attenuate AGEs formation. Se nanoparticles (SeNPs) possessed significant antiglycative activity *in vitro* by reducing glucose bonding, fructosamine, glyoxal and fluorescent product formation in a dose-dependent manner. Moreover, SeNPs also reduced the formation of AGEs upon protein exposure to fructosamine and hydroxyl radical (Yu et al., 2015). Similarly, SeNPs were shown to reduce D-ribose-induced β-lactoglobulin glycation with decreased AGEs formation resulting from a decrease in glycation sites (Du et al., 2020).

In agreement with modeling systems, laboratory *in vivo* studies demonstrated a significant decrease in AGEs levels upon Se treatment in diabetic animals. Se treatment (1 μg/kg bw as sodium selenite) in STZ-diabetic Sprague–Dawley rats significantly reduced blood glucose and HbA1c, hepatic oxidative stress markers, as well as down-regulated RAGE and NF-κB mRNA expression, altogether resulting in improved liver histology (Pillai et al., 2012). Se polysaccharide from sweet corncob (50, 100, or 200 mg/kg daily for 3 weeks) also improved glycemic control, lipid spectrum, and AGEs levels in parallel with reducing pancreatic and hepatic cell damage in mice with STZ-induced diabetes. These changes were significantly more pronounced in comparison to non-selenized polysaccharide from sweet corncob, being indicative of the role of Se in the observed effects (Wang et al., 2022). Although AGEs treatment did not reduce tissue Se content in Sprague-Dawley rats, it decreased glutathione peroxidase (GPX) 1 expression due to DNA methyltransferase (DNMT) 1 and 2 overexpression associated with oxidative stress, cardiomyocyte apoptosis and impaired cardiac functioning, whereas sodium selenite treatment at dosage of 0.05 mg/kg body weight for 21 days significantly suppressed AGE-induced effects (Zhu et al., 2022).

In spontaneously hypertensive rats (SHR) Se supplementation in the form of selenomethionine (0.25 μg Se/g chow) reduced lipid peroxidation and AGEs levels and eNOS expression in the aortic wall, whereas GPX1 activity, both in blood and aortic tissue, was elevated following Se treatment (Ruseva et al., 2015). These findings are in agreement with the results of *in vitro* studies demonstrating that the treatment of HUVEC cells with 100 nmol/L Se significantly reduced AGE-induced COX-2, p38, and P-selectin expression (Li et al., 2011).

It is notable that maternal Se deficiency in C57BL/6 mice resulted in increased renal AGEs accumulation, as well as mitochondrial dysfunction characterized by reduced Complex II and Complex IV protein levels, although no significant alteration of selenoprotein mRNA expression pattern was observed (Neal et al., 2021), thus supporting the role of physiological Se levels in counteracting AGEs accumulation and toxicity.

Consistent with the observation of Se-induced reduction of AGEs levels, modulation of RAGE signaling was demonstrated upon Se treatment. Specifically, in mice with D-galactose-induced neurotoxicity, treatment with Se-enriched *Auricularia auricular* aqueous extract significantly reduced cognitive deficits and cerebral oxidative stress, as well as decreased hippocampal RAGE, phosphorylated c-Jun N-terminal kinases (p-JNK), Erk, P38, as well as the inhibitor of the nuclear factor kappa B (NF-κB) alpha (IκBα) and p-NF-κB protein expression with subsequent decrease in Tumor necrosis factor α (TNFα), interleukin (IL) 1β and IL-6 levels (Wang et al., 2019). Similarly, in STZ-HFD murine model of diabetes Se-enriched *Auricularia auricular* aqueous extract treatment significantly reduced liver damage, lipid metabolism dysregulation, oxidative stress, and inflammation due to down-regulation of RAGE, p-c-Jun, p-ERK, and p-P38 protein expression (Wei et al., 2020). In diabetic rats SeNPs, especially in combination with sildenafil, significantly reduced kidney histopathology scores, improved kidney function, down-regulated diabetes-induced kidney RAGE, High mobility group box 1 (HMGB1) protein, NF-κB, monocyte chemoattractant protein-1 (MCP-1), TNFα, and IL-1β levels, as well as decreased caspase 3 expression with a resulting antiapoptotic effect (El-Azab et al., 2022).

Existing data demonstrate that Se is involved in regulation of insulin signaling, although the association between Se status and diabetes appears to be U-shaped with a higher risk at both low and high exposure levels (Vinceti et al., 2015). Correspondingly, maternal alterations of Se supply including Se deficiency and excess were both associated with insulin resistance (Ojeda et al., 2019).

Given the involvement of Se in the regulation of physiological insulin signaling, a number of studies demonstrated the beneficial effects of Se in the improvement of insulin resistance and diabetes risk. Specifically, 200 μg/day Se supplementation in diabetic patients significantly reduced insulin resistance and hyperinsulinemia (Farrokhian et al., 2016). Correspondingly, Se intake was inversely associated with insulin resistance at total Se intake < 1.6 μg/kg/day (Wang et al., 2017), and Se deficiency was associated with insulin resistance in hepatitis C patients (Himoto et al., 2011). Laboratory studies demonstrated that combined treatment with Se and insulin was effective in the improvement of insulin sensitivity through up-regulation of PI3K and GLUT4 protein levels (Xu et al., 2011). It has been shown that Se-induced IRS/PI3K/Akt signaling resulting in increased GLUT1 translocation may be independent of insulin receptor activation in 3T3-L1 adipocytes (Heart and Sung, 2003). Correspondingly, Se deficiency-induced thioredoxin suppression was found to be associated with a decrease in IR, IRS-1/2, PI3K/pPI3K and Akt mRNA and protein expression altogether resulting in impaired insulin signal transduction (Yang et al., 2017). In parallel with insulin signaling regulation, Se exerts beneficial effects on β-cell functioning and insulin production (Campbell et al., 2008).

In contrast, Vinceti et al. (2018) demonstrated that Se exposure is associated with increased risk of diabetes mellitus type 2 (DM2) across a wide range of exposure doses (Vinceti et al., 2018), demonstrating the role of excessive Se exposure in the alteration of insulin signaling. Correspondingly, supranutritional Se intake resulted in insulin resistance due to inhibition of insulin-induced Akt phosphorylation, as well as reduced IRS-1 and PI3K expression (Stahel et al., 2017) which may be mediated by increased PTP-1B activity (Mueller et al., 2008). In addition, an *in silico* study demonstrated that substitution of insulin receptor Cys residues by SeCys results in destabilization of insulin-insulin receptor complex affecting downstream insulin signaling (Behar et al., 2020). Adverse effects of excessive Se exposure on insulin signaling may be associated with dysregulation of selenoprotein expression (Zhou et al., 2018). Specifically, high Se exposure was associated with increased glutathione peroxidase (GPX) 1 activity attenuating H_2O_2-induced phosphatase and tensin homolog (PTEN) inhibition, resulting in increased PTEN activity (Zhou et al., 2013). Selenoprotein P was also found to be higher in diabetic and prediabetic subjects, also being significantly associated with insulin resistance (Yang et al., 2011). Selenoprotein P promotes reductive stress, thus altering protein tyrosine phosphatase 1B (PTP1B) oxidative modification and inactivation, and resulting in sustained PTP-1B activity and reduced insulin signaling (Takamura, 2020). Aberrant expression of other selenoproteins including GPX3, selenoproteins S, M, T may also contribute to insulin resistance through a number of mechanisms including inflammation and endoplasmic reticulum stress, to name a few (Zhou et al., 2018).

Therefore, the existing data demonstrate that Se may attenuate AGEs formation in glycation systems, whereas physiological and therapeutic levels of Se counteract AGEs accumulation and RAGE activation with a subsequent inflammatory response, resulting in reduced tissue damage, as evidenced in animal models of diabetes mellitus. This association has been confirmed by an observation of increased AGEs production upon Se deficiency, as well as AGE-induced alterations in selenoprotein functioning, thus being indicative of a mutual relationship between AGE toxicity and Se metabolism. An essential role of Se in the modulation of insulin signaling cascade may also contribute to the indirect antiglycative effect of Se. In contrast to the strong evidence of the contribution of Se overload to insulin resistance and diabetes, indications of the impact of supraphysiological or toxic doses of Se on AGE/RAGE signaling are lacking and require future experimental attention.

2.3 Copper (Cu)

Cu is an essential metal that possesses high toxicity in high doses mainly due to its redox activity (Chen et al., 2020). Given the role of Cu overload in oxidative stress, as well AGE formation (Sajithlal et al., 1998), it is proposed that the AGE/RAGE pathway may be involved in the adverse effects of Cu dysregulation.

Physiological Cu concentrations exert antiglycative activity against MGO-induced HSA glycation whereas high Cu levels promote protein glycation, being indicative of the role of both Cu deficiency and excess in AGE production (Ramirez Segovia et al., 2017). The results of an earlier study demonstrated that Cu significantly increased AGEs formation in glucose-exposed collagen in a dose-dependent manner, although the incubation of the protein with Cu alone did not induce AGEs formation (Sajithlal et al., 1998). Cu was shown to promote glucose or fructose-induced glycation of synthetic β amyloid peptide ultimately leading to the formation of covalently cross-linked high-molecular-mass β-amyloid oligomers as a result of glucoxidation (Loske et al., 2000). Cu-induced ascorbate autooxidation was also considered as a potential glyoxal source, thus promoting AGEs formation from BSA glycation in a cell-free medium (Shangari et al., 2007). Promotion of HSA glycoxidation in presence of Cu also resulted in increased AGEs production and significant cytotoxicity of the modified protein in motor neurons (Marques et al., 2017). In addition, Cu ameliorated the inhibitory effect of aminoguanidine on protein glycation and AGEs formation (Jakus et al., 2001).

Given the role of high Cu in AGEs formation, it has been also proposed that the antiglycative activity of multiple AGE inhibitors may be mediated by its Cu-chelating activity rather than carbonyl-trapping activity, especially at micromolar levels (Price et al., 2001). Nonetheless, it has also been argued that Cu may not promote protein glycation *in vivo* due to a relatively high protein level that may chelate Cu (Birlouez-Aragon et al., 1996).

In turn, certain studies demonstrated that AGE overaccumulation may disrupt Cu metabolism. Specifically, it has been demonstrated that MGO induces ceruloplasmin protein aggregation with the release of Cu ions from the molecule, resulting in altered Cu transport and further aggravation of protein modification (Kang, 2006). Exposure of HSA to MGO also resulted in a significant reduction in thiol content and Cu release, indicative of decreased Cu-binding capacity and potentially contributing to increased free Cu levels (Penezić et al., 2015).

The results from glycation system models are supported by *in vivo* and *in vitro* studies of biological systems demonstrating increased AGE/RAGE signaling. Specifically, co-incubation of human amylin with Cu increased formation of peptide oligomers and fibrils that possessed significant cytotoxicity to neuroblastoma cells through RAGE and p75-NGFR signaling (Caruso et al., 2017). In turn, triethylene tetramine dihydrochloride (trientine), a selective Cu chelator, significantly reduced synaptic loss, diminished brain AGEs levels with subsequent inhibition of RAGE and NF-κB mRNA and protein expression, and decreased amyloid β (Aβ) accumulation due to beta-site amyloid precursor protein cleaving enzyme 1 (BACE1) down-regulation in APP/PS1 mice used as a transgenic Alzheimer's disease model (Wang et al., 2013). Correspondingly, trientine reduced MGO and

3-deoxyglucosone, as well as 8-hydroxy-2′-deoxyguanosine (8-OHdG) levels and semicarbazide-sensitive amine oxidase (SSAO) activity in lens from diabetic rats (Hamada et al., 2005). Notably, despite a significant increase in total antioxidant capacity due to Cu treatment in STZ-diabetic rats, Cu exposure was associated with increased protein glycation and lipid peroxidation (Civelek et al., 2010).

In contrast, dietary Cu deficiency (0.4 mg/kg diet) in male mice resulted in a significant increase in serum fructosamine, pentosidine, and blood glycated hemoglobin levels in comparison to animals fed a Cu-adequate diet (6.4 mg Cu/kg) (Saari, 1994; Saari and Dahlen, 1999), being indicative of the role of both sub- and supra-physiological Cu levels in AGEs formation and toxicity.

The observed association between Cu overload and AGEs formation and RAGE activation corresponds to the existing data on the role of this metal in insulin resistance. Tan and Soma Roy (2021) demonstrated a significant association between dietary Cu intake and insulin resistance only at high intake levels (Tan and Soma Roy, 2021). The results of meta-analysis demonstrated that DM2 is associated with significantly higher circulating Cu levels (Qiu et al., 2017). It is also noteworthy that DM2 is characterized by a significant increase in circulating Cu levels without a significant increase in ceruloplasmin concentrations (Skalnaya et al., 2017), being indicative of increased free Cu fraction (Squitti et al., 2017). Therefore, diabetes is associated with severe alteration of Cu handling (Lowe et al., 2017). In turn, an increase in free catalytically active Cu levels may promote insulin resistance and diabetogenesis through development of oxidative stress (Tanaka et al., 2009). In contrast, certain studies demonstrated that Cu deficiency may be also associated with insulin resistance, diabetes and its complications (Cui et al., 2022). Correspondingly, Cu was also shown to possess insulin mimetic activity through up-regulation of PI3K/Akt pathway (Barthel et al., 2007), and to increase glucose-stimulated insulin release in Cohen diabetic rats (Weksler-Zangen et al., 2013). Therefore, both Cu deficiency and excess may be associated with altered insulin signaling and subsequent hyperglycemia that may contribute to AGEs upon dysregulated Cu metabolism.

The existing data demonstrate the significant role of Cu in protein glycation, AGEs formation, and RAGE stimulation, that is supported by antiglycative effects of Cu chelation. Cu-induced modulation of AGE/RAGE cascade was also demonstrated in *in vivo* models, being indicative of its role in Cu neurotoxicity. Single indications of AGEs overaccumulation upon Cu deficiency allow the possibility of a bimodal effect of Cu on AGEs toxicity, when both deficiency and excess of the metal contribute to excessive glycation.

2.4 Iron (Fe)

Fe is the most abundant trace metal in organisms, involved in a plethora of physiological functions due to its redox activity. Therefore, Fe overload is associated with adverse health effects due to an increase in catalytically-active Fe pool and subsequent development of oxidative stress. Redox activity of Fe was also shown to contribute to protein glycation and AGEs formation (Takagi et al., 1995). In patients with β-thalassemia major characterized by Fe overload serum Fe and ferritin levels directly correlated with pentosidine, whereas concentration of non-transferrin-bound

Fe was directly associated with circulating carboxymethyl-lysine and pentosidine levels (Mirlohi et al., 2018).

In vitro studies demonstrated that Fe promoted glucose-induced glycosylation of collagen with the formation of AGEs being cytotoxic for both human umbilical vein endothelial cells (HUVECs) and primary human monocytes that was shown to be dose-dependent (Xiao et al., 2007). However, Galiniak et al. (2015) did not reveal any effect of Fe chelation in reducing protein glycation *in vitro*. Moreover, antiglycative agents did not possess any significant Fe-binding activity, being indicative of insignificant role of Fe in protein glycation (Galiniak et al., 2015). It has been demonstrated that the rate of hemoglobin (Hg) glycation in non-diabetic subjects is higher in patients with anemia, reflecting the role of Fe deficiency as a factor of protein glycation (Shanthi et al., 2013).

Despite contradictory indications of the role of Fe in the formation of AGEs and protein glycation in glycation systems, *in vivo* studies demonstrate that Fe overload is associated with AGEs overproduction. Specifically, we have also demonstrated that ferric citrate supplementation to HFD-fed rats significantly increased hepatic and testicular AGEs and carboxymethyllysine (CML) content in a dose-dependent manner, as well as increased testicular RAGE and p38 MAPK levels with subsequent elevation of TNFα levels and oxidative stress that may contribute to impaired testosterone synthesis. Correlation analysis also demonstrated increased testicular Fe accumulation and the number of AGE-positive cells (Chen et al., 2020). Administration of ferric Fe (0.5–3 g/kg diet) to STZ/nicotinamide-diabetic mice along with hepatic Fe accumulation increased liver steatosis score and inflammation, also promoted accumulation of AGEs and specifically CML in liver at least partially due to down-regulation of GloI content. Increased AGE levels were accompanied by increased RAGE, NF-κB, and p38 MAPK expression, whereas that of HMGB1 was decreased in Fe-supplemented rats (Chao et al., 2017). Consistent with the significant association between Fe and AGEs accumulation in diabetic animals, Fe overload led to increased AGE/RAGE signaling, considered a potential mechanism in diabetic retinopathy (Ciudin et al., 2010).

Consistent with the role of Fe overload in AGEs accumulation, a number of studies demonstrated a tight relationship between Fe and RAGE signaling in *in vivo* models. It has been demonstrated that intracerebral hemorrhage associated with elevated brain Fe content is associated with RAGE expression in brain tissues around the hemorrhage, as also seen in rats infused with $FeCl_2$ into basal ganglia, reflecting the key role of Fe in RAGE up-regulation in hemorrhagic stroke. Fe-induced increase in RAGE signaling was also shown to be associated with NF-κB protein overexpression and subsequent up-regulation of IL-1β, IL-6, IL-8R, COX-2, and inducible nitric oxide synthase (iNOS) mRNA expression, that were reversed by RAGE-specific antagonist (Yang et al., 2015). Alcohol-exposed mice were characterized by alcoholic liver disease with hepatic Fe overload and increased RAGE protein expression, whereas RAGE knockout prevented Fe overload and alcohol-induced injury, reflecting the role of RAGE signaling in Fe metabolism dysregulation. These findings were also supported by the observed direct correlation between serum RAGE and serum Fe, transferrin, ferritin, and hepcidin levels in patients with alcoholic liver disease (Li et al., 2023).

It is also noteworthy that heme may be considered as a ligand for RAGE with micromolar affinity and its interaction with the receptor is Fe^{3+}-dependent, being also competitive with AGEs for receptor binding. Moreover, intraperitoneal heme injection was shown to up-regulate pulmonary TNFα and IL-1β gene expression in RAGE-dependent manner with down-regulation of ERK1/2 and Akt protein levels as clearly demonstrated in knockout mice (May et al., 2021).

Dysregulation of Fe homeostasis is associated with cell death through ferroptosis that was also influenced by AGEs. Specifically, AGE exposure induced ferroptosis in engineered cardiac tissues, whereas ferroptosis inhibitors ferrostatin-1 and deferoxamine prevented AGE-induced cell damage. Similar effect was observed in mice with STZ/HFD-induced DM2 (Wang et al., 2022). Concomitantly, AGEs treatment as well as incubation with serum of diabetic patients with osteoporosis increased apoptosis and ferroptosis in hFOB1.19 osteoblast cell line that were reversed by treatment with Fe chelator deferoxamine (DFO) (Ge et al., 2022). Exposure of mouse embryonic osteoblasts to MGO also induced a ferroptosis-specific pattern characterized by increased cellular Fe accumulation, reduced GPX4 expression, and ROS overproduction, resulting in reduced cell viability, alkaline phosphatase activity, and decreased number of mineralized nodules in the cells. In turn, co-treatment with N-acetylcysteine or ferrostatin-1 reduced MGO toxicity (Feng et al., 2022). AGE exposure also induced Fe accumulation and ferroptosis in rat kidney epithelial cell line, while ferrostatin-1 ameliorated ferroptosis-associated gene expression and lipid peroxidation (Wang et al., 2022).

Fe metabolism is tightly interrelated with carbohydrate metabolism (Fillebeen et al., 2020) and alterations of Fe status were shown to be associated with impaired insulin signaling and hyperglycemia. Specifically, markers of Fe metabolism including circulating ferritin and transferrin were found to be associated with muscle, liver, and adipose tissue insulin resistance during a 7-year follow up in CODAM study (Wlazlo et al., 2015). Heme Fe intake is associated with higher risk of DM2 (Rajpathak et al., 2006). At the same time, several studies demonstrated that Fe deficiency may be also tightly associated with altered carbohydrate metabolism (Soliman et al., 2017), thus being indicative of the role of both Fe deficiency and overload in glucometabolic disorders.

Development of hepatic Fe overload upon high fat high carbohydrate diet (HFHCD) preceded insulin resistance and liver steatosis being indicative of its role in altered carbohydrate and lipid metabolism dysregulation (Tsuchiya et al., 2013). Adipose tissue Fe accumulation was associated with reduced IRS-1, pAkt, and total Akt protein expression along with a down-regulation of adiponectin gene expression and protein levels (Dongiovanni et al., 2013; Ma et al., 2017). Reduction of IRS-1 and IRS-2 was responsible for inhibition of insulin-induced Akt activation upon Fe exposure in primary hepatocytes (Varghese et al., 2018). In primary cultured neurons Fe(II) treatment significantly reduced phosphorylation insulin receptor tyrosine, IRS-1 and PI3K (Wan et al., 2019). Fe treatment also promoted IRS1 phosphorylation at Serine 307 residue in STZ/nicotinamide-induced diabetes, being indicative of the role of Fe overload in promotion of insulin resistance (Liu et al., 2016). Inhibitory effect of Fe on insulin signaling may be mediated by increased ROS production, autophagy (Sung et al., 2019), JNK activation (Cui et al., 2019), as well as other mechanisms. Correspondingly, Fe chelation by deferoxamine significantly up-regulated InsR mRNA and protein levels with an increase in Akt, forkhead box protein O1 (FOXO1), and glycogen synthase kinase-3 β (Gsk3β) phosphorylation altogether resulting in GluT1, glucokinase, phosphofructokinase (PFK) and glyceraldehyde-3-phosphate dehydrogenase (GAPDH) activation in hepatocytes (Dongiovanni et al., 2008). Similar effects of Fe chelation were observed in brain (Mechlovich et al., 2014). In skeletal muscles Fe restriction was associated with increased Irs-1 and Glut4 gene expression (Mehdad et al., 2015).

It is also noteworthy that Fe metabolism is tightly related to β-cell functioning (Backe et al., 2016), whereas Fe metabolism dysregulation with Fe overload is associated with oxidative stress and β-cell apoptosis (Hansen et al., 2012). Fe overexposure was shown to affect insulin secretion by MIN6 β-cells due to alteration of exocytosis mechanisms (Blesia et al., 2021), whereas dietary Fe restriction or Fe chelation in ob/ob mice significantly improved glycemic control and β-cell function (Cooksey et al., 2010).

Although the results from *in vitro* glycation system models are rather contradictory, *in vivo* and *in vitro* studies in biological systems demonstrate that Fe overload is directly associated with AGEs accumulation that may be associated with Fe-induced insulin resistance and hyperglycemia. Activation of RAGE signaling was shown to contribute to Fe-induced NF-κB activation and subsequent production of proinflammatory cytokines and tissue damage. It is also demonstrated that altered Fe handling and ferroptosis could be considered as the potential mechanisms of AGEs toxicity. In contrast to Fe overload, the evidence of involvement of AGE/RAGE pathway in adverse effects of Fe deficiency is insufficient.

2.5 Manganese (Mn)

Mn is an essential redox-active metal that is involved in a variety of metabolic processes due to its function as a cofactor for multiple enzymes including Mn-superoxide dismutase (Mn-SOD), arginase, and glutamine synthetase to name a few (Chen et al., 2018). However, Mn overaccumulation is associated with adverse health effects including neurotoxicity (Tinkov et al., 2021). In contrast to other more abundant redox metals, such as Fe and Cu, the role of AGEs in Mn toxicity has been studied to a lesser extent.

An *in vitro* study demonstrated that Mn similarly to Fe promoted glyoxal and glucosone formation while decreasing MGO formation in peritoneal dialysis fluids, although the total level of glucose degradation products was increased by Mn (Gensberger-Reigl et al., 2020). In contrast, in BAEC cells Mn treatment significantly prevented AGE formation and cytotoxicity along with increasing cellular Zn concentrations and AGE-induced NF-κB mRNA expression, whereas protein expression was diminished (Zhuang et al., 2012). An earlier study demonstrated that Mn treatment was capable of restoring erythrocyte GloI activity inhibited by S-p-bromobenzylglutathione (Aronsson et al., 1981) which may contribute to reduction of AGEs formation by detoxication of reactive carbonyls. In contrast, human GloII was not capable of binding Mn (Limphong et al., 2009).

Certain studies demonstrated the potential interplay between Mn and RAGE activation. We have demonstrated that Mn neurotoxicity in dopaminergic neurons may be dependent on RAGE expression, whereas its toxicity to serotoninergic neurons was independent of RAGE (Lawes et al., 2020). It is also notable that Mn is essential for binding certain ligands like αX I-domain of β2 integrins to RAGE, and the effect of Mn for receptor binding is higher than that for Mg (Buyannemekh et al., 2017). In contrast, in Mn-exposed bovine aortic endothelial (BAEC) cells Mn treatment tended to reduce RAGE mRNA and protein expression without reaching a level of significance (Zhuang et al., 2012).

In view of the insufficient data on the propensity of Mn to interfere with AGE metabolism, the role of Mn in the development of hyperglycemia through modulation of insulin production and signaling may indirectly indicate high risk of AGEs overaccumulation upon excessive Mn exposure. The existing epidemiological data demonstrate a J-shaped (Yang et al., 2020) relationship between Mn exposure and insulin resistance and DM2, being indicative of the essential role of Mn in carbohydrate metabolism in physiological concentrations and its impairment upon both Mn deficiency and overload. Such effects may be also mediated by the role of Mn in Mn-SOD functioning that was shown to be associated with both improved (Boden et al., 2012) and altered (Han et al., 2016) insulin sensitivity.

Mn exposure significantly reduced glucose-induced insulin response in INS-1 832/13 cells with the effect comparable to that of iAs (Beck et al., 2019). At the same time, maternal Mn deficiency significantly promoted hyperglycemia and hyperinsulinemia in Wistar/NIN (WNIN) rat offspring fed a HFD (Ganeshan et al., 2011). Such an effect may be mediated by Mn-supplementation-induced improvement of insulin secretion by pancreatic Langerhans islets (Lee et al., 2013). In Huntington's disease cell models, Mn treatment significantly increased insulin-like growth factor receptor (IGFR)/IR-dependent AKT phosphorylation with the increase in glucose uptake (Bryan et al., 2020).

Taken together, recent findings do not provide sufficient insight into the relationship between Mn exposure and AGE/RAGE pathway activation. The existing data demonstrate both positive and negative impact of Mn treatment on AGEs accumulation and RAGE signaling, in agreement with the diverse effects of Mn on glucose metabolism regulation upon deficiency and toxicity. Further studies addressing the dose-dependency of the effects of Mn are required to better elucidate the patterns and particular mechanism linking sub- and supraphysiological Mn levels to AGE toxicity.

2.6 Chromium (Cr)

Biological activity of Cr strongly depends on its particular form when trivalent Cr (Cr(III)) is involved the in regulation of insulin signaling cascade, and hexavalent Cr (Cr(VI)) is considered a potent carcinogen (Monga et al., 2022). The association between chromium exposure markers and AGE/RAGE signaling also appears to be dependent on the particular form of the metal. In shipyard welders exposed to metal fumes and polyaromatic hydrocarbons (PAH), urinary Cr levels were found to be associated with serum RAGE levels in correlation and adjusted regression models (Lai et al., 2020), although lack of Cr speciation does not allow to specify the mode of its interaction with RAGE.

In turn, laboratory studies clearly demonstrated that Cr(III) may attenuate AGE accumulation and RAGE activation, whereas Cr(VI) may up-regulate AGE/RAGE signaling. Specifically, Cr(III) picolinate treatment was shown to reduce circulating AGEs levels and improve glycemic control due to increased insulin and nitric oxide (NO) production in a rat model of diabetes mellitus induced by STZ and HFD feeding (Huang et al., 2014). At the same time, in a db/db diabetic mice Cr(III) picolinate (5, 10, 100 or 250 mg/kg) did not reduce renal AGEs accumulation, although a high-dose treatment slightly improved glycemic control (Mozaffari et al., 2012). *Inonotus obliquus*-derived chromium (III) complex (3.0 mg/mL) was also shown to reduce circulating AGEs levels in hepatic L02 cells more effectively than aminoguanidine (0.4 mg/mL) (Wang et al., 2018). It is also notable that protein glycation is associated with alteration of Cr(III) transport and functioning. Specifically, glycation of human serum transferrin by incubation for 14 days at 37°C with or without 1.0 M glucose resulted in altered Cr(III) transport and subsequent decreased delivery to the tissues (Deng et al., 2016). In contrast to Cr(III), analysis of hepatopancreatic transcriptome from exposed to Cr(VI) (10 μl $K_2Cr_2O_7$ or 4 mg/L) for 24 hours revealed up-regulation of genes involved in the AGE-RAGE signaling pathway (Zhang et al., 2018).

Distinct associations of Cr(III) and Cr(VI) with AGE accumulation and toxicity may be mediated by the influence of Cr species on insulin signaling and regulation of systemic glucose levels. Cr(III) has been recognized as a glucose tolerance factor in 1950s (Schwarz and Mertz, 1959). Since then the understanding of the mechanisms of hypoglycemic effects of Cr(III) has been significantly evolved, although the existing data still demonstrate its role in modulation of insulin signaling (Vincent and Brown, 2019).

Laboratory data demonstrated that in obese animals Cr(III) significantly increased insulin-induced phosphorylation of IRS-1 and up-regulated PI3K activity, whereas PTP-1B activity was diminished (Wang et al., 2006). In addition, Cr(III) significantly increased insulin receptor tyrosine kinase activity thus promoting insulin signal transduction (Davis and Vincent, 1997). Improved IRS-1/PI3K/Akt signaling upon Cr treatment was shown to up-regulate GluT4 translocation and increased glucose uptake (Hua et al., 2012). At the same time, another study demonstrated that Cr(III)-induced increase in GluT4 translocation was independent of IR/IRS-1/PI3K/Akt pathway being mediated by improvement of membrane fluidity (Chen et al., 2006). Cr(III)-induced increase in IRS-1/PI3K/AKT/GSK-3β signaling was also responsible for improvement of cognitive functions in Alzheimer's disease (AD)-like dementia induced by intracerebroventricular STZ injection (Akhtar et al., 2020). p38MAPK activation induced by Cr(III) was also shown to be important for Cr-induced increase in GluT4 mRNA and protein expression in L6 skeletal muscle cells (Feng et al., 2018) and adipocytes (Feng et al., 2018b). Correspondingly, maternal Cr restriction resulted in insulin resistance though inhibition of

IRS-1/PI3K/Akt pathway at least partially due to up-regulation of PTP-1B altogether resulting in increased gluconeogenesis as evidenced by increased phosphoenolpyruvate carboxykinase (PEPCK) and glucose-6-phosphatase (G6Pase) activity (Zhang et al., 2016). It has been proposed that adipose tissue Cr deficiency is associated with altered adipocyte insulin sensitivity (Tinkov et al., 2015).

In turn, Cr(VI) was shown to possess an opposite effect on insulin sensitivity resulting in insulin resistance; at least partially through ROS-mediated inhibition of insulin receptor (IR) tyrosine and Akt phosphorylation, as well as other pathways (Ge et al., 2008). In addition, *in utero* Cr(VI) exposure promoted a decrease in insulin receptor and IRS-1 protein levels in liver and skeletal muscle along with a decrease in Glut4 content, whereas Glut2 protein levels were found to be elevated (Shobana et al., 2017).

In agreement with the effects of Cr forms on carbohydrate metabolism and oxidative stress, the existing data demonstrated that Cr(III) exerted an inhibitory effect on AGEs formation upon hyperglycemic conditions in diabetes mellitus models, whereas Cr(VI) appeared to stimulate AGE/RAGE pathway, although the existing data are insufficient for deriving sound conclusions.

2.7 Cobalt (Co)

Biological activity of inorganic Co is mediated by its role as hypoxia mimetic due to up-regulation of HIF-1 signaling, whereas Co overload may be associated with dysregulation of this pathway, as well as oxidative stress, mitochondrial dysfunction (Leyssens et al., 2017), and altered handling of Fe and other metals (Skalny et al., 2021b).

Non-toxic doses of Co were considered as the potential inhibitor of AGEs toxicity due to the role of the metal role as hypoxia inducible factor (HIF-1) mimetic (Miyata and Dan, 2008). Treatment with $CoCl_2$ (0.2 mM in drinking water) in parallel with up-regulation of HIF and HIF-target gene expression significantly reduced kidney damage, renal transforming growth factor (TGFβ) and connective tissue growth factor (CTGF) expression, NADPH-oxidase levels as well AGE accumulation as assessed by kidney pentosidine content in hypertensive, type 2 diabetic rats with nephropathy (SHR/NDmcr-cp) (Ohtomo et al., 2008). Correspondingly, *in vitro* studies demonstrated that Co is capable of reducing AGEs formation and amyloid transformation in BSA molecules incubated with glucose for 24 hours at 60°C (Litvinov et al., 2021).

In contrast to single indications of Co-induced decrease in AGEs production, laboratory studies demonstrate that Co(II) may significantly stimulate RAGE signaling. Specifically, $CoCl_2$ treatment similarly to hypoxia (1% O_2) increased HIF-1α as well as RAGE protein and mRNA expression in pancreatic tumor cells. It has been also demonstrated that hypoxia-induced RAGE expression is dependent on NF-κB p65 expression, whereas suppression of HIF-1a did not have a significant impact on RAGE expression (Kang et al., 2014). $CoCl_2$-induced RAGE signaling was shown to contribute to NF-κB activation further increasing RAGE expression in Panc-1 and pancreatic adenocarcinoma cell lines (Taneja et al., 2020). It has been demonstrated that Co-mimicked hypoxia increased the release of HMGB1 which binds RAGE and Toll-like receptor (TLR-4) with downstream activation of PI3K/Akt, p38 MAPK, and NF-κB cascades (Chang et al., 2017).

The influence of Co on insulin signaling mediates its effect on circulating glucose levels, as it is an essential determinant of AGEs formation. Both Co deficiency and excess were found to be associated with DM2 incidence, being indicative of a U-shaped relationship between Co intake/exposure and insulin resistance with subsequent diabetes (Cao et al., 2021). However, certain studies demonstrated a dose-dependent increase in the prevalence ratios of insulin resistance with increasing urinary Co (Hu et al., 2022). In contrast, blood Co levels were characterized by an inverse association with HOMA-IR (Chen et al., 2021). Laboratory data also demonstrate diverse effects of Co treatment in insulin signaling and glucose metabolism. Specifically, $CoCl_2$ treatment significantly reduced hyperglycemia and improved glucose utilization after glucose load in STZ-diabetic rats (Vasudevan et al., 2007). An increase in insulin receptor and GluT1 mRNA and protein expression was observed in $CoCl_2$-exposed HepG2 hepatocytes (Dongiovanni et al., 2008). Correspondingly, using $CoCl_2$ for mimicking hypoxia in exercised rats significantly increased skeletal muscle glucose uptake by increasing GluT1 protein expression as well as hexokinase and PFK activity indicative of stimulated glycolysis (Saxena et al., 2012). Co exerted hypoglycemic effect through down-regulation of PEPCK mRNA expression and subsequent suppression of gluconeogenesis (Saker et al., 1998).

In contrast, certain studies demonstrated adverse effects of Co on insulin signal transduction. $CoCl_2$ exposure significantly reduced insulin-induced insulin receptor and IRS-1 phosphorylation in a HIF-dependent manner (Regazzetti et al., 2009). In addition, $CoCl_2$ increased Ser793 IRS-1 phosphorylation in HEK293 cells stably expressing IRS-1, being indicative of its association with insulin resistance (Tzatsos and Tsichlis, 2007). At the same time, toxic effects of $CoCl_2$ in the hippocampal neuronal cell line (Hou et al., 2023) as well as chemoresistant glioblastoma cells (Lee et al., 2021) were associated with reduced PI3K/Akt phosphorylation.

Taken together, the existing limited data demonstrate that Co may attenuate AGEs formation, being in agreement with hypoglycemic effects of non-toxic doses of Co. At the same time, Co overexposure is associated with up-regulation of RAGE signaling subsequent NF-κB signaling activation, that may underlie proinflammatory effects of the metal.

3. Toxic metals and AGE/RAGE signaling

In contrast to essential metals, the toxic ones are not involved in cellular functions, and their exposure even at low levels may exert significant adverse health effects. The key mechanisms of heavy metal toxicity include inhibition of enzymes and metabolic pathways, oxidative stress, mitochondrial dysfunction, and inflammation. The role of AGE/RAGE signaling in heavy metal toxicity appears possible but yet successfully highlighted.

3.1 Arsenic (As)

Based on the high incidence of As overload affecting more than 200 million people worldwide and a broad toxicity of this metalloid (Nurchi et al., 2020), As has been ranked as the number one most toxic substance in 2022 by the Agency for toxic substances and disease registry (ATSDR) (ATSDR, 2022). As toxicity is generally mediated by its ability to induce oxidative stress (Hu et al., 2020) that is known to be interrelated with AGEs formation and toxicity (Nowotny et al., 2015).

Existing studies have demonstrated that As promotes glycation of RNase by mimicking phosphate effects (Watkins et al., 1987). Moreover, catalytic constant of arsenate was found to be twofold higher than that for phosphate and carbonate (Gil et al., 2004). However, data from biological systems are insufficient. Only one study in Swiss albino mice demonstrated that sodium arsenite exposure (10 mg/kg body weight for 90 days) induced increased kidney AGE levels and RAGE expression associated with kidney apoptosis, inflammation, and altered insulin signaling (Dutta et al., 2018).

Epidemiological and laboratory studies indicated a significant association between As exposure and RAGE activation, although the character of this association was distinct. Specifically, chronic As exposure was associated with reactive spirometric pattern in children, as well as reduced soluble RAGE (sRAGE) level and higher matrix metalloproteinase (MMP) 9 concentration in sputum. Moreover, sputum sRAGE levels were found to be inversely correlated with urinary dimethylarsinic (DMA), monomethylarsonic (MMA) and dimethylarsinic acid (DMA) percentage in the studied cohort. The authors propose that As-induced decrease in sRAGE levels prevents its binding to AGEs and other RAGE-specific ligands, thus promoting inflammatory signaling due to higher rage of membrane RAGE activation (Olivas-Calderón et al., 2015). At the same time, RAGE expression in human pharyngeal epithelial cells exposed to road dust collected from 10 Chinese cities did not correlate significantly with RAGE expression (Tung et al., 2021).

Experimental studies demonstrate that oral exposure to sodium arsenite (2.5–10 mg/kg) results in lung inflammation and fibrosis with a concomitant increase in lung RAGE and HMBG1, as well as PI3K, p-AKT, IL-1β, IL-18, and MMP-9 expression, being indicative of the role of HMGB1/RAGE pathway in As-induced pneumofibrosis (Wang et al., 2021). In contrast, sputum RAGE levels were found to be inversely associated with urinary As levels, being in agreement with the observation of As-induced decrease in bronchoalveolar lavage RAGE expression in C57B16 mice exposed to 0, 10 or 50 ppb As in drinking water for 4 weeks (Lantz et al., 2007).

In the nervous system inorganic arsenic exposure (3 ppm with drinking water) from the fetal period to 4 months old resulted in behavioral deficits associated with increased Aβ(1−42) levels due to increased BACE1 activity and amyloid precursor protein (APP) expression, in parallel with the increase in both membrane and soluble RAGE levels that may contribute to Aβ intake from the bloodstream (Niño et al., 2018). It has been proposed that As-induced RAGE expression may be mediated by astrocyte activation (Niño et al., 2022).

Recent findings on the association between As exposure and AGE accumulation with RAGE signaling are in line with the earlier demonstrated role of As in insulin resistance and diabetes mellitus with subsequent hyperglycemia (Navas-Acien et al., 2008; Zhou et al., 2022). Correspondingly, laboratory studies revealed a plethora of mechanisms that may underlie diabetogenic effects of As exposure. As inhibited insulin-induced glucose uptake due to down-regulation of IR/IRS-1/PI3K/Akt pathway and reduced expression and translocation of glucose transporters GluT4 (Walton et al., 2004; Xue et al., 2011; Li et al., 2021), GluT1, GluT3 (Niyomchan et al., 2018), or GluT2 (Xu et al., 2022). Alteration of PKB/Akt due to inhibition of 3-phosphoinositide-dependent kinase-1 (PDK-1) activity was shown to be critical for arsenite and methylarsonous-induced insulin resistance, while PI3K or PTEN activity were not affected (Paul et al., 2007). Insulin resistance may be induced by other mechanisms including p38 MAPK signaling (Bazuine et al., 2003), ROS overproduction, mitophagy, and inflammasome activation (Jia et al., 2020). However, the impact of As on insulin signaling cascade was shown to be dependent on the period of exposure, when long-term (24 h) exposure reduced insulin-induced Akt signaling, whereas short-term treatment up-regulated Akt phosphorylation in HepG2 human hepatoma cells (Hamann et al., 2014). In neurons As also impaired insulin signaling through down-regulation of insulin receptor tyrosine kinase activity, a decrease in IRS1 levels, and inhibition of PI3K activity (Wisessaowapak et al., 2021), whereas insulin attenuated As neurotoxicity through up-regulation of PI3K/Akt/sirtuin 1 (SIRT1) pathway (Niyomchan et al., 2015).

Diabetogenic effects of As exposure were shown to be associated with both pancreatic β-cell dysfunction and enhanced gluconeogenesis (Liu et al., 2014). Specifically, sodium arsenite reduced insulin mRNA expression and basal insulin secretion, as well as impair glucose-induced insulin secretion by pancreatic β-cells (Díaz-Villaseñor et al., 2006).

Despite being rather limited, the existing data demonstrate that As exposure may promote protein glycation and AGEs accumulation *in vitro* and *in vivo*, as well as modulate RAGE activity in kidneys, lungs, and brain resulting in nephrotoxicity, lung fibrosis, and cerebral Aβ accumulation (Figure 2). Increased AGEs accumulation may be associated with As-induced hyperglycemia due to both insulin resistance and altered insulin production in β-cells.

The existing data demonstrate that RAGE activation is associated with Aβ accumulation upon exposure to As (Niño et al., 2018), Pb (Shen et al., 2020), Hg (Kim et al., 2014), and Al (Weng et al., 2020; Wongpun et al., 2022) due to the role of RAGE in the uptake of Aβ from the bloodstream. In turn, reduced LRP1 expression was associated with decreased Aβ efflux from Hg and Al-exposed neuronal cells, further aggravating Aβ intracellular aggregation. The observed up-regulation of APP and BACE1 expression in response to As and Al exposure may be mediated by RAGE-dependent NF-κB activation, that is known to up-regulate APP and BACE1 expression in parallel with inflammatory responses (Snow et al., 2016). AGE/RAGE signaling may also promote amyloidogenic APP processing through increased expression of cathepsin B that is known to be essential for Aβ formation (Batkulwar et al., 2018). Reduction of soluble RAGE upon As exposure

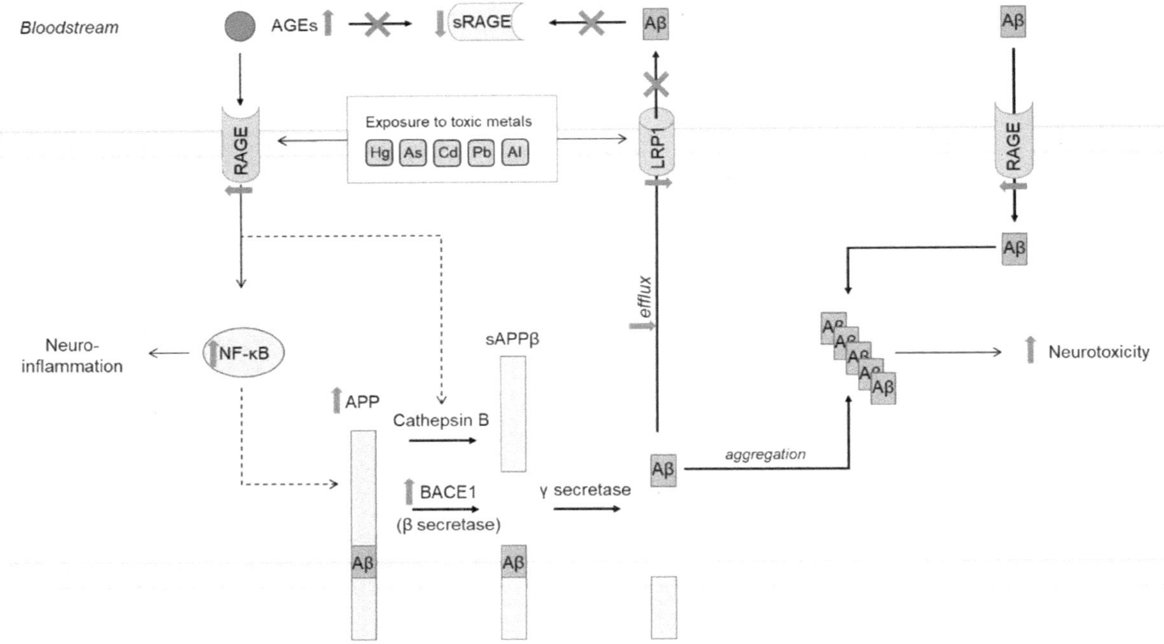

Figure 2. The proposed involvement of AGE/RAGE signaling in metal-induced Aβ accumulation and toxicity.

may aggravate the effects of metals on membrane-bound RAGE activation due to decreased binding of sRAGE to RAGE-activating ligands AGEs and Aβ. It is also noteworthy, that the proposed mechanism may be characteristic for Cu neurotoxicity as demonstrated in studies with Cu chelator trientine (Wang et al., 2013). Solid lines are indicative of direct interaction between the proteins, thin lines are indicative of regulatory effects, thin dotted line demonstrates the pathway that was not associated with metal exposure but may be hypothetically involved.

3.2 Lead (Pb)

Pb is ranked as the second most toxic substance (ATSDR, 2022) that possesses significant neurotoxicity as well as other adverse health effects. Despite being banned for use as a component of petrol additive, global burden of Pb exposure and its potential impact on human health cannot be underestimated.

An increasing body of evidence has demonstrated that Pb toxicity may be associated with AGEs overproduction. Specifically, serum Pb levels in metal-exposed workers were characterized by a positive association with circulating AGEs levels (Wang et al., 2018). Laboratory studies also demonstrated the association between Pb exposure and AGEs accumulation. Exposure to Pb (300 mg/L Pb acetate solution for 10 weeks) to high-fat-fed mice significantly increased brain AGEs accumulation along with neuroinflammation altogether resulting in a more profound cognitive dysfunction as compared to both high-fat diet-fed and Pb-exposed groups (Huang et al., 2020). Exposure to 250 ppm Pb with drinking water for 28 days resulted in significant nephrotoxicity that was accompanied by a significant increase in renal content MGO, an AGEs precursor, whereas treatment with metformin, insulin sensitizer, significantly reduced Pb toxicity (Huang et al., 2017), being indicative of the role of altered insulin signaling in the association between Pb exposure and AGE accumulation.

Increased AGEs accumulation upon Pb exposure may also result from its impaired detoxication. Specifically, Pb and Cd exposure were shown to inhibit hepatic and cerebral GloI activity, although GloII was up-regulated in response to metal treatment *in vivo* (Winter and Piskorska, 1984).

In parallel with Pb-induced increase in AGE accumulation, metal exposure may be associated with up-regulation of RAGE signaling. Specifically, oral exposure of Sprague-Dawley rats to 14–27 mg/kg Pb (as Pb acetate) for 5 days a week for 4–8 weeks resulted in a significant accumulation of Aβ in the cerebrospinal fluid (CSF) and choroid plexus in parallel with an increase in RAGE expression and its translocation from the cytoplasm to apical microvilli resulting in altered efflux of Aβ40 from CSF to the bloodstream (Shen et al., 2020).

The role of Pb exposure in AGE formation and accumulation with subsequent RAGE activation may be also dependent of Pb-induced hyperglycemia. Multiple epidemiological (Wang et al., 2022) and laboratory (Faulk et al., 2014) studies revealed a significant association between Pb exposure markers and hyperglycemia due to insulin resistance. It has been demonstrated that Pb-induced insulin resistance may be mediated by inhibition of hepatic (Wang et al., 2022) and renal (Balaji et al., 2021) PI3K/AKT signaling. The latter may be mediated by IRS-1 phosphorylation as observed in the hippocampus and the cortex of Pb-exposed rats (Liu et al., 2022). Impaired PI3K/AKT signaling was also shown to be responsible for Pb-induced inhibition of GluT4 translocation and the resulting decrease in glucose uptake by hippocampal neurons (Zhao et al., 2021). Yun et al. (2019) demonstrated that Pb acetate exposure significantly altered insulin signaling pathway in rat hippocampus by down-regulating insulin receptor (Insr), Irs1, PI3K and Akt2 gene expression, whereas GluT1 and GluT3 gene and protein expressions were found to be increased in a dose-dependent manner (Yun et al., 2019).

Laboratory *in vivo* and *in vitro* studies demonstrate that Pb exposure is associated with AGEs accumulation brain tissues with subsequent RAGE activation contributing to Aβ accumulation (Figure 2). Stimulation of AGEs formation appears to be associated with impaired MGO detoxication by glyoxalases, as well as hyperglycemia due to Pb-induced dysregulation of IR/IRS1/PI3K/Akt/GluT cascade of insulin signaling.

3.3 Mercury (Hg)

Hg is the third most toxic compound (ATSDR, 2022) that exists in several chemical forms possessing significant toxicity to brain and other organs although the toxicity patterns appear to be species-specific. Although significant success in reducing Hg emissions has been achieved within Minamata Convention on Mercury, Hg exposure still affects a high number of subjects through diet and occupational exposure (Basu et al., 2018). In addition to a broad spectrum of mechanisms underlying Hg toxicity, certain studies demonstrated the role of Hg exposure in accumulation of AGEs and protein glycation. Specifically, thimerosal (an alkylmercury source) was shown to promote hemoglobin glycation in presence of high glucose and fructose levels with up to sevenfold increase in AGEs levels (de Magalhães Silva et al., 2020). Correspondingly, prenatal methylmercury (MeHg) exposure significantly affected expression of genes involved in AGE/RAGE signaling, although this effect was observed in 48–72 hours post-fertilization and at the highest tested MeHg dose of 60 μg/L (Wang et al., 2023).

Consistent with the observed Hg-induced accumulation of AGEs, prenatal MeHg exposure resulted in a significant decrease in hippocampal GloI and GloII, glutathione reductase, and brain-derived neurotrophic factor (BDNF) expression (Karpova et al., 2014), being indicative of reduced detoxication of reactive carbonyls (e.g., glyoxal and MGO) that are considered as AGEs precursors. At the same time, exposure of isolated human erythrocytes to $HgCl_2$ was associated with increased GloI activity along with induction of oxidative and nitrosative stress (Ahmad and Mahmood, 2019), although such effect may be considered as a compensatory response to oxidative and carbonyl stress.

A number of studies demonstrated that Hg exposure is associated with up-regulation of RAGE signaling that may be involved in toxic effects of metal exposure. Kim et al. (2014) demonstrated that MeHg-induced increase in RAGE expression in brain capillary endothelium may be involved in altered Aβ transport and its subsequent hippocampal accumulation in Wistar rats (Kim et al., 2014). Oral prenatal and postnatal MeHg exposure (0.5 mg/kg body weight/day) resulted in liver damage and a concomitant neurotoxicity with behavioral alterations. The latter were shown to be associated with MeHg-induced up-regulation of Akt/GSK3β/mammalian target of rapamycin (mTOR) signaling and tau hyperphosphorylation in parallel with increased RAGE expression in the occipital cortex (Rosa-Silva et al., 2020). In contrast, another study demonstrated that prenatal MeHg exposure in rats resulted in a significant decrease in hippocampal RAGE levels being indicative of RAGE/Erk1/2 signaling inhibition, as well as down-regulation of PI3K/Akt/mTOR pathway in parallel with activation of p38MAPK signaling, altogether underlying behavioral and anxiety disorders (Heimfarth et al., 2018).

The proposed role of Hg exposure in increasing AGEs formation and toxicity is also supported by observation of the significant association between Hg exposure markers and insulin resistance (Lee et al., 2017; Kim et al., 2015). However, the particular mechanisms underlying interference of inorganic Hg (iHg) and MeHg with insulin signaling and carbohydrate metabolism are still to be elucidated. Certain studies demonstrated that Hg exposure may affect the insulin signal transduction cascade. MeHg-induced neurotoxicity was shown to be associated with inhibition of PI3K/Akt signaling (Wei et al., 2023), as observed both in neuronal cells (Chung et al., 2019) and astrocytes (Pierozan et al., 2017). Alteration of PI3K/Akt signaling was also associated with decreased insulin secretion in Hg-exposed pancreatic β-cells (Chen et al., 2006). Combined exposure to 2 mg Hg and Cd reduced IRS1 and GluT4 mRNA expression in adipose tissue, whereas G6Pase was up-regulated (Camsari et al., 2017). $HgCl_2$ was shown to increase GluT1, but not GluT4-mediated increase in glucose transport in adipocytes (Barnes and Kircher, 2005). Hg-induced alterations in adiponectin production may also contribute to adipose tissue and systemic insulin resistance (Chauhan et al., 2019). At the same time, iHg-induced up-regulation of PI3K, Akt, and PTEN gene expression was observed in kidneys (Kadry et al., 2022). In view of inconsistent data on the mechanisms of Hg-induced disruption of insulin signaling, Roy et al. (2017) stated insufficient understanding of cause-effect relationships between Hg exposure and diabetes (Roy et al., 2017).

In contrast, more evidence exists on the role of Hg exposure in beta cell dysfunction and altered insulin production (Schumacher et al., 2017). Specifically, both iHg and MeHg exposure was shown to induce oxidative stress with subsequent apoptosis and necrosis in β-cells (Chen et al., 2006; Chen et al., 2010). Apoptotic β cell death was also associated with mitochondrial dysfunction and endoplasmic reticulum stress induced by MeHg exposure (Yang et al., 2022). Correspondingly, high doses of MeHg were shown to inhibit insulin production through down-regulation of proinsulin production (Chida et al., 2022).

Limited evidence also supports the notion that Hg exposure is associated with increased AGEs accumulation, at least in part, due to alteration of reactive carbonyl detoxication by glyoxalases, as well as the role of the metals in hyperglycemia. In turn, modulation of RAGE signaling upon Hg exposure may be associated with neurotoxic effects of the metal.

3.4 Cadmium (Cd)

Cd is a toxic metal ranked as the seventh most toxic substance according to ATSDR 2022 Substance Priority List (ATSDR, 2022) having a toxic effect on organs including kidney, live, lungs through a plethora of molecular mechanisms (Lee and Thévenod, 2020), although contribution of AGEs to Cd toxicity is still insufficiently studied.

In vitro glycation system models demonstrated that AGEs accumulation upon Cd exposure may be associated with promotion of AGEs formation from its precursors, glyoxal and MGO. Suhartono et al. (2014) demonstrated that Cd is capable of promoting formation of MGO and H_2O_2 during BSA glycation upon high-glucose conditions (Suhartono et al., 2014a). The authors also demonstrated the role of Cd exposure in ovarian MGO accumulation in female rats (Husna et al., 2014). At the same time, another study by this group failed to reveal any significant increase in MGO accumulation in kidneys of Cd-exposed rats, although stimulation of H_2O_2 production, malondialdehyde (MDA) and advanced oxidation protein products (AOPP) was observed (Suhartono et al., 2014b). It has been demonstrated that exposure of purified high-density lipoprotein 3 (HDL3) to 12 and 24 μM $CdCl_2$ increased AGEs levels nearly 2- and 4-fold, respectively (Kim et al., 2017).

Only single studies demonstrating the impact of Cd exposure on AGEs accumulation and toxicity in biological systems exist. Specifically, intraperitoneal $CdCl_2$ injection (5 μg/kg/body weight) induced hyperglycemia, as well as cerebral oxidative stress, neuroinflammation with increased TNFα and IL-1 level, as well as elevated AGEs accumulation, that were reduced by whey protein and *Brassicaceae* extract treatment (Al-Malki et al., 2017). Acute oral exposure to Cd (6.3 mg/kg, single dose) in rats significantly increased liver and kidney protein glycosylation as well as induced oxidative stress associated with tissue damage (Ramamurthy et al., 2016). At the same time, an earlier study demonstrated that Cd pretreatment significantly inhibited AGEs-induced iNOS expression and nitrite release in RAW 264.7 Cells (Sumi and Ignarro, 2004).

Inconsistent data exist on the role of Cd exposure in RAGE signaling. Urinary Cd as well as Cr levels were shown to be independently associated with serum RAGE levels in metal- and PAH-exposed shipyard welders, whereas circulating AGE concentration was interrelated only with urinary nickel (Ni) levels (Lai et al., 2020). At the same time, our recent study demonstrated that Cd neurotoxicity to dopaminergic and serotoninergic neurons did not depend on RAGE expression, indicative of RAGE-independent mechanisms in Cd-induced neurotoxicity (Lawes et al., 2020).

Despite inconsistent data on the role of Cd exposure in formation of AGEs, multiple epidemiological, animal and cellular studies demonstrated that Cd may be involved in insulin resistance and diabetogenesis (Tinkov et al., 2017; Buha et al., 2020) characterized by hyperglycemia that promotes protein glycation.

Briefly, Cd may affect insulin signal transduction through a number of mechanisms. Specifically, Cd exposure significantly reduced phosphorylation of insulin receptor, IRS1/2 and Akt with subsequent inhibition of GluT4 expression and translocation (Han et al., 2003; Sarmiento-Ortega et al., 2022; Gasser et al., 2022; Chen et al., 2021). Cd is also capable of decreasing the number of insulin receptors on adipocyte membrane (Ficková et al., 2003). At the same time, another study demonstrated Cd-induced increase in insulin receptor and IRS1 phosphorylation in liver, whereas Akt phosphorylation was decreased (Treviño et al., 2015). Reduction of insulin secretion upon Cd exposure may be observed prior development of insulin resistance (Li et al., 2019), also contributing to hyperglycemia. These findings establish that Cd-induced toxicity in β-cells is associated with inflammation and lipid accumulation (Hong et al., 2021), oxidative stress and ferroptosis (Hong et al., 2022).

Taken together, the existing data demonstrate that Cd is capable of promoting protein glycation and AGEs formation in *in vitro* glycation systems, as well as increasing the risk of hyperglycemia, which that may contribute to increased formation of AGEs associated with inflammation, oxidative stress, and kidney, liver, and brain damage. Therefore, among Group 12 metals Cd and Hg characterized by lower reactivity were shown to promote AGEs formation, whereas more reactive Zn possessed antiglycative effects.

3.5 Aluminium (Al)

Increasing production of Al and wide application of Al-containing materials in the everyday life has significantly increased the risk of chronic Al toxicity (Exley, 2016) with neurotoxicity being the most significant health hazard (Skalny et al., 2021c).

The existing data demonstrate that interference of Al with AGE/RAGE signaling may significantly contribute to Al neurotoxicity. In a $AlCl_3$/D-galactose-induced model of Alzheimer's disease the increase in cerebral Al level was associated with elevated brain AGEs and oxidative stress markers content along with altered memory and learning abilities, impaired neurotransmitter metabolism, and amyloid accumulation (Wei et al., 2017). Consistent with these observations, an *in vitro* study by Ferretti et al. (2004) demonstrated that Al as well as Fe significantly increased lipid peroxidation in glycated HDL incubated for 24–72 hours with 50 mM glucose, being indicative of the role of protein glycation in prooxidant effects induced by metals (Ferretti et al., 2004).

The ability of Al to modulate RAGE signaling in terms of neurotoxicity was also investigated. In Sprague Dawley rats daily oral administration of 110 mg/kg body weight $AlCl_3$ impaired memory and learning ability together with accumulation of $A\beta_{1-42}$ due to increased APP, BACE1, and Tau-5 protein expression, whereas low-density lipoprotein cholesterol receptor related protein 1 (LRP-1) protein expression was reduced. $AlCl_3$ exposure also induced neuroinflammation with increased RAGE protein expression, P65 and IκBα protein phosphorylation and subsequent up-regulation of NF-κB-targeted proinflammatory cytokine production (Weng et al., 2020). Similarly, $AlCl_3$ and D-galactose treatment also increased RAGE expression and subsequent NF-κB activation, increased amyloidogenesis due to BACE1 activation, promoted tau phosphorylation and induced synaptic dysfunction with neuronal apoptosis (Wongpun et al., 2022). At the same time, certain studies did not detect significant up-regulation of RAGE expression upon $AlCl_3$ and D-galactose exposure (Luo et al., 2009; Sun et al., 2009; Promyo et al., 2020).

The interplay between Al exposure and AGE accumulation is expected to be at least partially mediated by the role of Al in hyperglycemia due to altered insulin signaling cascade. Specifically, $AlCl_3$ exposure also inhibited brain glucose uptake through a decrease in GLuT1 and GluT3 protein levels, while metformin treatment ameliorated these effects (Song et al., 2022) being indicative of the role of altered insulin sensitivity in $AlCl_3$ neurotoxicity. This suggestion is consistent with the observation of effective improvement of cognitive function in Al-exposed rats by combined treatment with insulin and glucose (Nampoothiri et al., 2017). $AlCl_3$ was also shown to reduce hippocampal insulin, GLP-1, and GluT4 levels, Akt phosphorylation, increase IRS-1^{S307} phosphorylation (Saleh et al., 2021), whereas tyrosine phosphorylation of IRS-1 was reduced (Attia et al., 2020). $AlCl_3$ and D-galactose also down-regulated PI3K and Akt mRNA and protein expression in the hippocampus in a model of AD (Li et al., 2016). In turn, systemic insulin resistance following $AlCl_3$ exposure was shown to be associated with both pancreatic damage and the reduction of skeletal muscle GLUT4 mRNA and protein expression (Wei et al., 2018).

Therefore, the effects of Al exposure may include increased AGEs accumulation, while Al-induced insulin resistance could significantly contribute to the promotion of protein glycation and AGEs formation. Activation of RAGE signaling upon Al was shown to be involved in neuroinflammation and Aβ accumulation, although the results on Al-induced RAGE activation are rather inconsistent. At the same time, one can propose that activation of AGE/RAGE signaling may be considered as one of the mechanisms of Al neurotoxicity.

4. Conclusions

Taken together, existing data demonstrate that both toxic and essential metals may interfere with the formation and accumulation of AGEs and subsequent RAGE signaling. Specifically, toxic metals as well as high doses of essential metals (Fe, Cu) promote protein glycation and AGEs formation, whereas Zn and Se reduce the rate of AGEs formation in glycation models and biological systems. The effect of metal on AGEs formation may be mediated by modulation of oxidative stress, direct interaction of the metal with protein molecules, as well as regulation of glyoxalases which are involved in detoxication of reactive carbonyls (glyoxal, MGO) involved in AGEs formation. It is also noteworthy that both essential and toxic metals have a significant impact on glucose homeostasis through modulation of insulin signaling through IR/IRS/PI3K/Akt pathway and glucose uptake by GluTs, as well as regulation of glycolysis and gluconeogenesis. The contribution of metals to the formation of high-glucose conditions may therefore play a significant role in the formation of AGEs (Figure 3).

Despite being somewhat inconsistent, existing data demonstrate that both exposure to toxic metals and essential metal overload may promote AGEs formation through oxidative stress and lipid peroxidation, leading to increased generation of reactive carbonyls considered as AGEs precursors. Inhibition of reactive carbonyl detoxication by glyoxalases due to enzyme inactivation may also contribute to increased AGEs production. Both reduced insulin production and sensitivity upon metal exposure result in hyperglycemia which plays an essential role in promotion of protein glycation and AGEs formation. Metal-induced AGEs accumulation is also

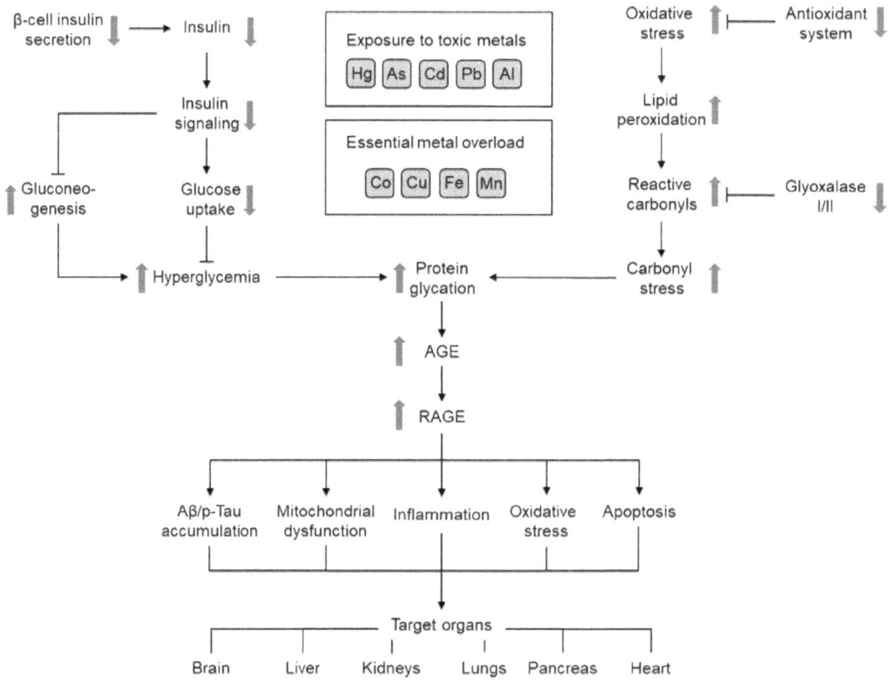

Figure 3. The proposed mechanism of AGEs formation and accumulation with subsequent activation of RAGE signaling.

accompanied by activation of RAGE. Increased AGE/RAGE signaling in upon metal overload target tissues is associated with inflammation, oxidative stress, apoptosis, mitochondrial dysfunction, accumulation of Aβ and phosphorylated tau protein, being indicative of the potential role of AGEs in metal toxicity.

In vivo and *in vitro* studies demonstrated that metal exposure is associated with an increase in AGE accumulation in models of diabetes, obesity, neurotoxicity and neurodegeneration, nephrotoxicity, and liver damage. Metal-induced AGEs overaccumulation was found to be associated with inflammation, oxidative stress, mitochondrial dysfunction, Aβ and phosphorylated tau accumulation. In addition to direct AGEs-induced toxicity, these effects may be at least partially mediated by stimulation of RAGE signaling, as demonstrated by up-regulation of RAGE/NF-κB inflammatory pathway.

Finally, an alternative mechanism for stimulation of AGEs formation and accumulation upon metal exposure involves metal-induced dysregulation of insulin production and/or sensitivity with the ensuing development of insulin resistance and sustained hyperglycemia. This suggestion is supported by the observation of ameliorating effects of insulin on AGEs/RAGE-related toxicity and neurodegeneration (Pinkas and Aschner, 2017). It is also noteworthy that AGEs may have a significant impact on physiological effects of metals. Specifically, protein glycation was shown to affect metal affinity for protein molecules, thus reducing metal binding and transport to the target cells.

Taken together, the existing data demonstrate that stimulation of AGE formation and subsequent RAGE activation should be considered as a potential mechanism for both metal toxicity and protective effects of essential metal(loids) like Zn and Se. Therefore, targeting the AGE/RAGE pathway with antiglycative agents should be considered as a potential strategy for reducing metal toxicity, although further detailed studies are required in support of this hypothesis.

Acknowledgements

The study was performed with support of the Russian Ministry of Science and Higher Education, Project No. FENZ-2023-0004.

References

Aaseth, J., Skalny, A.V., Roos, P.M., Alexander, J., Aschner, M. and Tinkov, A.A. 2021. Copper, iron, selenium and lipo-glycemic dysmetabolism in Alzheimer's disease. Int. J. Mol. Sci. 22(17): 9461. doi: 10.3390/ijms22179461.

Abd El-Khalik, S.R., Nasif, E., Arakeep, H.M. and Rabah, H. 2022. The prospective ameliorative role of zinc oxide nanoparticles in STZ-induced diabetic nephropathy in rats: mechanistic targeting of autophagy and regulating Nrf2/TXNIP/NLRP3 inflammasome signaling. Biol. Trace Elem. Res. 200(4): 1677–1687. doi: 10.1007/s12011-021-02773-4.

Abdelkader, H., Mustafa, W.W., Alqahtani, A.M., Alsharani, S., Al Fatease, A. and Alany, R.G. 2022. Glycation-induced age-related illnesses, antiglycation and drug delivery strategies. J. Pharm Pharmacol. 74(11): 1546–1567. doi: 10.1093/jpp/rgac051.

Adamopoulos, C., Farmaki, E., Spilioti, E., Kiaris, H., Piperi, C. and Papavassiliou, A.G. 2014. Advanced glycation end-products induce endoplasmic reticulum stress in human aortic endothelial cells. Clin. Chem. Lab. Med. 52(1): 151–60. doi: 10.1515/cclm-2012-0826.

Agency for Toxic Substances and Disease Registry (ATSDR). 2023. The ATSDR 2022 Substance Priority List. https://www.atsdr.cdc.gov/spl/index.html#2022spl Accessed January 2023.

Ahmad, S. and Mahmood, R. 2019. Mercury chloride toxicity in human erythrocytes: enhanced generation of ROS and RNS, hemoglobin oxidation, impaired antioxidant power, and inhibition of plasma membrane redox system. Environ Sci. Pollut. Res. Int. 26(6): 5645–5657. doi: 10.1007/s11356-018-04062-5.

Akhtar, A., Dhaliwal, J., Saroj, P., Uniyal, A., Bishnoi, M. and Sah, S.P. 2020. Chromium picolinate attenuates cognitive deficit in ICV-STZ rat paradigm of sporadic Alzheimer's-like dementia via targeting neuroinflammatory and IRS-1/PI3K/AKT/GSK-3β pathway. Inflammopharmacology 28(2): 385–400. doi: 10.1007/s10787-019-00681-7.

Al-Malki, A.L., Barbour, E.K., Ea, H., Moselhy, S.S., ALZahrani, A.H.S. and Kumosani, T.A. 2017. Signaling pathways regulated by Brassicaceae extract inhibit the formation of advanced glycated end products in rat brain. Afr. J. Tradit. Complement Altern. Med. 14(2): 234–240. doi: 10.21010/ajtcam.v14i2.25.

Anandan, S., Mahadevamurthy, M., Ansari, M.A., Alzohairy, M.A., Alomary, M.N., Farha Siraj, S. et al. 2019. Biosynthesized ZnO-NPs from Morus indica attenuates methylglyoxal-induced protein glycation and RBC damage: *In-vitro, in-vivo* and molecular docking study. Biomolecules 9(12): 882. doi: 10.3390/biom9120882.

Aronsson, A.C., Sellin, S., Tibbelin, G. and Mannervik, B. 1981. Probing the active site of glyoxalase I from human erythrocytes by use of the strong reversible inhibitor S-p-bromobenzylglutathione and metal substitutions. Biochem. J. 197(1): 67–75. doi: 10.1042/bj1970067.

Attia, H., Albuhayri, S., Alaraidh, S., Alotaibi, A., Yacoub, H., Mohamad, R. et al. 2020. Biotin, coenzyme Q10, and their combination ameliorate aluminium chloride-induced Alzheimer's disease via attenuating neuroinflammation and improving brain insulin signaling. J. Biochem. Mol. Toxicol. e22519. doi: 10.1002/jbt.22519.

Backe, M.B., Moen, I.W., Ellervik, C., Hansen, J.B. and Mandrup-Poulsen, T. 2016. Iron regulation of pancreatic beta-cell functions and oxidative stress. Annu Rev. Nutr. 36: 241–73. doi: 10.1146/annurev-nutr-071715-050939.

Bahmani, F., Kia, M., Soleimani, A., Mohammadi, A.A. and Asemi, Z. 2016. The effects of selenium supplementation on biomarkers of inflammation and oxidative stress in patients with diabetic nephropathy: a randomised, double-blind, placebo-controlled trial. Br J. Nutr. 116(7): 1222–1228. doi: 10.1017/S0007114516003251.

Balaji, K., Vijayakumar, J., Sankaran, P.K., Senthilkumar, S., Vijayaraghavan, R., Selvaraj, J. et al. 2021. Molecular studies on the nephroprotective potential of celastrus paniculatus against lead-acetate-induced nephrotoxicity in experimental rats: role of the PI3K/AKT signaling pathway. Molecules 2021 Nov 2; 26(21): 6647. doi: 10.3390/molecules26216647. PMID: 34771053; PMCID: PMC8587739.

Barman, S., Pradeep, S.R. and Srinivasan, K. 2018. Zinc supplementation alleviates the progression of diabetic nephropathy by inhibiting the overexpression of oxidative-stress-mediated molecular markers in streptozotocin-induced experimental rats. J. Nutr. Biochem. 54: 113–129.

Barman, S. and Srinivasan, K. 2019a. Zinc supplementation ameliorates diabetic cataract through modulation of crystallin proteins and polyol pathway in experimental rats. Biol. Trace Elem. Res. 187(1): 212–223. doi: 10.1007/s12011-018-1373-3.

Barman, S. and Srinivasan, K. 2019b. Ameliorative effect of zinc supplementation on compromised small intestinal health in streptozotocin-induced diabetic rats. Chem. Biol. Interact. 307: 37–50. doi: 10.1016/j.cbi.2019.04.018.

Barnes, D.M. and Kircher, E.A. 2005. Effects of mercuric chloride on glucose transport in 3T3-L1 adipocytes. Toxicol. *In Vitro* 19(2): 207–14. doi: 10.1016/j.tiv.2004.08.005.

Barthel, A., Ostrakhovitch, E.A., Walter, P.L., Kampkötter, A. and Klotz, L.O. 2007. Stimulation of phosphoinositide 3-kinase/Akt signaling by copper and zinc ions: mechanisms and consequences. Arch Biochem Biophys. 463(2): 175–82. doi: 10.1016/j.abb.2007.04.015.

Basu, N., Horvat, M., Evers, D.C., Zastenskaya, I., Weihe, P. and Tempowski, J. 2018. A state-of-the-science review of mercury biomarkers in human populations worldwide between 2000 and 2018. Environ Health Perspect. 126(10): 106001. doi: 10.1289/EHP3904.

Batkulwar, K., Godbole, R., Banarjee, R., Kassaar, O., Williams, R.J. and Kulkarni, M.J. 2018. Advanced glycation end products modulate amyloidogenic APP processing and tau phosphorylation: a mechanistic link between glycation and the development of Alzheimer's disease. ACS Chem. Neurosci. 9(5): 988–1000. doi: 10.1021/acschemneuro.7b00410.

Bazuine, M., Ouwens, D.M., Gomes de Mesquita, D.S. and Maassen, J.A. 2003. Arsenite stimulated glucose transport in 3T3-L1 adipocytes involves both Glut4 translocation and p38 MAPK activity. Eur. J. Biochem. 270(19): 3891–903. doi: 10.1046/j.1432-1033.2003.03771.x.

Beck, R., Chandi, M., Kanke, M., Stýblo, M. and Sethupathy, P. 2019. Arsenic is more potent than cadmium or manganese in disrupting the INS-1 beta cell microRNA landscape. Arch Toxicol. 93(11): 3099–3109. doi: 10.1007/s00204-019-02574-8.

Behar, A., Dennouni-Medjati, N., Harek, Y., Dali-Sahi, M., Belhadj, M. and Meziane, F.Z. 2020. Selenium overexposure induces insulin resistance: *In silico* study. Diabetes Metab Syndr. 14(6): 1651–1657. doi: 10.1016/j.dsx.2020.08.005.

Birlouez-Aragon, I., Tessier, F., Mompeyssin, V. and Baciuska, J. 1996. Lack of effect of copper on advanced Maillard reaction and glucose autoxidation at physiological concentrations of albumin. Redox Rep. 2(2): 127–32. doi: 10.1080/13510002.1996.11747039.

Blesia, V., Patel, V.B., Al-Obaidi, H., Renshaw, D. and Zariwala, M.G. 2021. Excessive iron induces oxidative stress promoting cellular perturbations and insulin secretory dysfunction in MIN6 beta cells. Cells 10(5): 1141. doi: 10.3390/cells10051141.

Boden, M.J., Brandon, A.E., Tid-Ang, J.D., Preston, E., Wilks, D., Stuart, E. et al. 2012. Overexpression of manganese superoxide dismutase ameliorates high-fat diet-induced insulin resistance in rat skeletal muscle. Am. J. Physiol. Endocrinol. Metab. 303(6): 798–805. doi: 10.1152/ajpendo.00577.2011.

Bryan, M.R., Nordham, K.D., Rose, D.I.R., O'Brien, M.T., Joshi, P., Foshage, A.M. et al. 2020. Manganese acts upon insulin/IGF receptors to phosphorylate AKT and increase glucose uptake in Huntington's disease cells. Mol. Neurobiol. 57(3): 1570–1593. doi: 10.1007/s12035-019-01824-1.

Buha, A., Đukić-Ćosić, D., Ćurčić, M., Bulat, Z., Antonijević, B., Moulis, J.M. et al. 2020. Emerging links between cadmium exposure and insulin resistance: human, animal, and cell study data. Toxics 8(3): 63. doi: 10.3390/toxics8030063.

Buyannemekh, D. and Nham, S.U. 2017. Characterization of αX I-domain binding to receptors for advanced glycation end products (RAGE). Mol Cells 40(5): 355–362. doi: 10.14348/molcells.2017.0021.

Cameron, A.D., Ridderström, M., Olin, B. and Mannervik, B. 1999. Crystal structure of human glyoxalase II and its complex with a glutathione thiolester substrate analogue. Structure 7(9): 1067–1078.

Campbell, S.C., Aldibbiat, A., Marriott, C.E., Landy, C., Ali, T., Ferris, W.F. et al. 2008. Selenium stimulates pancreatic beta-cell gene expression and enhances islet function. FEBS Lett. 582(15): 2333–7. doi: 10.1016/j.febslet.2008.05.038. Epub 2008 Jun 4. PMID: 18538137.

Camsari, C., Folger, J.K., McGee, D., Bursian, S.J., Wang, H., Knott, J.G. et al. 2017. Effects of periconception cadmium and mercury co-administration to mice on indices of chronic diseases in male offspring at maturity. Environ Health Perspect. 125(4): 643–650. doi: 10.1289/EHP481.

Cao, B., Fang, C., Peng, X., Li, X., Hu, X., Xiang, P. et al. 2021. U-shaped association between plasma cobalt levels and type 2 diabetes. Chemosphere 267: 129224. doi: 10.1016/j.chemosphere.2020.129224.

Caruso, G., Distefano, D.A., Parlascino, P., Fresta, C.G., Lazzarino, G., Lunte, S.M. et al. 2017. Receptor-mediated toxicity of human amylin fragment aggregated by short- and long-term incubations with copper ions. Mol. Cell Biochem. 425(1-2): 85–93. doi: 10.1007/s11010-016-2864-1.

Chang, Y.C., Lin, C.W., Hsieh, M.C., Wu, H.J., Wu, W.S., Wu, W.C. et al. 2017. High mobility group B1 up-regulates angiogenic and fibrogenic factors in human retinal pigment epithelial ARPE-19 cells. Cell Signal. 40: 248–257. doi: 10.1016/j.cellsig.2017.09.019.

Chao, K.C., Chen, S.H., Chang, C.C., Lee, Y.C., Wang, C.M. and Chang, J.S. 2017. Effects of ferric citrate supplementation on advanced glycation end products in a rat model of streptozotocin/nicotinamide-induced diabetes. Mol. Nutr. Food Res. 61(5). doi: 10.1002/mnfr.201600753.

Chauhan, S., Dunlap, K. and Duffy, L.K. 2019. Effects of methylmercury and theaflavin digallate on adipokines in mature 3T3-L1 adipocytes. Int. J. Mol. Sci. 20(11): 2755. doi: 10.3390/ijms20112755.

Chen, G., Liu, P., Pattar, G.R., Tackett, L., Bhonagiri, P., Strawbridge, A.B. et al. 2006. Chromium activates glucose transporter 4 trafficking and enhances insulin-stimulated glucose transport in 3T3-L1 adipocytes via a cholesterol-dependent mechanism. Mol. Endocrinol. 20(4): 857–70. doi: 10.1210/me.2005-0255.

Chen, H., Li, P., Shen, Z., Wang, J. and Diao, L. 2021. Protective effects of selenium yeast against cadmium-induced necroptosis through miR-26a-5p/PTEN/PI3K/AKT signaling pathway in chicken kidney. Ecotoxicol. Environ. Saf. 220: 112387. doi: 10.1016/j.ecoenv.2021.112387.

Chen, J., Jiang, Y., Shi, H., Peng, Y., Fan, X. and Li, C. 2020. The molecular mechanisms of copper metabolism and its roles in human diseases. Pflugers Arch. 472(10): 1415–1429. doi: 10.1007/s00424-020-02412-2.

Chen, P., Bornhorst, J. and Aschner, M. 2018. Manganese metabolism in humans. Front. Biosci. (Landmark Ed). 23: 1655–1679. doi: 10.2741/4665.

Chen, S.H., Yuan, K.C., Lee, Y.C., Shih, C.K., Tseng, S.H., Tinkov, A.A. et al. 2020. Iron and advanced glycation end products: emerging role of iron in androgen deficiency in obesity. Antioxidants (Basel) 9(3): 261. doi: 10.3390/antiox9030261.

Chen, Y., Huang, H., He, X., Duan, W. and Mo, X. 2021. Sex differences in the link between blood cobalt concentrations and insulin resistance in adults without diabetes. Environ Health Prev. Med. 26(1): 42. doi: 10.1186/s12199-021-00966-w.

Chen, Y.W., Huang, C.F., Tsai, K.S., Yang, R.S., Yen, C.C., Yang, C.Y. et al. 2006. The role of phosphoinositide 3-kinase/Akt signaling in low-dose mercury-induced mouse pancreatic beta-cell dysfunction *in vitro* and *in vivo*. Diabetes 55(6): 1614–24. doi: 10.2337/db06-0029.

Chen, Y.W., Huang, C.F., Tsai, K.S., Yang, R.S., Yen, C.C., Yang, C.Y. et al. 2006. Methylmercury induces pancreatic beta-cell apoptosis and dysfunction. Chem. Res. Toxicol. 19(8): 1080–5. doi: 10.1021/tx0600705.

Chen, Y.W., Yang, C.Y., Huang, C.F., Hung, D.Z., Leung, Y.M. and Liu, S.H. 2009. Heavy metals, islet function and diabetes development. Islets 1(3): 169–76. doi: 10.4161/isl.1.3.9262.

Chen, Y.W., Huang, C.F., Yang, C.Y., Yen, C.C., Tsai, K.S. and Liu, S.H. 2010. Inorganic mercury causes pancreatic beta-cell death via the oxidative stress-induced apoptotic and necrotic pathways. Toxicol. Appl. Pharmacol. 243(3): 323–31. doi: 10.1016/j.taap.2009.11.024.

Chida, D., Toyama, T., Chiba, T., Kaneko, T., Arisawa, K. and Saito, Y. 2022. Effects of the interplay between selenocystine and methylmercury on their cytotoxicity and glucose-driven insulin secretion from mouse insulinoma cells. BPB Reports 5(4): 74–79.

Chung, Y.P., Yen, C.C., Tang, F.C., Lee, K.I., Liu, S.H., Wu, C.C. et al. 2019. Methylmercury exposure induces ROS/Akt inactivation-triggered endoplasmic reticulum stress-regulated neuronal cell apoptosis. Toxicology 425: 152245. doi: 10.1016/j.tox.2019.152245.

Ciudin, A., Hernández, C. and Simó, R. 2010. Iron overload in diabetic retinopathy: a cause or a consequence of impaired mechanisms? Exp. Diabetes Res. 2010: 714108. doi: 10.1155/2010/714108.

Civelek, S., Gelişgen, R., Andican, G., Seven, A., Küçük, S.H., Ozdoğan, M. et al. 2010. Advanced glycation end products and antioxidant status in nondiabetic and streptozotocin induced diabetic rats: effects of copper treatment. Biometals 23(1): 43–9. doi: 10.1007/s10534-009-9265-9.

Cooksey, R.C., Jones, D., Gabrielsen, S., Huang, J., Simcox, J.A., Luo, B. et al. 2010. Dietary iron restriction or iron chelation protects from diabetes and loss of beta-cell function in the obese (ob/ob lep-/-) mouse. Am. J. Physiol. Endocrinol. Metab. 298(6): 1236–43. doi: 10.1152/ajpendo.00022.2010.

Cui, R., Choi, S.E., Kim, T.H., Lee, H.J., Lee, S.J., Kang, Y. et al. 2019. Iron overload by transferrin receptor protein 1 regulation plays an important role in palmitate-induced insulin resistance in human skeletal muscle cells. FASEB J. Feb; 33(2): 1771–1786. doi: 10.1096/fj.201800448R.

Cui, X., Wang, Y., Liu, H., Shi, M., Wang, J. and Wang, Y. 2022. The molecular mechanisms of defective copper metabolism in diabetic cardiomyopathy. Oxid. Med. Cell Longev. 2022: 5418376. doi: 10.1155/2022/5418376.

Davis, C.M. and Vincent, J.B. 1997. Chromium oligopeptide activates insulin receptor tyrosine kinase activity. Biochemistry 36(15): 4382–5. doi: 10.1021/bi963154t.

de Magalhães Silva, M., de Araújo Dantas, M.D., da Silva Filho, R.C., Dos Santos Sales, M.V., de Almeida Xavier, J., Leite, A.C.R. et al. 2020. Toxicity of thimerosal in biological systems: Conformational changes in human hemoglobin, decrease of oxygen binding capacity, increase of protein glycation and amyloid's formation. Int. J. Biol. Macromol. 154: 661–671. doi: 10.1016/j.ijbiomac.2020.03.156.

Deng, G., Dyroff, S.L., Lockart, M., Bowman, M.K. and Vincent, J.B. 2016. The effects of the glycation of transferrin on chromium binding and the transport and distribution of chromium *in vivo*. J. Inorg. Biochem. 164: 26–33. doi: 10.1016/j.jinorgbio.2016.08.008.

Díaz-Villaseñor, A., Sánchez-Soto, M.C., Cebrián, M.E., Ostrosky-Wegman, P. and Hiriart, M. 2006. Sodium arsenite impairs insulin secretion and transcription in pancreatic beta-cells. Toxicol. Appl. Pharmacol. 214(1): 30–4. doi: 10.1016/j.taap.2005.11.015.

Dongiovanni, P., Valenti, L., Ludovica Fracanzani, A., Gatti, S., Cairo, G. and Fargion, S. 2008. Iron depletion by deferoxamine up-regulates glucose uptake and insulin signaling in hepatoma cells and in rat liver. Am. J. Pathol. 172(3): 738–47. doi: 10.2353/ajpath.2008.070097.

Dongiovanni, P., Ruscica, M., Rametta, R., Recalcati, S., Steffani, L., Gatti, S. et al. 2013. Dietary iron overload induces visceral adipose tissue insulin resistance. Am. J. Pathol. 182(6): 2254–63. doi: 10.1016/j.ajpath.2013.02.019.

Du, P.C., Tu, Z.C., Wang, H. and Hu, Y.M. 2020. Mechanism of selenium nanoparticles inhibiting advanced glycation end products. J. Agric. Food Chem. 68(39): 10586–10595. doi: 10.1021/acs.jafc.0c03229.

Dutta, S., Saha, S., Mahalanobish, S., Sadhukhan, P. and Sil, P.C. 2018. Melatonin attenuates arsenic induced nephropathy via the regulation of oxidative stress and inflammatory signaling cascades in mice. Food Chem. Toxicol. 118: 303–316. doi: 10.1016/j.fct.2018.05.032.

El-Azab, M.F., Al-Karmalawy, A.A., Antar, S.A., Hanna, P.A., Tawfik, K.M. and Hazem, R.M. 2022. A novel role of Nano selenium and sildenafil on streptozotocin-induced diabetic nephropathy in rats by modulation of inflammatory, oxidative, and apoptotic pathways. Life Sci. 303: 120691. doi: 10.1016/j.lfs.2022.120691.

Exley, C. 2016. The toxicity of aluminium in humans. Morphologie 100(329): 51–5. doi: 10.1016/j.morpho.2015.12.003.

Farrokhian, A., Bahmani, F., Taghizadeh, M., Mirhashemi, S.M., Aarabi, M.H., Raygan, F. et al. 2016. Selenium supplementation affects insulin resistance and serum hs-CRP in patients with Type 2 diabetes and coronary heart disease. Horm Metab Res. 48(4): 263–8. doi: 10.1055/s-0035-1569276.

Faulk, C., Barks, A., Sánchez, B.N., Zhang, Z., Anderson, O.S., Peterson, K.E. et al. 2014. Perinatal lead (Pb) exposure results in sex-specific effects on food intake, fat, weight, and insulin response across the murine life-course. PLoS One 9(8): e104273. doi: 10.1371/journal.pone.0104273.

Feng, W., Ding, Y., Zhang, W., Chen, Y., Li, Q., Wang, W. et al. 2018a. Chromium malate alleviates high-glucose and insulin resistance in L6 skeletal muscle cells by regulating glucose uptake and insulin sensitivity signaling pathways. Biometals 31(5): 891–908. doi: 10.1007/s10534-018-0132-4.

Feng, W., Liu, Y., Fei, F., Chen, Y., Ding, Y., Yan, M. et al. 2018b. Improvement of high-glucose and insulin resistance of chromium malate in 3T3-L1 adipocytes by glucose uptake and insulin sensitivity signaling pathways and its mechanism. RSC Adv. 9(1): 114–127. doi: 10.1039/c8ra07470d.

Feng, Y., Yang, D., Zhi, X., Deng, H., Zhang, W., Wang, R. et al. 2022. Role of interaction between reactive oxygen species and ferroptosis pathway in methylglyoxal-induced injury in mouse embryonic osteoblasts. Nan Fang Yi Ke Da Xue Xue Bao 42(1): 108–115. doi: 10.12122/j.issn.1673-4254.2022.01.13.

Ferretti, G., Bacchetti, T., Marchionni, C. and Dousset, N. 2004. Effect of non-enzymatic glycation on aluminium-induced lipid peroxidation of human high density lipoproteins (HDL). Nutr. Metab Cardiovasc Dis. 14(6): 358–65. doi: 10.1016/s0939-4753(04)80026-7.

Ficková, M., Eybl, V., Kotyzová, D., Micková, V., Möstbök, S. and Brtko, J. 2003. Long lasting cadmium intake is associated with reduction of insulin receptors in rat adipocytes. Biometals 16(4): 561–6. doi: 10.1023/a:1023485130767. PMID: 12779241.

Fillebeen, C., Lam, N.H., Chow, S., Botta, A., Sweeney, G. and Pantopoulos, K. 2020. Regulatory connections between iron and glucose metabolism. Int. J. Mol. Sci. 21(20): 7773. doi: 10.3390/ijms21207773.

Fukunaka, A. and Fujitani, Y. 2018. Role of zinc homeostasis in the pathogenesis of diabetes and obesity. Int. J. Mol. Sci. 2018 Feb 6; 19(2): 476. doi: 10.3390/ijms19020476. PMID: 29415457; PMCID: PMC5855698.

Galiniak, S., Bartosz, G. and Sadowska-Bartosz, I. 2015. Is iron chelation important in preventing glycation of bovine serum albumin *in vitro*? Cell Mol. Biol. Lett. 20(4): 562–70. doi: 10.1515/cmble-2015-0033.

Ganeshan, M., Sainath, P.B., Padmavathi, I.J., Venu, L., Kishore, Y.D., Kumar, K.A. et al. 2011. Maternal manganese restriction increases susceptibility to high-fat diet-induced dyslipidemia and altered adipose function in WNIN male rat offspring. Exp. Diabetes Res. 2011: 486316. doi: 10.1155/2011/486316.

Gasser, M., Lenglet, S., Bararpour, N., Sajic, T., Wiskott, K., Augsburger, M. et al. 2022. Cadmium acute exposure induces metabolic and transcriptomic perturbations in human mature adipocytes. Toxicology 470: 153153. doi: 10.1016/j.tox.2022.153153.

Ge, W., Jie, J., Yao, J., Li, W., Cheng, Y. and Lu, W. 2022. Advanced glycation end products promote osteoporosis by inducing ferroptosis in osteoblasts. Mol. Med. Rep. 25(4): 140. doi: 10.3892/mmr.2022.12656.

Ge, X., Liu, Z., Qi, W., Shi, X. and Zhai, Q. 2008. Chromium (VI) induces insulin resistance in 3T3-L1 adipocytes through elevated reactive oxygen species generation. Free Radic. Res. 42(6): 554–63. doi: 10.1080/10715760802155113. PMID: 18569013.

Gensberger-Reigl, S., Auditore, A., Huppert, J. and Pischetsrieder, M. 2021. Metal cations promote α-dicarbonyl formation in glucose-containing peritoneal dialysis fluids. Glycoconj J. 38(3): 319–329. doi: 10.1007/s10719-020-09964-6.

Gil, H., Vásquez, B., Peña, M. and Uzcategui, J. 2004. Effect of buffer carbonate and arsenate on the kinetics of glycation of human hemoglobin. J. Phys. Org. Chem. 17(6-7): 537–540.

Grădinaru, D., Margină, D., Ungurianu, A., Nițulescu, G., Pena, C.M., Ionescu-Tîrgoviște, C. et al. 2021. Zinc status, insulin resistance and glycoxidative stress in elderly subjects with type 2 diabetes mellitus. Exp. Ther. Med. 22(6): 1393. doi: 10.3892/etm.2021.10829.

Hamada, Y., Nakashima, E., Naruse, K., Nakae, M., Naiki, M., Fujisawa, H. et al. 2005. A copper chelating agent suppresses carbonyl stress in diabetic rat lenses. J. Diabetes Complications 19(6): 328–34. doi: 10.1016/j.jdiacomp.2005.08.002.

Hamann, I., Petroll, K., Hou, X., Anwar-Mohamed, A., El-Kadi, A.O. and Klotz, L.O. 2014. Acute and long-term effects of arsenite in HepG2 cells: modulation of insulin signaling. Biometals 27(2): 317–32. doi: 10.1007/s10534-014-9714-y.

Han, J.C., Park, S.Y., Hah, B.G., Choi, G.H., Kim, Y.K., Kwon, T.H. et al. 2003. Cadmium induces impaired glucose tolerance in rat by down-regulating GLUT4 expression in adipocytes. Arch Biochem. Biophys. 413(2): 213–20. doi: 10.1016/s0003-9861(03)00120-6.
Han, Y.H., Buffolo, M., Pires, K.M., Pei, S., Scherer, P.E. and Boudina, S. 2016. Adipocyte-specific deletion of manganese superoxide dismutase protects from diet-induced obesity through increased mitochondrial uncoupling and biogenesis. Diabetes 65(9): 2639–51. doi: 10.2337/db16-0283.
Hansen, J.B., Tonnesen, M.F., Madsen, A.N., Hagedorn, P.H., Friberg, J., Grunnet, L.G. et al. 2012. Divalent metal transporter 1 regulates iron-mediated ROS and pancreatic β cell fate in response to cytokines. Cell Metab. 16(4): 449–61. doi: 10.1016/j.cmet.2012.09.001.
Heart, E. and Sung, C.K. 2003. Insulin-like and non-insulin-like selenium actions in 3T3-L1 adipocytes. J. Cell Biochem. 88(4): 719–31. doi: 10.1002/jcb.10395.
Hegab, Z., Gibbons, S., Neyses, L. and Mamas, M.A. 2012. Role of advanced glycation end products in cardiovascular disease. World J. Cardiol. 4(4): 90–102. doi: 10.4330/wjc.v4.i4.90.
Heimfarth, L., Delgado, J., Mignori, M.R., Gelain, D.P., Moreira, J.C.F. and Pessoa-Pureur, R. 2018. Developmental neurotoxicity of the hippocampus following *in utero* exposure to methylmercury: impairment in cell signaling. Arch. Toxicol. 92(1): 513–527. doi: 10.1007/s00204-017-2042-6.
Himoto, T., Yoneyama, H., Kurokohchi, K., Inukai, M., Masugata, H., Goda, F. et al. 2011. Selenium deficiency is associated with insulin resistance in patients with hepatitis C virus-related chronic liver disease. Nutr. Res. 31(11): 829–35. doi: 10.1016/j.nutres.2011.09.021.
Hong, H., Xu, Y., Xu, J., Zhang, J., Xi, Y., Pi, H. et al. 2021. Cadmium exposure impairs pancreatic β-cell function and exaggerates diabetes by disrupting lipid metabolism. Environ Int. 149: 106406. doi: 10.1016/j.envint.2021.106406.
Hong, H., Lin, X., Xu, Y., Tong, T., Zhang, J., He, H. et al. 2022. Cadmium induces ferroptosis mediated inflammation by activating Gpx4/Ager/p65 axis in pancreatic β-cells. Sci. Total Environ. 849: 157819. doi: 10.1016/j.scitotenv.2022.157819.
Hou, Y., Zhang, Y., Jiang, S., Xie, N., Zhang, Y., Meng, X. et al. 2023. Salidroside intensifies mitochondrial function of CoCl2-damaged HT22 cells by stimulating PI3K-AKT-MAPK signaling pathway. Phytomedicine 109: 154568. doi: 10.1016/j.phymed.2022.154568.
Hu, J., Cao, J., Xu, Q. and Lu, M. 2022. Dose-response relationships between urinary cobalt concentrations and obesity, insulin resistance, and metabolic-related disorders in the general population. Environ Sci. Pollut. Res. Int. 29(20): 29682–29688. doi: 10.1007/s11356-021-17861-0.
Hu, Y., Li, J., Lou, B., Wu, R., Wang, G., Lu, C. et al. 2020. The role of reactive oxygen species in arsenic toxicity. Biomolecules 10(2): 240. doi: 10.3390/biom10020240.
Hua, Y., Clark, S., Ren, J. and Sreejayan, N. 2012. Molecular mechanisms of chromium in alleviating insulin resistance. J. Nutr. Biochem. 23(4): 313–9. doi: 10.1016/j.jnutbio.2011.11.001.
Huang, S., Peng, W., Jiang, X., Shao, K., Xia, L., Tang, Y. et al. 2014. The effect of chromium picolinate supplementation on the pancreas and macroangiopathy in type II diabetes mellitus rats. J. Diabetes Res. 2014: 717219. doi: 10.1155/2014/717219.
Huang, Y. and Yun, K. 2020. Study on effects and mechanism of lead and high-fat diet on cognitive function and central nervous system in mice. World Neurosurg. 138: 758–763. doi: 10.1016/j.wneu.2020.01.165.
Huang, Y.S., Li, Y.C., Tsai, P.Y., Lin, C.E., Chen, C.M., Chen, S.M. et al. 2017. Accumulation of methylglyoxal and d-lactate in Pb-induced nephrotoxicity in rats. Biomed. Chromatogr. 31(5). doi: 10.1002/bmc.3869.
Husna, A.H., Ramadhani, E.A., Eva, D.T., Yulita, A.F. and Suhartono, E. 2014. The role formation of methylglyoxal, carbonyl compound, hydrogen peroxide and advance oxidation protein product induced cadmium in ovarian rat. Int. J. Chem. Eng. Appl. 5(4): 319–23.
Iqbal, S., Qais, F.A., Alam, M.M. and Naseem, I. 2018. Effect of glycation on human serum albumin-zinc interaction: a biophysical study. J. Biol. Inorg. Chem. 23(3): 447–458. doi: 10.1007/s00775-018-1554-8.
Jakus, V., Bauerová, K. and Rietbrock, N. 2001. Effect of aminoguanidine and copper(II) ions on the formation of advanced glycosylation end products. *In vitro* study on human serum albumin. Arzneimittelforschung 51(4): 280–3. doi: 10.1055/s-0031-1300038.

Jansen, J., Rosenkranz, E., Overbeck, S., Warmuth, S., Mocchegiani, E., Giacconi, R. et al. 2012. Disturbed zinc homeostasis in diabetic patients by *in vitro* and *in vivo* analysis of insulinomimetic activity of zinc. J. Nutr. Biochem. 23(11): 1458–66. doi: 10.1016/j.jnutbio.2011.09.008.

Jia, X., Qiu, T., Yao, X., Jiang, L., Wang, N., Wei, S. et al. 2020. Arsenic induces hepatic insulin resistance via mtROS-NLRP3 inflammasome pathway. J. Hazard Mater. 399: 123034. doi: 10.1016/j.jhazmat.2020.123034.

Kadry, M.O. and Abdel Megeed, R.M. 2022. Ubiquitous toxicity of Mercuric Chloride in target tissues and organs: Impact of Ubidecarenone and liposomal-Ubidecarenone STAT 5A/PTEN/PI3K/AKT signaling pathways. J. Trace Elem. Med. Biol. 74: 127058. doi: 10.1016/j.jtemb.2022.127058.

Kandarakis, S.A., Piperi, C., Topouzis, F. and Papavassiliou, A.G. 2014. Emerging role of advanced glycation-end products (AGEs) in the pathobiology of eye diseases. Prog. Retin. Eye Res. 42: 85–102. doi: 10.1016/j.preteyeres.2014.05.002.

Kang, J.H. 2006. Oxidative modification of human ceruloplasmin by methylglyoxal: an *in vitro* study. J. Biochem. Mol. Biol. 39(3): 335–8. doi: 10.5483/bmbrep.2006.39.3.335.

Kang, R., Hou, W., Zhang, Q., Chen, R., Lee, Y.J., Bartlett, D.L. et al. 2014. RAGE is essential for oncogenic KRAS-mediated hypoxic signaling in pancreatic cancer. Cell Death Dis. 5(10): e1480. doi: 10.1038/cddis.2014.445.

Karpova, N.N., Lindholm, J.S., Kulesskaya, N., Onishchenko, N., Vahter, M., Popova, D. et al. 2014. TrkB overexpression in mice buffers against memory deficits and depression-like behavior but not all anxiety- and stress-related symptoms induced by developmental exposure to methylmercury. Front. Behav. Neurosci. 8: 315. doi: 10.3389/fnbeh.2014.00315.

Kheirouri, S., Alizadeh, M. and Maleki, V. 2018. Zinc against advanced glycation end products. Clin. Exp. Pharmacol. Physiol. 45(6): 491–498. doi: 10.1111/1440-1681.12904.

Kim, D.K., Park, J.D. and Choi, B.S. 2014. Mercury-induced amyloid-beta (Aβ) accumulation in the brain is mediated by disruption of Aβ transport. J. Toxicol. Sci. 39(4): 625–35. doi: 10.2131/jts.39.625.

Kim, J.Y., Kim, S.J., Bae, M.A., Kim, J.R. and Cho, K.H. 2018. Cadmium exposure exacerbates severe hyperlipidemia and fatty liver changes in zebrafish via impairment of high-density lipoproteins functionality. Toxicol *In Vitro* 47: 249–258. doi: 10.1016/j.tiv.2017.11.007.

Kim, K.N., Park, S.J., Choi, B. and Joo, N.S. 2015. Blood mercury and insulin resistance in nondiabetic koreans (KNHANES 2008–2010). Yonsei Med. J. 56(4): 944–50. doi: 10.3349/ymj.2015.56.4.944.

Kuzan, A. 2021. Toxicity of advanced glycation end products (Review). Biomed Rep. May; 14(5): 46. doi: 10.3892/br.2021.1422.

Lai, C.H., Chou, C.C., Chuang, H.C., Lin, G.J., Pan, C.H. and Chen, W.L. 2020. Receptor for advanced glycation end products in relation to exposure to metal fumes and polycyclic aromatic hydrocarbon in shipyard welders. Ecotoxicol. Environ. Saf. 202: 110920. doi: 10.1016/j.ecoenv.2020.110920.

Lantz, R.C., Lynch, B.J., Boitano, S., Poplin, G.S., Littau, S., Tsaprailis, G. et al. 2007. Pulmonary biomarkers based on alterations in protein expression after exposure to arsenic. Environ Health Perspect. 115(4): 586–91. doi: 10.1289/ehp.9611.

Lawes, M., Pinkas, A., Frohlich, B.A., Iroegbu, J.D., Ijomone, O.M. and Aschner, M. 2020. Metal-induced neurotoxicity in a RAGE-expressing *C. elegans* model. Neurotoxicology 80: 71–75. doi: 10.1016/j.neuro.2020.06.013.

Lee, S.H., Jouihan, H.A., Cooksey, R.C., Jones, D., Kim, H.J., Winge, D.R. et al. 2013. Manganese supplementation protects against diet-induced diabetes in wild type mice by enhancing insulin secretion. Endocrinology 154(3): 1029–38. doi: 10.1210/en.2012-1445.

Lee, S.H., Choi, B., Park, S.J., Kim, Y.S. and Joo, N.S.. 2017. The cut-off value of blood mercury concentration in relation to insulin resistance. J. Obes. Metab. Syndr. 26(3): 197–203. doi: 10.7570/jomes.2017.26.3.197.

Lee, W.K. and Thévenod, F. 2020. Cell organelles as targets of mammalian cadmium toxicity. Arch Toxicol. 94(4): 1017–1049. doi: 10.1007/s00204-020-02692-8.

Lee, Y.W., Cherng, Y.G., Yang, S.T., Liu, S.H., Chen, T.L. and Chen, R.M. 2021. Hypoxia induced by cobalt chloride triggers autophagic apoptosis of human and mouse drug-resistant glioblastoma cells through targeting the PI3K-AKT-mTOR signaling pathway. Oxid. Med. Cell Longev. 2021: 5558618. doi: 10.1155/2021/5558618.

Leyssens, L., Vinck, B., Van Der Straeten, C., Wuyts, F. and Maes, L. 2017. Cobalt toxicity in humans—A review of the potential sources and systemic health effects. Toxicology 387: 43–56. doi: 10.1016/j.tox.2017.05.015.

Li, H., Kang, T., Qi, B., Kong, L., Jiao, Y., Cao, Y. et al. 2016. Neuroprotective effects of ginseng protein on PI3K/Akt signaling pathway in the hippocampus of D-galactose/$AlCl_3$ inducing rats model of Alzheimer's disease. J. Ethnopharmacol. 179: 162–9. doi: 10.1016/j.jep.2015.12.020.

Li, W., Wu, L., Sun, Q., Yang, Q., Xue, J., Shi, M. et al. 2021. MicroRNA-191 blocking the translocation of GLUT4 is involved in arsenite-induced hepatic insulin resistance through inhibiting the IRS1/AKT pathway. Ecotoxicol. Environ. Saf. Jun 1; 215: 112130. doi: 10.1016/j.ecoenv.2021.112130. Epub 2021 Mar 18. PMID: 33743404.

Li, X., Li, M., Xu, J., Zhang, X., Xiao, W. and Zhang, Z. 2019. Decreased insulin secretion but unchanged glucose homeostasis in cadmium-exposed male C57BL/6 mice. J. Toxicol. 2019: 8121834. doi: 10.1155/2019/8121834.

Li, Y., Qin, M., Zhong, W., Liu, C., Deng, G., Yang, M. et al. 2023. RAGE promotes dysregulation of iron and lipid metabolism in alcoholic liver disease. Redox Biol. 59: 102559. doi: 10.1016/j.redox.2022.102559.

Li, Y.B., Han, J.Y., Jiang, W. and Wang, J. 2011. Selenium inhibits high glucose-induced cyclooxygenase-2 and P-selectin expression in vascular endothelial cells. Mol. Biol. Rep. 38(4): 2301–6. doi: 10.1007/s11033-010-0362-1.

Limphong, P., McKinney, R.M., Adams, N.E., Bennett, B., Makaroff, C.A., Gunasekera, T. et al. 2009. Human glyoxalase II contains an Fe(II)Zn(II) center but is active as a mononuclear Zn(II) enzyme. Biochemistry 48(23): 5426–34. doi: 10.1021/bi9001375.

Litvinov, R.A., Gontareva, A.V., Usmiyanova, L.E. and Klimenko, D.R. 2021. Influence of certain d-metals on formation of advanced glycation end products, aggregation and amyloid transformation of albumin in glycation reaction. Pharm Pharmacol. 9(4): 306–317.

Liu, K.L., Chen, P.Y., Wang, C.M., Chen, W.Y., Chen, C.W., Owaga, E. et al. 2016. Dose-related effects of ferric citrate supplementation on endoplasmic reticular stress responses and insulin signalling pathways in streptozotocin-nicotinamide-induced diabetes. Food Funct. 7(1): 194–201. doi: 10.1039/c5fo01252j.

Liu, R., Bai, L., Liu, M., Wang, R., Wu, Y., Li, Q. et al. 2022. Combined exposure of lead and high-fat diet enhanced cognitive decline via interacting with CREB-BDNF signaling in male rats. Environ. Pollut. 304: 119200. doi: 10.1016/j.envpol.2022.119200.

Liu, S., Guo, X., Wu, B., Yu, H., Zhang, X. and Li, M. 2014. Arsenic induces diabetic effects through beta-cell dysfunction and increased gluconeogenesis in mice. Sci. Rep. 4: 6894. doi: 10.1038/srep06894.

Loske, C., Gerdemann, A., Schepl, W., Wycislo, M., Schinzel, R., Palm, D. et al. 2000. Transition metal-mediated glycoxidation accelerates cross-linking of beta-amyloid peptide. Eur. J. Biochem. 267(13): 4171–8. doi: 10.1046/j.1432-1327.2000.01452.x.

Lowe, J., Taveira-da-Silva, R. and Hilário-Souza, E. 2017. Dissecting copper homeostasis in diabetes mellitus. IUBMB Life 69(4): 255–262. doi: 10.1002/iub.1614.

Lu, J., Stewart, A.J., Sadler, P.J., Pinheiro, T.J. and Blindauer, C.A. 2008. Albumin as a zinc carrier: properties of its high-affinity zinc-binding site. Biochem. Soc. Trans. 36(6): 1317–1321.

Luevano-Contreras, C., Garay-Sevilla, M.E. and Chapman-Novakofski, K. 2013. Role of dietary advanced glycation end products in diabetes mellitus. J. Evid. Based Integr. Med. 18(1): 50–66.

Luo, Y., Niu, F., Sun, Z., Cao, W., Zhang, X., Guan, D. et al. 2009. Altered expression of Abeta metabolism-associated molecules from D-galactose/AlCl(3) induced mouse brain. Mech Ageing Dev. 130(4): 248–52. doi: 10.1016/j.mad.2008.12.005.

Ma, X., Pham, V.T., Mori, H., MacDougald, O.A., Shah, Y.M. and Bodary, P.F. 2017. Iron elevation and adipose tissue remodeling in the epididymal depot of a mouse model of polygenic obesity. PLoS One 12(6): e0179889. doi: 10.1371/journal.pone.0179889.

Maret, W. 2017. Zinc in pancreatic islet biology, insulin sensitivity, and diabetes. Prev. Nutr. Food Sci. 22(1): 1–8. doi: 10.3746/pnf.2017.22.1.1.

Marques, C.M.S., Nunes, E.A., Lago, L., Pedron, C.N., Manieri, T.M., Sato, R.H. et al. 2017. Generation of Advanced Glycation End-Products (AGEs) by glycoxidation mediated by copper and ROS in a human serum albumin (HSA) model peptide: reaction mechanism and damage in motor neuron cells. Mutat. Res. Genet. Toxicol. Environ. Mutagen. 824: 42–51. doi: 10.1016/j.mrgentox.2017.10.005.

May, O., Yatime, L., Merle, N.S., Delguste, F., Howsam, M., Daugan, M.V. et al. 2021. The receptor for advanced glycation end products is a sensor for cell-free heme. FEBS J. 288(11): 3448–3464. doi: 10.1111/febs.15667.

Mechlovich, D., Amit, T., Bar-Am, O., Mandel, S., Youdim, M.B. and Weinreb, O. 2014. The novel multi-target iron chelator, M30 modulates HIF-1α-related glycolytic genes and insulin signaling pathway in the frontal cortex of APP/PS1 Alzheimer's disease mice. Curr. Alzheimer Res. 11(2): 119–27. doi: 10.2174/1567205010666131212112529.

Mehdad, A., Campos, N.A., Arruda, S.F. and Siqueira, E.M. 2015. Iron deprivation may enhance insulin receptor and Glut4 transcription in skeletal muscle of adult rats. J. Nutr. Health Aging 19(8): 846–54. doi: 10.1007/s12603-015-0541-9.

Mirlohi, M.S., Yaghooti, H., Shirali, S., Aminasnafi, A. and Olapour, S. 2018. Increased levels of advanced glycation end products positively correlate with iron overload and oxidative stress markers in patients with β-thalassemia major. Ann. Hematol. 97(4): 679–684. doi: 10.1007/s00277-017-3223-3.

Miyata, T. and Dan, T. 2008. Inhibition of advanced glycation end products (AGEs): an implicit goal in clinical medicine for the treatment of diabetic nephropathy? Diabetes Res. Clin. Pract. 82S1: S25–9. doi: 10.1016/j.diabres.2008.09.012.

Monga, A., Fulke, A.B. and Dasgupta, D. 2022. Recent developments in essentiality of trivalent chromium and toxicity of hexavalent chromium: Implications on human health and remediation strategies. J. Hazard Mater Adv. 100113.

Moulahoum, H., Ghorbanizamani, F., Timur, S. and Zihnioglu, F. 2020. Zinc enhances carnosine inhibitory effect against structural and functional age-related protein alterations in an albumin glycoxidation model. Biometals 33(6): 353–364. doi: 10.1007/s10534-020-00254-0.

Mozaffari, M.S., Baban, B., Abdelsayed, R., Liu, J.Y., Wimborne, H. and Rodriguez, N.W. 2012. Abebe. Renal and glycemic effects of high-dose chromium picolinate in db/db mice: assessment of DNA damage. J. Nutr. Biochem. 23(8): 977–85. doi: 10.1016/j.jnutbio.2011.05.004.

Mueller, A.S., Klomann, S.D., Wolf, N.M., Schneider, S., Schmidt, R., Spielmann, J. et al. 2008. Redox regulation of protein tyrosine phosphatase 1B by manipulation of dietary selenium affects the triglyceride concentration in rat liver. J. Nutr. 138(12): 2328–36. doi: 10.3945/jn.108.089482.

Münch, G., Schicktanz, D., Behme, A., Gerlach, M., Riederer, P., Palm, D. et al. 1999. Amino acid specificity of glycation and protein–AGE crosslinking reactivities determined with a dipeptide SPOT library. Nat. Biotechnol. 17(10): 1006–1010.

Nampoothiri, M., Ramalingayya, G.V., Kutty, N.G., Krishnadas, N. and Rao, C.M. 2017. Insulin combined with glucose improves spatial learning and memory in aluminum chloride-induced dementia in rats. J. Environ. Pathol. Toxicol. Oncol. 36(2): 159–169. doi: 10.1615/JEnvironPatholToxicolOncol.2017020185.

Navas-Acien, A., Silbergeld, E.K., Pastor-Barriuso, R. and Guallar, E. 2008. Arsenic exposure and prevalence of type 2 diabetes in US adults. JAMA 300(7): 814–22. doi: 10.1001/jama.300.7.814.

Neal, E.S., Hofstee, P., Askew, M.R., Kent, N.L., Bartho, L.A., Perkins, A.V. et al. 2021. Maternal selenium deficiency in mice promotes sex-specific changes to urine flow and renal expression of mitochondrial proteins in adult offspring. Physiol. Rep. 9(6): e14785. doi: 10.14814/phy2.14785.

Niño, S.A., Martel-Gallegos, G., Castro-Zavala, A., Ortega-Berlanga, B., Delgado, J.M., Hernández-Mendoza, H. et al. 2018. Chronic arsenic exposure increases Aβ(1-42) production and receptor for advanced glycation end products expression in rat brain. Chem. Res. Toxicol. 31(1): 13–21. doi: 10.1021/acs.chemrestox.7b00215.

Niño, S.A., Chi-Ahumada, E., Carrizales, L., Estrada-Sánchez, A.M., Gonzalez-Billault, C., Zarazúa, S. et al. 2022. Life-long arsenic exposure damages the microstructure of the rat hippocampus. Brain Res. 1775: 147742. doi: 10.1016/j.brainres.2021.147742.

Niyomchan, A., Watcharasit, P., Visitnonthachai, D., Homkajorn, B., Thiantanawat, A. and Satayavivad, J. 2015. Insulin attenuates arsenic-induced neurite outgrowth impairments by activating the PI3K/Akt/SIRT1 signaling pathway. Toxicol. Lett. 236(3): 138–44. doi: 10.1016/j.toxlet.2015.05.008.

Niyomchan, A., Visitnonthachai, D., Suntararuks, S., Ngamsiri, P., Watcharasit, P. and Satayavivad, J. 2018. Arsenic impairs insulin signaling in differentiated neuroblastoma SH-SY5Y cells. Neurotoxicology 66: 22–31. doi: 10.1016/j.neuro.2018.03.004.

Nowotny, K., Jung, T., Höhn, A., Weber, D. and Grune, T. 2015. Advanced glycation end products and oxidative stress in type 2 diabetes mellitus. Biomolecules 5(1): 194–222. doi: 10.3390/biom5010194.

Nurchi, V.M., Djordjevic, A.B., Crisponi, G., Alexander, J., Bjørklund, G. and Aaseth, J. 2020. Arsenic toxicity: molecular targets and therapeutic agents. Biomolecules 10(2): 235. doi: 10.3390/biom10020235.

Ohtomo, S., Nangaku, M., Izuhara, Y., Takizawa, S., Strihou, C. and Miyata, T. 2008. Cobalt ameliorates renal injury in an obese, hypertensive type 2 diabetes rat model. Nephrol. Dial Transplant. 23(4): 1166–72. doi: 10.1093/ndt/gfm715.

Ojeda, M.L., Nogales, F., Membrilla, A. and Carreras, O. 2019. Maternal selenium status is profoundly involved in metabolic fetal programming by modulating insulin resistance, oxidative balance and energy homeostasis. Eur. J. Nutr. 58(8): 3171–3181. doi: 10.1007/s00394-018-1861-4.

Olivas-Calderón, E., Recio-Vega, R., Gandolfi, A.J., Lantz, R.C., González-Cortes, T., Gonzalez-De Alba, C. et al. 2015. Lung inflammation biomarkers and lung function in children chronically exposed to arsenic. Toxicol. Appl. Pharmacol. 287(2): 161–167. doi: 10.1016/j.taap.2015.06.001.

Paul, D.S., Harmon, A.W., Devesa, V., Thomas, D.J. and Stýblo, M. 2007. Molecular mechanisms of the diabetogenic effects of arsenic: inhibition of insulin signaling by arsenite and methylarsonous acid. Environ. Health Perspect. 115(5): 734–42. doi: 10.1289/ehp.9867.

Penezić, A.Z., Jovanović, V.B., Pavićević, I.D., Aćimović, J.M. and Mandić, L.M. 2015. HSA carbonylation with methylglyoxal and the binding/release of copper(II) ions. Metallomics 7(10): 1431–8. doi: 10.1039/c5mt00159e.

Pierozan, P., Biasibetti, H., Schmitz, F., Ávila, H., Fernandes, C.G., Pessoa-Pureur, R. et al. 2017. Neurotoxicity of methylmercury in isolated astrocytes and neurons: the cytoskeleton as a main target. Mol. Neurobiol. 54(8): 5752–5767. doi: 10.1007/s12035-016-0101-2.

Pillai, S.S., Sugathan, J.K. and Indira, M. 2012. Selenium downregulates RAGE and NFκB expression in diabetic rats. Biol. Trace Elem. Res. 149(1): 71–7. doi: 10.1007/s12011-012-9401-1.

Pinkas, A. and Aschner, M. 2016. Advanced glycation end-products and their receptors: related pathologies, recent therapeutic strategies, and a potential model for future neurodegeneration studies. Chem. Res. Toxicol. 29(5): 707–14. doi: 10.1021/acs.chemrestox.6b00034.

Pinkas, A. and Aschner, M. 2017. AGEs/RAGE-related neurodegeneration: daf-16 as a mediator, insulin as an ameliorant, and *C. elegans* as an expedient research model. Chem. Res. Toxicol. 30(1): 38–42. doi: 10.1021/acs.chemrestox.6b00264.

Price, D.L., Rhett, P.M., Thorpe, S.R. and Baynes, J.W. 2001. Chelating activity of advanced glycation end-product inhibitors. J. Biol. Chem. Dec 276(52): 48967–72. doi: 10.1074/jbc.M108196200.

Promyo, K., Iqbal, F., Chaidee, N. and Chetsawang, B. 2020. Aluminum chloride-induced amyloid β accumulation and endoplasmic reticulum stress in rat brain are averted by melatonin. Food Chem. Toxicol. 146: 111829. doi: 10.1016/j.fct.2020.111829.

Qiu, Q., Zhang, F., Zhu, W., Wu, J. and Liang, M. 2017. Copper in diabetes mellitus: a meta-analysis and systematic review of plasma and serum studies. Biol. Trace Elem. Res. 177(1): 53–63. doi: 10.1007/s12011-016-0877-y.

Rabbani, N. and Thornalley, P.J. 2018. Advanced glycation end products in the pathogenesis of chronic kidney disease. Kidney Int. 93(4): 803–813. doi: 10.1016/j.kint.2017.11.034.

Rajpathak, S., Ma, J., Manson, J., Willett, W.C. and Hu, F.B. 2006. Iron intake and the risk of type 2 diabetes in women: a prospective cohort study. Diabetes Care. 29(6): 1370–6. doi: 10.2337/dc06-0119.

Ramamurthy, C.H., Subastri, A., Suyavaran, A., Subbaiah, K.C., Valluru, L. and Thirunavukkarasu, C. 2016. Solanum torvum Swartz. fruit attenuates cadmium-induced liver and kidney damage through modulation of oxidative stress and glycosylation. Environ. Sci. Pollut. Res. Int. 23(8): 7919–29. doi: 10.1007/s11356-016-6044-3.

Ramirez Segovia, A.S., Wrobel, K., Acevedo Aguilar, F.J., Corrales Escobosa, A.R. and Wrobel, K. 2017. Effect of Cu(ii) on *in vitro* glycation of human serum albumin by methylglyoxal: a LC-MS-based proteomic approach. Metallomics 9(2): 132–140. doi: 10.1039/c6mt00235h.

Regazzetti, C., Peraldi, P., Grémeaux, T., Najem-Lendom, R., Ben-Sahra, I., Cormont, M. et al. 2009. Hypoxia decreases insulin signaling pathways in adipocytes. Diabetes 58(1): 95–103. doi: 10.2337/db08-0457.

Ridderström, M., Cameron, A.D., Jones, T.A. and Mannervik, B. 1998. Involvement of an active-site Zn^{2+} ligand in the catalytic mechanism of human glyoxalase I. J. Biol. Chem. 273(34): 21623–8. doi: 10.1074/jbc.273.34.21623.

Rosa-Silva, H.T.D., Panzenhagen, A.C., Schmidtt, V., Alves Teixeira, A., Espitia-Pérez, P., de Oliveira Franco, Á. et al. 2020. Hepatic and neurobiological effects of foetal and breastfeeding and adulthood exposure to methylmercury in Wistar rats. Chemosphere 244: 125400. doi: 10.1016/j.chemosphere.2019.125400.

Roy, C., Tremblay, P.Y. and Ayotte, P. 2017. Is mercury exposure causing diabetes, metabolic syndrome and insulin resistance? A systematic review of the literature. Environ. Res. 156: 747–760. doi: 10.1016/j.envres.2017.04.038.

Roy, C.N., Semba, R.D., Sun, K., Bandinelli, S., Varadhan, R., Patel, K.V. et al. 2012. Circulating selenium and carboxymethyl-lysine, an advanced glycation endproduct, are independent predictors of anemia in older community-dwelling adults. Nutrition 28(7-8): 762–6. doi: 10.1016/j.nut.2011.11.005.

Ruseva, B., Atanasova, M., Tsvetkova, R., Betova, T., Mollova, M., Alexandrova, M. et al. 2015. Effect of selenium supplementation on redox status of the aortic wall in young spontaneously hypertensive rats. Oxid. Med. Cell Longev. 2015: 609053. doi: 10.1155/2015/609053.

Saari, J.T. 1994. Implication of nonenzymatic glycosylation as a mode of damage in dietary copper deficiency. Nutr. Res. 14(11): 1689–1699.

Saari, J.T. and Dahlen, G.M. 1999. Early and advanced glycation end-products are increased in dietary copper deficiency. J. Nutr. Biochem. 10(4): 210–4. doi: 10.1016/s0955-2863(98)00100-4.

Sajithlal, G.B., Chithra, P. and Chandrakasan, G. 1998. The role of metal-catalyzed oxidation in the formation of advanced glycation end products: an *in vitro* study on collagen. Free Radic Biol. Med. 25(3): 265–9. doi: 10.1016/s0891-5849(98)00035-5.

Saker, F., Ybarra, J., Leahy, P., Hanson, R.W., Kalhan, S.C. and Ismail-Beigi, F. 1998. Glycemia-lowering effect of cobalt chloride in the diabetic rat: role of decreased gluconeogenesis. Am. J. Physiol. 274(6): E984–91. doi: 10.1152/ajpendo.1998.274.6.E984.

Saleh, R.A., Eissa, T.F., Abdallah, D.M., Saad, M.A. and El-Abhar, H.S. 2021. Peganum harmala enhanced GLP-1 and restored insulin signaling to alleviate AlCl3-induced Alzheimer-like pathology model. Sci. Rep. 11(1): 12040. doi: 10.1038/s41598-021-90545-4.

Sarmiento-Ortega, V.E., Moroni-González, D., Díaz, A., Eduardo, B. and Samuel, T. 2022. Oral subacute exposure to cadmium LOAEL dose induces insulin resistance and impairment of the hormonal and metabolic liver-adipose axis in wistar rats. Biol. Trace Elem. Res. 200(10): 4370–4384. doi: 10.1007/s12011-021-03027-z.

Saxena, S., Shukla, D. and Bansal, A. 2012. Augmentation of aerobic respiration and mitochondrial biogenesis in skeletal muscle by hypoxia preconditioning with cobalt chloride. Toxicol. Appl. Pharmacol. 264(3): 324–34. doi: 10.1016/j.taap.2012.08.033.

Schumacher, L. and Abbott, L.C. 2017. Effects of methyl mercury exposure on pancreatic beta cell development and function. J. Appl. Toxicol. 37(1): 4–12. doi: 10.1002/jat.3381.

Schwarz, K. and Mertz, W. 1959. Chromium(III) and the glucose tolerance factor. Arch. Biochem. Biophys. 85: 292–5. doi: 10.1016/0003-9861(59)90479-5.

Sengani, M., Chakraborty, S., Balaji, M.P., Govindasamy, R., Alahmadi, T.A., Al Obaid, S. et al. 2023. Anti-diabetic efficacy and selective inhibition of methyl glyoxal, intervention with biogenic Zinc oxide nanoparticle. Environ. Res. 216(Pt2): 114475. doi: 10.1016/j.envres.2022.114475.

Shan, W., Qi, J., Li, C. and Nie, X. 2019. Agonism of GPR39 displays protective effects against advanced glycation end-product (AGE)-induced degradation of extracellular matrix in human SW1353 cells. Arch. Biochem. Biophys. 677: 108164. doi: 10.1016/j.abb.2019.108164.

Shangari, N., Chan, T.S., Chan, K., Huai Wu, S. and O'Brien, P.J. 2007. Copper-catalyzed ascorbate oxidation results in glyoxal/AGE formation and cytotoxicity. Mol. Nutr. Food Res. Apr; 51(4): 445–55. doi: 10.1002/mnfr.200600109.

Shanthi, B., Revathy, C., Manjula Devi, A.J. and Subhashree. 2013. Effect of iron deficiency on glycation of haemoglobin in nondiabetics. J. Clin. Diagn. Res. 7(1): 15–7. doi: 10.7860/JCDR/2012/4881.2659.

Shen, X., Xia, L., Liu, L., Jiang, H., Shannahan, J., Du, Y. et al. 2020. Altered clearance of beta-amyloid from the cerebrospinal fluid following subchronic lead exposure in rats: Roles of RAGE and LRP1 in the choroid plexus. J. Trace Elem. Med. Biol. 61: 126520. doi: 10.1016/j.jtemb.2020.126520.

Shobana, N., Aruldhas, M.M., Tochhawng, L., Loganathan, A., Balaji, S., Kumar, M.K. et al. 2017. Transient gestational exposure to drinking water containing excess hexavalent chromium modifies insulin signaling in liver and skeletal muscle of rat progeny. Chem. Biol. Interact. 277: 119–128. doi: 10.1016/j.cbi.2017.09.003.

Skalnaya, M.G., Skalny, A.V. and Tinkov, A.A. 2017. Serum copper, zinc, and iron levels, and markers of carbohydrate metabolism in postmenopausal women with prediabetes and type 2 diabetes mellitus. J. Trace Elem. Med. Biol. 43: 46–51. doi: 10.1016/j.jtemb.2016.11.005.

Skalny, A.V., Aschner, M. and Tinkov, A.A. 2021a. Zinc. Adv. Food Nutr. Res. 96: 251–310. doi: 10.1016/bs.afnr.2021.01.003.

Skalny, A.V., Gluhcheva, Y., Ajsuvakova, O.P., Pavlova, E., Petrova, E., Rashev, P. et al. 2021b. Perinatal and early-life cobalt exposure impairs essential metal metabolism in immature ICR mice. Food Chem. Toxicol. 149: 111973. doi: 10.1016/j.fct.2021.111973.

Skalny, A.V., Aschner, M., Jiang, Y., Gluhcheva, Y.G., Tizabi, Y., Lobinski, R. et al. 2021c. Molecular mechanisms of aluminum neurotoxicity: Update on adverse effects and therapeutic strategies. Adv. Neurotoxicol. 5: 1–34. doi: 10.1016/bs.ant.2020.12.001.

Snow, W.M. and Albensi, B.C. 2016. Neuronal gene targets of NF-κB and their dysregulation in Alzheimer's disease. Front. Mol. Neurosci. 9: 118. doi: 10.3389/fnmol.2016.00118.

Soliman, A.T., De Sanctis, V., Yassin, M. and Soliman, N. 2017. Iron deficiency anemia and glucose metabolism. Acta Biomed. Apr 28; 88(1): 112–118. doi: 10.23750/abm.v88i1.6049.

Song, Y., Liu, Z., Zhu, X., Hao, C., Hao, W., Wu, S. et al. 2022. Metformin alleviates the cognitive impairment caused by aluminum by improving energy metabolism disorders in mice. Biochem Pharmacol. 202: 115140. doi: 10.1016/j.bcp.2022.115140.

Squitti, R., Mendez, A.J., Simonelli, I. and Ricordi, C. 2017. Diabetes and Alzheimer's Disease: Can elevated free copper predict the risk of the disease? J. Alzheimers Dis. 56(3): 1055–1064. doi: 10.3233/JAD-161033.

Stahel, P., Kim, J.J., Cieslar, S.R., Warrington, J.M., Xiao, C. and Cant, J.P. 2017. Supranutritional selenium intake from enriched milk casein impairs hepatic insulin sensitivity via attenuated IRS/PI3K/AKT signaling and decreased PGC-1α expression in male Sprague-Dawley rats. J. Nutr. Biochem. 41: 142–150. doi: 10.1016/j.jnutbio.2016.12.012.

Suhartono, E., Triawanti, T., Leksono and Djati, M.S. 2014a. The role of cadmium in proteins glycation by glucose: formation of methylglyoxal and hydrogen peroxide *in vitro*. J. Med. Biol. Eng. 3(1): 59–62.

Suhartono, E., Leksono, A.S. and Djati, M.S. 2014b. Oxidative stress and kidney glycation in rats exposed cadmium. Int. J. Eng. Res. Appl. 5(6): 497.

Sumi, D. and Ignarro, L.J. 2004. Regulation of inducible nitric oxide synthase expression in advanced glycation end product-stimulated raw 264.7 cells: the role of heme oxygenase-1 and endogenous nitric oxide. Diabetes 53(7): 1841–50. doi: 10.2337/diabetes.53.7.1841.

Sun, W., Yang, J., Wang, W., Hou, J., Cheng, Y., Fu, Y. et al. 2018. The beneficial effects of Zn on Akt-mediated insulin and cell survival signaling pathways in diabetes. J. Trace Elem. Med. Biol. 46: 117–127. doi: 10.1016/j.jtemb.2017.12.005.

Sun, Z.Z., Chen, Z.B., Jiang, H., Li, L.L., Li, E.G. and Xu, Y. 2009. Alteration of Aβ metabolism-related molecules in predementia induced by $AlCl_3$ and D-galactose. AGE (Dordr) 31(4): 277–84. doi: 10.1007/s11357-009-9099-y.

Sung, H.K., Song, E., Jahng, J.W.S., Pantopoulos, K. and Sweeney, G. 2019. Iron induces insulin resistance in cardiomyocytes via regulation of oxidative stress. Sci. Rep. 9(1): 4668. doi: 10.1038/s41598-019-41111-6.

Taccioli, C., Wan, S.G., Liu, C.G., Alder, H., Volinia, S., Farber, J.L. et al. 2009. Zinc replenishment reverses overexpression of the proinflammatory mediator S100A8 and esophageal preneoplasia in the rat. Gastroenterology 136(3): 953–66. doi: 10.1053/j.gastro.2008.11.039.

Takagi, Y., Kashiwagi, A., Tanaka, Y., Asahina, T., Kikkawa, R. and Shigeta, Y. 1995. Significance of fructose-induced protein oxidation and formation of advanced glycation end product. J. Diabetes Complications 9(2): 87–91. doi: 10.1016/1056-8727(94)00022-g.

Takamura, T. 2020. Hepatokine selenoprotein P-mediated reductive stress causes resistance to intracellular signal transduction. Antioxid. Redox Signal. 33(7): 517–524. doi: 10.1089/ars.2020.8087.

Tan, P.Y. and Soma Roy, M. 2021. Dietary copper and selenium are associated with insulin resistance in overweight and obese Malaysian adults. Nutr. Res. 93: 38–47. doi: 10.1016/j.nutres.2021.06.008.

Tanaka, A., Kaneto, H., Miyatsuka, T., Yamamoto, K., Yoshiuchi, K., Yamasaki, Y. et al. 2009. Role of copper ion in the pathogenesis of type 2 diabetes. Endocr. J. 56(5): 699–706. doi: 10.1507/endocrj.k09e-051.

Taneja, S., Vetter, S. and Leclerc, E. 2020. Effect of hypoxia on RAGE signaling in pancreatic cancer. The FASEB Journal 34(S1): 1–1.

Tarwadi, K.V. and Agte, V.V. 2011. Effect of micronutrients on methylglyoxal-mediated *in vitro* glycation of albumin. Biol. Trace Elem. Res. 143(2): 717–25. doi: 10.1007/s12011-010-8915-7.

Tinkov, A.A., Sinitskii, A.I., Popova, E.V., Nemereshina, O.N., Gatiatulina, E.R., Skalnaya, M.G. et al. 2015. Alteration of local adipose tissue trace element homeostasis as a possible mechanism of obesity-related insulin resistance. Med. Hypotheses 85(3): 343–7. doi: 10.1016/j.mehy.2015.06.005.

Tinkov, A.A., Filippini, T., Ajsuvakova, O.P., Aaseth, J., Gluhcheva, Y.G., Ivanova, J.M. et al. 2017. The role of cadmium in obesity and diabetes. Sci. Total Environ. 601-602: 741–755. doi: 10.1016/j.scitotenv.2017.05.224.

Tinkov, A.A., Paoliello, M.M.B., Mazilina, A.N., Skalny, A.V., Martins, A.C., Voskresenskaya, O.N. et al. 2021. Molecular targets of manganese-induced neurotoxicity: a five-year update. Int. J. Mol. Sci. 22(9): 4646. doi: 10.3390/ijms22094646.

Tóbon-Velasco, J.C., Cuevas, E. and Torres-Ramos, M.A. 2014. Receptor for AGEs (RAGE) as mediator of NF-κB pathway activation in neuroinflammation and oxidative stress. CNS Neurol. Disord. Drug Targets 13(9): 1615–26. doi: 10.2174/1871527313666140806144831.

Treviño, S., Waalkes, M.P., Flores Hernández, J.A., León-Chavez, B.A., Aguilar-Alonso, P. and Brambila, E. 2015. Chronic cadmium exposure in rats produces pancreatic impairment and insulin resistance in multiple peripheral tissues. Arch Biochem. Biophys. 583: 27–35. doi: 10.1016/j.abb.2015.07.010.

Tsuchiya, H., Ebata, Y., Sakabe, T., Hama, S., Kogure, K. and Shiota, G. 2013. High-fat, high-fructose diet induces hepatic iron overload via a hepcidin-independent mechanism prior to the onset of liver steatosis and insulin resistance in mice. Metabolism 62(1): 62–9. doi: 10.1016/j.metabol.2012.06.008.

Tung, N.T., Ho, K.F., Niu, X., Sun, J., Shen, Z., Wu, F. et al. 2021. Loss of E-cadherin due to road dust PM2.5 activates the EGFR in human pharyngeal epithelial cells. Environ. Sci. Pollut. Res. Int. 28(38): 53872–53887. doi: 10.1007/s11356-021-14469-2.

Tupe, R.S. and Agte, V.V. 2010. Role of zinc along with ascorbic acid and folic acid during long-term *in vitro* albumin glycation. Br. J. Nutr. 103(3): 370–7. doi: 10.1017/S0007114509991929.

Tzatsos, A. and Tsichlis, P.N. 2007. Energy depletion inhibits phosphatidylinositol 3-kinase/Akt signaling and induces apoptosis via AMP-activated protein kinase-dependent phosphorylation of IRS-1 at Ser-794. J. Biol. Chem. 282(25): 18069–18082. doi: 10.1074/jbc.M610101200.

Varghese, J., James, J., Vaulont, S., Mckie, A. and Jacob, M. 2018. Increased intracellular iron in mouse primary hepatocytes *in vitro* causes activation of the Akt pathway but decreases its response to insulin. Biochim. Biophys Acta Gen. Subj. 1862(9): 1870–1882. doi: 10.1016/j.bbagen.2018.05.022.

Vasudevan, H. and McNeill, J.H. 2007. Chronic cobalt treatment decreases hyperglycemia in streptozotocin-diabetic rats. Biometals 20(2): 129–34. doi: 10.1007/s10534-006-9020-4.

Vincent, J.B. and Brown, S. 2019. Introduction: a history of chromium studies (1955–2007). pp. 1–58. In The Nutritional Biochemistry of Chromium (III). Elsevier.

Vinceti, M., Grioni, S., Alber, D., Consonni, D., Malagoli, C., Agnoli, C. et al. 2015. Toenail selenium and risk of type 2 diabetes: the ORDET cohort study. J. Trace Elem. Med. Biol. 29: 145–50. doi: 10.1016/j.jtemb.2014.07.017.

Vinceti, M., Filippini, T. and Rothman, K.J. 2018. Selenium exposure and the risk of type 2 diabetes: a systematic review and meta-analysis. Eur. J. Epidemiol. 33(9): 789–810. doi: 10.1007/s10654-018-0422-8.

Vlassara, H. and Striker, G.E. 2013. Advanced glycation endproducts in diabetes and diabetic complications. Endocrinol. Metab Clin. North Am. 42(4): 697–719. doi: 10.1016/j.ecl.2013.07.005.

Voziyan, P.A., Khalifah, R.G., Thibaudeau, C., Yildiz, A., Jacob, J., Serianni, A.S. et al. 2003. Modification of proteins *in vitro* by physiological levels of glucose: pyridoxamine inhibits conversion of Amadori intermediate to advanced glycation end-products through binding of redox metal ions. J. Biol. Chem. 278(47): 46616–46624.

Walton, F.S., Harmon, A.W., Paul, D.S., Drobná, Z., Patel, Y.M. and Styblo, M. 2004. Inhibition of insulin-dependent glucose uptake by trivalent arsenicals: possible mechanism of arsenic-induced diabetes. Toxicol. Appl. Pharmacol. 198(3): 424–33. doi: 10.1016/j.taap.2003.10.026.

Wan, W., Cao, L., Kalionis, B., Murthi, P., Xia, S. and Guan, Y. 2019. Iron deposition leads to hyperphosphorylation of tau and disruption of insulin signaling. Front. Neurol. 10: 607. doi: 10.3389/fneur.2019.00607.

Wang, B., Zhang, W., Chen, C., Chen, Y., Xia, F., Wang, N. et al. 2022. Lead exposure and impaired glucose homeostasis in Chinese adults: A repeated measures study with 5 years of follow-up. Ecotoxicol. Environ. Saf. 243: 113953. doi: 10.1016/j.ecoenv.2022.113953.

Wang, C., Gao, X., Santhanam, R.K., Chen, Z., Chen, Y., Xu, L. et al. 2018. Effects of polysaccharides from Inonotus obliquus and its chromium (III) complex on advanced glycation end-products formation, α-amylase, α-glucosidase activity and H_2O_2-induced oxidative damage in hepatic L02 cells. Food Chem. Toxicol. 116(PtB): 335–345. doi: 10.1016/j.fct.2018.04.047.

Wang, C.Y., Xie, J.W., Xu, Y., Wang, T., Cai, J.H., Wang, X. et al. 2013. Trientine reduces BACE1 activity and mitigates amyloidosis via the AGE/RAGE/NF-κB pathway in a transgenic mouse model of Alzheimer's disease. Antioxid. Redox Signal. 19(17): 2024–39. doi: 10.1089/ars.2012.5158.

Wang, J., Zhang, T., Liu, X., Fan, H. and Wei, C. 2019. Aqueous extracts of se-enriched Auricularia auricular attenuates D-galactose-induced cognitive deficits, oxidative stress and neuroinflammation via suppressing RAGE/MAPK/NF-κB pathway. Neurosci. Lett. 704: 106–111. doi: 10.1016/j.neulet.2019.04.002.

Wang, M.Z., Cai, Y.F., Fang, Q.J., Liu, Y.L., Wang, J., Chen, J.X. et al. 2022. Inhibition of ferroptosis of renal tubular cells with total flavones of Abelmoschus manihot alleviates diabetic tubulopathy. Anat. Rec. (Hoboken). doi: 10.1002/ar.25123.

Wang, N., Sheng, Z., Zhou, S., Jiang, F. and Zhang, Z. 2022. Chronic lead exposure exacerbates hepatic glucolipid metabolism disorder and gut microbiota dysbiosis in high-fat-diet mice. Food Chem Toxicol. 170: 113451. doi: 10.1016/j.fct.2022.113451.

Wang, R., Guo, S., Kang, B. and Yang, L. 2023. Toxicogenomic signatures associated with methylmercury induced developmental toxicity in the zebrafish embryos. Chemosphere 313: 137380. doi: 10.1016/j.chemosphere.2022.137380.

Wang, S., Cao, J., Jia, W., Guo, W., Yan, S., Wang, Y. et al. 2020. Single molecule observation of hard–soft-acid–base (HSAB) interaction in engineered Mycobacterium smegmatis porin A (MspA) nanopores. Chem. Sci. 11(3): 879–887.

Wang, W., Zheng, F. and Zhang, A. 2021. Arsenic-induced lung inflammation and fibrosis in a rat model: Contribution of the HMGB1/RAGE, PI3K/AKT, and TGF-β1/SMAD pathways. Toxicol. Appl. Pharmacol. 432: 115757. doi: 10.1016/j.taap.2021.115757.

Wang, X., Wu, W., Zheng, W., Fang, X., Chen, L., Rink, L. et al. 2019. Zinc supplementation improves glycemic control for diabetes prevention and management: a systematic review and meta-analysis of randomized controlled trials. Am. J. Clin. Nutr. 110(1): 76–90. doi: 10.1093/ajcn/nqz041.

Wang, X., Chen, X., Zhou, W., Men, H., Bao, T., Sun, Y. et al. 2022. Ferroptosis is essential for diabetic cardiomyopathy and is prevented by sulforaphane via AMPK/NRF2 pathways. Acta Pharm Sin B. 12(2): 708–722. doi: 10.1016/j.apsb.2021.10.005.

Wang, Y., Lin, M., Gao, X., Pedram, P., Du, J., Vikram, C. et al. 2017. High dietary selenium intake is associated with less insulin resistance in the Newfoundland population. PLoS One 12(4): e0174149. doi: 10.1371/journal.pone.0174149.

Wang, Y., Wang, C.C., Chen, W.L., Wang, G.C., Chiang, S.T., Liaw, F.Y. et al. 2018. The association between metal concentration in human body and serum advanced glycation end-products (ages) among metal workers. Occup. Environ. Med. 75: A136.

Wang, Z.Q., Zhang, X.H., Russell, J.C., Hulver, M. and Cefalu, W.T. 2006. Chromium picolinate enhances skeletal muscle cellular insulin signaling *in vivo* in obese, insulin-resistant JCR:LA-cp rats. J. Nutr. 136(2): 415–20. doi: 10.1093/jn/136.2.415.

Wang, Z., Wang, X., Xiu, W. and Ma, Y. 2022. Characteristics of selenium polysaccharide from sweet corncob and its effects on non-enzymatic glycosylation *in vivo*. Appl. Biol. Chem. 65(1): 10.

Watkins, N.G., Neglia-Fisher, C.I., Dyer, D.G., Thorpe, S.R. and Baynes, J.W. 1987. Effect of phosphate on the kinetics and specificity of glycation of protein. J. Biol. Chem. 262(15): 7207–12.

Wei, C., Wang, J., Duan, C., Fan, H. and Liu, X. 2020. Aqueous extracts of Se-enriched auricularia auricular exhibits antioxidant capacity and attenuate liver damage in high-fat diet/streptozotocin-induced diabetic mice. J. Med. Food. 23(2): 153–160. doi: 10.1089/jmf.2019.4416.

Wei, X., Wei, H., Yang, D., Li, D., Yang, X., He, M. et al. 2018. Effect of aluminum exposure on glucose metabolism and its mechanism in rats. Biol. Trace Elem. Res. 186(2): 450–456. doi: 10.1007/s12011-018-1318-x.

Wei, Y., Liu, D., Zheng, Y., Li, H., Hao, C. and Ouyang, W. 2017. Protective effects of kinetin against aluminum chloride and D-galactose induced cognitive impairment and oxidative damage in mouse. Brain Res. Bull. 134: 262–272. doi: 10.1016/j.brainresbull.2017.08.014.

Wei, Y., Ni, L., Pan, J., Li, X., Deng, Y., Xu, B., Yang, T., Sun, J. and Liu, W. 2023. Methylmercury promotes oxidative stress and autophagy in rat cerebral cortex: Involvement of PI3K/AKT/mTOR or AMPK/TSC2/mTOR pathways and attenuation by N-acetyl-L-cysteine. Neurotoxicol. Teratol. 95: 107137. doi: 10.1016/j.ntt.2022.107137.

Weksler-Zangen, S., Jörns, A., Tarsi-Chen, L., Vernea, F., Aharon-Hananel, G., Saada, A. et al. 2013. Dietary copper supplementation restores β-cell function of Cohen diabetic rats: a link between mitochondrial function and glucose-stimulated insulin secretion. Am. J. Physiol. Endocrinol. Metab. 304(10): E1023–34. doi: 10.1152/ajpendo.00036.2013.

Weng, M.H., Chen, S.Y., Li, Z.Y. and Yen, G.C. 2020. Camellia oil alleviates the progression of Alzheimer's disease in aluminum chloride-treated rats. Free Radic. Biol Med. 152: 411–421. doi: 10.1016/j.freeradbiomed.2020.04.004.

Winter, R. and Piskorska, D. 1984. Effect of low doses of lead and cadmium on glyoxalase I and glyoxalase II activities of the liver and brain of rats. Med. Pr. 35(3): 177–83.

Wisessaowapak, C., Watcharasit, P. and Satayavivad, J. 2021. Arsenic disrupts neuronal insulin signaling through increasing free PI3K-p85 and decreasing PI3K activity. Toxicol. Lett. 349: 40–50. doi: 10.1016/j.toxlet.2021.06.002.

Wlazlo, N., van Greevenbroek, M.M., Ferreira, I., Jansen, E.H., Feskens, E.J., van der Kallen, C.J. et al. 2015. Iron metabolism is prospectively associated with insulin resistance and glucose intolerance over a 7-year follow-up period: the CODAM study. Acta Diabetol. 52(2): 337–48. doi: 10.1007/s00592-014-0646-3.

Wongpun, J., Chanmanee, T., Tocharus, C., Chokchaisiri, R., Chantorn, S., Pabuprapap, W. et al. 2022. The effects of festidinol treatment on the D-galactose and aluminum chloride-induced Alzheimer-like pathology in mouse brain. Phytomedicine 98: 153925. doi: 10.1016/j.phymed.2022.153925.

Xiao, H., Cai, G. and Liu, M. 2007. Fe^{2+}-catalyzed non-enzymatic glycosylation alters collagen conformation during AGE-collagen formation *in vitro*. Arch. Biochem. Biophys. 468(2): 183–92. doi: 10.1016/j.abb.2007.08.035.

Xiong, M., Liu, L., Liu, Z. and Gao, H. 2015. Inhibitory effect of zinc on the advanced glycation end product-induced apoptosis of mouse osteoblastic cells. Mol. Med. Rep. 12(4): 5286–92. doi: 10.3892/mmr.2015.4088.

Xu, T.J., Yuan, B.X. and Zou, Y.M. 2011. Effect of combination of insulin and selenium on insulin signal transduction in cardiac muscle of STZ-induced diabetic rats. Yao Xue Xue Bao 46(3): 274–9.

Xu, Y., Gu, C., Wu, L., Ye, F., Li, W., Li, H. et al. 2022. Intrauterine exposure of mice to arsenite induces abnormal and transgenerational glycometabolism. Chemosphere 294: 133757. doi: 10.1016/j.chemosphere.2022.133757.

Xue, P., Hou, Y., Zhang, Q., Woods, C.G., Yarborough, K., Liu, H. et al. 2011. Prolonged inorganic arsenite exposure suppresses insulin-stimulated AKT S473 phosphorylation and glucose uptake in 3T3-L1 adipocytes: involvement of the adaptive antioxidant response. Biochem. Biophys. Res. Commun. 407(2): 360–5. doi: 10.1016/j.bbrc.2011.03.024.

Yamagishi, S.I. and Matsui, T. 2018. Role of hyperglycemia-induced advanced glycation end product (AGE) accumulation in atherosclerosis. Ann. Vasc. Dis. 11(3): 253–258. doi: 10.3400/avd.ra.18-00070.

Yang, C.Y., Liu, S.H., Su, C.C., Fang, K.M., Yang, T.Y., Liu, J.M. et al. 2022. Methylmercury induces mitochondria- and endoplasmic reticulum stress-dependent pancreatic β-cell apoptosis via an oxidative stress-mediated JNK signaling pathway. Int. J. Mol. Sci. 23(5): 2858. doi: 10.3390/ijms23052858.

Yang, F., Wang, Z., Zhang, J.H., Tang, J., Liu, X., Tan, L. et al. 2015. Receptor for advanced glycation end-product antagonist reduces blood-brain barrier damage after intracerebral hemorrhage. Stroke 46(5): 1328–36. doi: 10.1161/STROKEAHA.114.008336.

Yang, J., Hamid, S., Cai, J., Liu, Q., Xu, S. and Zhang, Z. 2017. Selenium deficiency-induced thioredoxin suppression and thioredoxin knock down disbalanced insulin responsiveness in chicken cardiomyocytes through PI3K/Akt pathway inhibition. Cell Signal. 38: 192–200. doi: 10.1016/j.cellsig.2017.07.012.

Yang, J., Yang, A., Cheng, N., Huang, W., Huang, P., Liu, N. et al. 2020. Sex-specific associations of blood and urinary manganese levels with glucose levels, insulin resistance and kidney function in US adults: National health and nutrition examination survey 2011–2016. Chemosphere 258: 126940. doi: 10.1016/j.chemosphere.2020.126940.

Yang, S.J., Hwang, S.Y., Choi, H.Y., Yoo, H.J., Seo, J.A., Kim, S.G. et al. 2011. Serum selenoprotein P levels in patients with type 2 diabetes and prediabetes: implications for insulin resistance, inflammation, and atherosclerosis. J. Clin. Endocrinol. Metab. 96(8): E1325–9. doi: 10.1210/jc.2011-0620.

Yu, S., Zhang, W., Liu, W., Zhu, W., Guo, R., Wang, Y. et al. 2015. The inhibitory effect of selenium nanoparticles on protein glycation *in vitro*. Nanotechnology 26(14): 145703. doi: 10.1088/0957-4484/26/14/145703.

Yun, S., Wu, Y., Niu, R., Feng, C. and Wang, J. 2019. Effects of lead exposure on brain glucose metabolism and insulin signaling pathway in the hippocampus of rats. Toxicol. Lett. 310: 23–30. doi: 10.1016/j.toxlet.2019.04.011.

Zhang, D., Liu, J., Qi, T., Ge, B., Wang, Z., Jiang, S. et al. 2018. Transcriptome analysis of hepatopancreas from the Cr (VI)-stimulated mantis shrimp (Oratosquilla oratoria) by illumina paired-end sequencing: assembly, annotation, and expression analysis. J. Agric Food Chem. 66(11): 2598–2606. doi: 10.1021/acs.jafc.7b05074.

Zhang, Q., Sun, X., Xiao, X., Zheng, J., Li, M., Yu, M., Ping, F. et al. 2016. Maternal chromium restriction leads to glucose metabolism imbalance in mice offspring through insulin signaling and Wnt signaling pathways. Int. J. Mol. Sci. 17(10): 1767. doi: 10.3390/ijms17101767.

Zhao, Z.H., Du, K.J., Wang, T., Wang, J.Y., Cao, Z.P., Chen, X.M. et al. 2021. Maternal lead exposure impairs offspring learning and memory via decreased GLUT4 membrane translocation. Front. Cell Dev. Biol. 9: 648261. doi: 10.3389/fcell.2021.648261.

Zhou, J., Huang, K. and Lei, X.G. 2013. Selenium and diabetes—evidence from animal studies. Free Radic. Biol. Med. 65: 1548–1556. doi: 10.1016/j.freeradbiomed.2013.07.012.

Zhou, M., Zhao, E. and Huang, R. 2022. Association of urinary arsenic with insulin resistance: Cross-sectional analysis of the National Health and Nutrition Examination Survey, 2015–2016. Ecotoxicol. Environ. Saf. 231: 113218. doi: 10.1016/j.ecoenv.2022.113218.

Zhou, J.C., Zhou, J., Su, L., Huang, K. and Lei, X.G. 2018. Selenium and diabetes. Selenium 317–344.

Zhu, H., Wang, X., Meng, X., Kong, Y., Li, Y., Yang, C. et al. 2022. Selenium supplementation improved cardiac functions by suppressing DNMT2-mediated GPX1 promoter DNA methylation in AGE-induced heart failure. Oxid. Med. Cell Longev. 2022: 5402997. doi: 10.1155/2022/5402997.

Zhuang, X., Pang, X., Zhang, W., Wu, W., Zhao, J., Yang, H. and Qu, W. 2012. Effects of zinc and manganese on advanced glycation end products (AGEs) formation and AGEs-mediated endothelial cell dysfunction. Life Sci. 90(3-4): 131–9. doi: 10.1016/j.lfs.2011.10.025.

Chapter 10

The Importance of Rabenstein's Reaction in the Toxicity of Methylmercury

What is Missing?

João B. T. Rocha,[1,2,*] *Marcelo Farina,*[3] *Pablo A. Nogara*[4] and *Michael Aschner*[5]

1. Introduction

The medicinal and industrial interest by mankind for mercury (Hg) goes back several centuries. Mercury can be ubiquitously found in the environment at different chemical states (mainly Hg^0, Hg^{2+} and organic forms, particularly monomethylmercury – CH_3Hg^+). The majority of Hg found in the biosphere results from anthropogenic sources and is of great concern for human health (Ahlmark, 1948; Clarkson, 1972; Aschner and Aschner, 1990; Clarkson and Magos, 2006; Clarkson et al., 2007; Scheuhammer et al., 2007; Bernhoft, 2012; Lombardi et al., 2012; Cooke et al., 2013; Driscoll et al., 2013; Jaishankar et al., 2014; Rice et al., 2014; Bjørklund et al., 2017; AMAP/UN Environment, 2019; UNEP, 2019; O'Connor et al., 2019; Ganguly et al., 2022; Schartup, 2022; Basu et al., 2023; Crespo-Lopez et al., 2023; Mason et al., 2023; Sonke et al., 2023). Its toxicity will vary depending on the chemical form considered, as well as on the intensity and the period/duration of the exposure (developmental or mature/acute or chronic). Concerning the chemical

[1] Departamento de Bioquímica, ICBS, Universidade Federal do Rio Grande do Sul, Porto Alegre, RS, Brazil.
[2] Departamento de Bioquímica e Biologia Molecular, CCNE, Universidade Federal de Santa Maria, RS, Brazil.
[3] Departamento de Bioquímica, ICB, Universidade Federal de Santa Catarina, Florianópolis, SC, Brazil.
[4] Instituto Federal de Educao, Ciencia e Tecnologia Sul-rio-grandense (IFSul), Bagé, RS, Brazil.
[5] Department of Molecular Pharmacology, Albert Einstein College of Medicine, Bronx, NY 10461, USA.
* Corresponding author: jbtrocha@yahoo.com.br

forms, the predominant electrophilic forms of mercury (Hg^{2+} and CH_3Hg^+) in the biosphere have strong affinity for thiol groups found in a diversity of biomolecules (for instance, thousands of thiol-containing proteins). Consequently, the toxicity of electrophilic forms of mercury (E^+Hg) will involve the inactivation or disruption of thiol-containing molecules. In addition to thiol-containing molecules, Hg^{2+} and CH_3Hg^+ (also known as $MeHg^+$) have an extraordinary affinity for a class of rare proteins containing the selenol (-SeH) group (the selenoproteins). The selenol group is an analogue of the thiol (-SH) group found in thiol-containing proteins (Carvalho et al., 2008; Oliveira et al., 2017; Branco and Carvalho, 2019; Nogara et al., 2019; Ajsuvakova et al., 2020).

In this chapter, we will discuss the importance of a particular type of exchange reaction between $MeHg^+$ and the -SH or -SeH groups in the disposition and toxicity of E^+Hg in human body. Indeed, despite of the strong affinity of E^+Hg for -SH or -SeH groups, the complexed MeHg-S(e)-R can easily be dislocated by a free -SH or -SeH group. Here, we will use the notation -S(e)- to represent both the sulfide- or selenide bond to Hg and R- to represents an organic moiety, for instance, cysteine or cysteine containing-proteins. Note that the chalcogen atom (S or Se) in the chalcogen-mercury bond are derived from a free -S(e)H group. For simplicity we will refer to LMM-SH (low molecular mass thiol-containing biomolecules) to denote mainly reduced cysteine (Cys) and glutathione (GSH), whereas HMM-SH (high molecular mass thiol-containing biomolecules) will refer to thiol-containing proteins. In view of the stronger affinity of Hg^{2+} and $MeHg^+$ for the -SeH groups found in a few selenoproteins, we will also refer to HMM-SeH (selenoproteins). The expression HMM-S(e)H denotes thiol- or selenol-containing proteins. The -SeH group is much more reactive than the equivalent -SH groups and, consequently, in contrast to -SH, the -SeH cannot be found in stable low molecular mass-selenol-containing molecules (Barbosa et al., 2017; Nogara et al., 2023).

The exchange or the displacement of a complexed MeHg-S(e)-R by a free -SH or -SeH group was first described by Prof. Dallas L. Rabenstein, while working in the Department of Chemistry of Alberta University, Canada in the early years of the 70ties. This type of exchange reactions occur very rapidly, though it may seem to be paradoxical. The puzzle is how can a free thiol or selenol group interact with a complex of MeHg-S(e)-R that is thermodynamically very stable and break the -S(e)-Hg bond? Importantly, Prof. Rabenstein determined the kinetics of different exchanges reactions between several R-S(e)-HgMe and R-S(e)H molecules and found that they are diffusion controlled. In view of the importance of this type of reaction in the toxicity and mobility of E^+Hg, we have named it as the Rabenstein's reaction in honor of Prof. Dallas L. Rabenstein (Rabenstein and Fairhurst, 1975; Rabenstein, 1978a,b; Rabenstein and Evans, 1978; Nogara et al., 2019; Ajsuvakova et al., 2020).

Another biogeochemically relevant form of mercury is elemental mercury (Hg^0), which is much less reactive than Hg^{2+} and $MeHg^+$. Despite its low chemical reactivity, Hg^0 can easily be transported in the biosphere and metabolized to the toxic Hg^{2+} in mammalian tissues. In fact, inorganic elemental Hg (Hg^0) can enter the human body via lungs and can be distribute to several tissues. In the blood or inside different types of tissues, Hg^0 can be oxidized to the toxic Hg^{2+} by catalase and by still unknown processes. It is important to emphasize that the majority (if not all) of the toxicity of Hg^0 is mediated by its metabolism to Hg^{2+}, and the differences in the toxicity of Hg^0 and Hg^{2+} are accounted for by differences in their mobility and distribution in the human body. Basically, the Hg^0 enters more easily in several tissues (for instance, the brain) than Hg^{2+}, where it is metabolized to Hg^{2+} and trapped. A rare case of acute renal failure was reported in the literature after Hg^0 swallowing by an individual; however, the critical route of exposure was probably the lungs, because the subject had also high blood Hg levels and inflammation of the lungs (pneumonitis, a common symptom after Hg^0 exposure) (Zalups and Lash, 1994; Zalups, 2000; Clarkson and Magos, 2006; Clarkson et al., 2007; Bernhoft, 2012; Park and Zheng, 2012; Katsuma et al., 2014).

After chronic exposure to Hg^0, the most affected organs are the kidney, followed by the brain (where Hg causes symptoms such as erethism, memory loss, tremor, impaired vision, motor incoordination, etc.) (Calabrese et al., 2018; Evans et al., 2018; Johnson-Arbor et al., 2021). In the 20th century, calomel (Hg^+ or Hg_2Cl_2) was used as teething powders and anthelmintics in children and normally was associated with a syndrome named acrodynia or Pink disease (the syndrome was associated with redness of the skin or erythroderma). In the middle of the 20th, the syndrome was considered an idiosyncratic response to mercury (for a historical minireview see Bjørklund, 1995). Although nowadays the Hg-associated acrodynia is not common, there are several recent case reports of children exposure to Hg^0 in their home environment (Mercer et al., 2012; Atti et al., 2020). Of clinical importance, the diagnosis of Hg intoxication can be frequently delayed by non-specific signs or confounding the symptoms with infectious illness (Atti et al., 2020; Young et al., 2020; Johnson-Arbor et al., 2021).

After high levels of acute exposure to Hg^0, the lungs are most damaged, exhibiting severe inflammation (pneumonitis) followed by massive fibrosis and eventually death. The most common symptoms are those associated with typical pneumonitis, i.e., shortness of the breath or dyspnea and cough (Zalups and Lash, 1994; Park and Zheng, 2012; Bernhoft, 2012; Farina et al., 2024).

Hg^{2+} is toxic primarily to renal tubules (proximal convoluted renal tubules), where Hg^{2+} accumulates largely in proximal tubular epithelial cells. High levels of exposure to Hg^{2+} may cause irreversible renal failure and death. Upon chronic exposure, the tubular lesion cause proteinuria, vacuolation of epithelial cells and cellular degeneration. The renal tropism of Hg^{2+} is related to the presence of transporters of Hg^{2+}-cysteine complexes in the kidneys (for a comprehensive and elegant review see Zalups and Lash, 1994; Zalups, 2000; Clarkson and Magos, 2006; Bernhoft, 2012; Bridges and Zalups, 2017). Acute and chronic exposure to low levels of Hg^{2+} can be toxic to the cardiovascular system and can also result in its accumulation in other tissues, such as brain (Takahata et al., 1970; Lemos et al., 2012; Park and Zheng, 2012; Furieri et al., 2011; Rizzetti et al., 2017; Ferreira et al., 2022; Ganguly et al., 2022; Unoki et al., 2022; Gao et al., 2023; Naija and Yalcin, 2023). The neurological effects of Hg^{2+} are usually less frequent than that of Hg^0 (likely due to the low entrance of Hg^{2+} into the brain), but the symptoms of exposure to both forms of Hg can overlap to some extent (Clarkson and Magos, 2006; Bernhoft, 2012). Exposure to inorganic mercury forms (Hg^0, Hg^{2+}) has been associated with autoimmunity (for review see Bjørklund et al., 2020).

Dimethylmercury (CH_3HgCH_3) and monomethylmercury (CH_3Hg^+ or $MeHg^+$) are extremely neurotoxic agents (Nierenberg et al., 1998; Aaseth et al., 2020; Wunder and Bhati, 2021; Woolf, 2022; Farina et al., 2024); however, the existence of CH_3HgCH_3 in the environment is very low in view of its chemical instability. In contrast to CH_3HgCH_3, CH_3Hg^+ can be found in living parts of the aquatic environment and at high concentrations in the meat of predatory fish (ppm level) and aquatic mammals, but is not free, it is bounded to LMM-SH and HMM-SH. In fact, CH_3Hg^+ is formed from the methylation of mercuric mercury inside bacterial cells and bioaccumulates along the aquatic web food chain. Though $MeHg^+$ can accumulate in various tissues (liver, kidneys, brain, etc.), the central nervous system is considered its main. The developing mammalian brain is highly susceptible to even non-toxic maternal doses of $MeHg^+$ and exposure to even low levels during gestation and lactation can have long-lasting neurocognitive effects in humans (Ahlmark, 1948; Clarkson, 1972; Clarkson and Magos, 2006; Clarkson et al., 2007; Farina et al., 2011, 2024). A summary of main target organs of Hg^0, Hg^{2+} and CH_3Hg^+ is presented in Figure 1. The schematic molecular targets of $MeHg^+$ (and other electrophilic mercury forms) are depicted in Figure 2. As briefly commented above, $MeHg^+$ has a high affinity for -SH and -SeH found in specific proteins and in a few LMM-SH, for instance, GSH and Cys. $MeHg^+$ has even a greater affinity for inorganic analogues of the thiol or selenol groups (in Figure 2 they are represented by H_2S and H_2Se for simplicity).

Mercury form	Acute	Chronic
Hg^0	lungs	lungs, kidneys, brain
Hg^{2+}	kidneys	kidneys, cardiovascular
CH_3Hg^+	brain	brain, cardiovascular

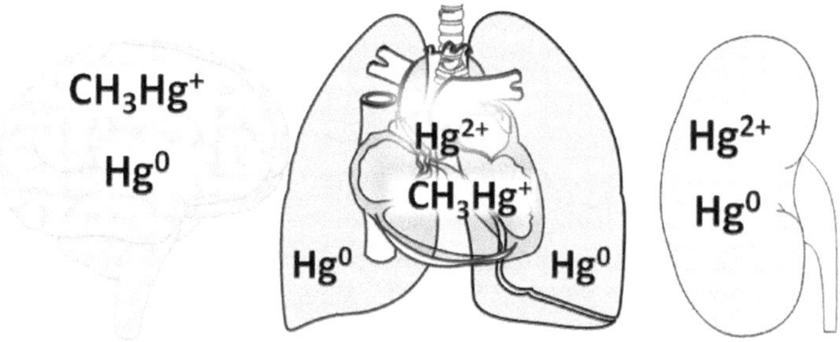

Figure 1. Schematic representation of target organs of mercury. Methylmercury (CH_3Hg^+) is chiefly neurotoxic but can also be associated with cardiovascular toxicity. Hg^0 is nephrotoxic and neurotoxic after chronic exposure. At high levels of exposure, Hg^0 can cause pneumonitis and even death by its pulmonary toxicity. Hg^{2+} is primarily toxic to the kidneys and heart (Houston, 2011; Park and Zheng, 2012; Bernhoft, 2012; Rice et al., 2014; Bjørklund et al., 2017; Ganguly et al., 2022; Farina et al., 2024).

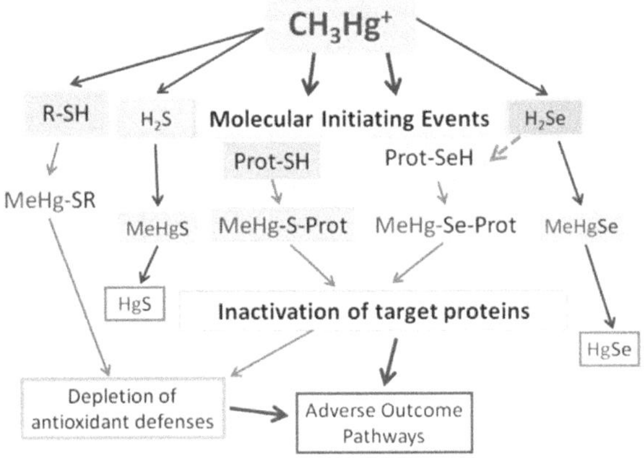

Figure 2. Schematic representation of $MeHg^+$ (CH_3Hg^+) interaction with its potential physiological targets. Adapted from Oliveira et al., 2017. The molecular initiating events (MIEs) of electrophilic mercury forms ($MeHg^+$ and Hg^{2+}) toxicity are mediated by the interaction of $MeHg^+$ with biomolecules containing thiol (-SH) or selenol (-SeH) groups. The knowledge about the specific molecules involved in the initiation of $MeHg^+$ toxicity is still elusive but probably involve the interaction of $MeHg^+$ (or Hg^{2+}) with several HMM-S(e)H (Prot-S(e)H in the figure), disruption of LMM-SH pool and H_2S and H_2Se. The interaction of mercury with these components is expected to activate adverse outcome pathways (AOPs) that ultimately will lead to neurotoxic effects in mammals (FitzGerald et al., 2022). Prot-SH and Prot-SeH are HMM-SH and HMM-SeH, respectively.

2. Molecular targets of electrophilic mercury forms

It is common to treat the cationic forms of mercury (or electrophilic mercury forms, e.g., $MeHg^+$ and Hg^{2+}) as the responsible for their toxicity; however, they have such strong affinity for -SH groups found in Cys (which can be found at mmolar range in reduced GSH and in thousands of proteins) that their occurrence as a free cationic atom or molecule are negligible inside living cells (Hughes, 1957; Rabenstein and Fairhurst, 1975; Rabenstein, 1978a,b; Rabenstein and Evans, 1978; Rabenstein and Reid, 1984; Rabenstein et al., 1982, 1983, 1986; Cheesman et al., 1988; Nogara et al., 2019).

In fact, the absorption, distribution, deposition, and toxicity of Hg in the human body will be dictated by the formation of complexes with thiol groups from free circulating Cys, GSH and thiol-containing proteins (for instance, albumin and hemoglobin) (Rabenstein, 1978; Nogara et al., 2019; Ajsuvakova et al., 2020). In view of its high affinity for thiol groups, in the meat of fish (fish muscle), $MeHg^+$ is found predominantly bound to different thiol-containing proteins (Figure 3 depicts schematically $MeHg^+$ in fish meat forming complexes with both actin, myosin and tropomyosin cysteinyl residues). After digestion in the human digestive tract, the CH_3Hg^+ is released as a complex with cysteine (Cys-S-HgMe complex). Here it is important to highlight that the processes depicted in Figure 3 are those involved in the transfer of $MeHg^+$ from contaminated fish meat to fish-eating vertebrates (including the human population). Indeed, in Figure 3 we are emphasizing the transport of MeHg-S-Cysteine complex, which can be transported by the neutral amino acid

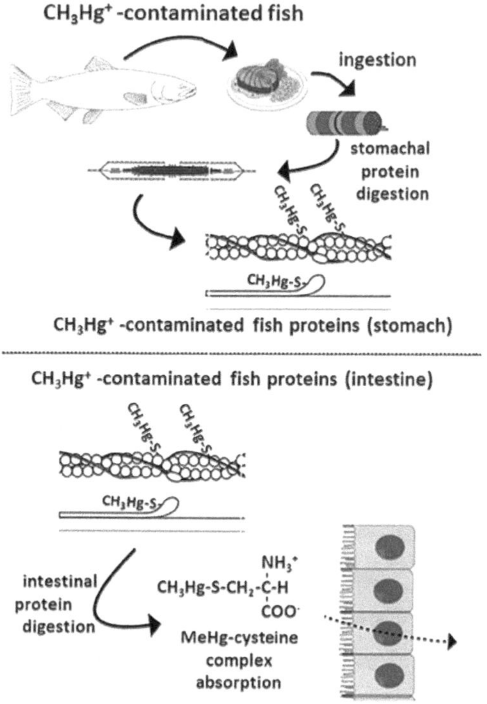

Figure 3. Schematic representation of MeHg-contaminated fish digestion. The MeHg found in the meat of contaminated fish is bound to thiol-containing proteins. In the figure, the myofibril, the contractile and regulatory proteins are indicated. MeHg is depicted bound to the myosin, actin, and tropomyosin. After digestion, the contaminated proteins will release the MeHg-Cysteine complex. The complex is then taken up by the enterocytes (represented by the dotted arrow).

transport as a mimic of methionine (Aschner and Clarkson, 1988a,b, 1989; Kerper et al., 1992; Simmons-Willis et al., 2002; Yin et al., 2008).

In view of the strong affinity of $MeHg^+$ for thiol- and selenol-groups no free or only negligible concentrations of free $MeHg^+$ are found in the intestinal tract or inside any human intracellular or extracellular media. The predominance of MeHg-S-R (where R- can be either Cys, GSH, lipoic acid, and any other LMM-SH or HMM-SH over MeHg-Se-R) are a consequence of the much higher relative concentration of thiol- over the selenol-groups containing molecules (see below in Section 3 for more details about their concentrations). In fact, the intestinal tract is expected to contain both LMM-SH or HMM-SH molecules but very low or negligible concentrations of either LMM-SeH molecules (if any) and very small quantities of HMM-SeH (selenoproteins).

The easy formation of MeHg-Cysteine complex is expected to facilitate the uptake of $MeHg^+$ either by the participation of transporter proteins or by passive diffusion. Accordingly, the octanol:water partition coefficient (K_{ow}) of the complex Cys-S-$HgCH_3$ is about 50 and much greater than that of CH_3HgCl. Theoretically, the cysteine-MeHg complex could be partitioned in the plasma membrane bilayer, get transported inside the enterocyte by passive diffusion (Figure 4). However, $MeHg^+$

in the chemical form of CH_3HgCl has a K_{ow} value of 1.7 in water and approximately 2.5 in physiological solution (Halbach, 1985; Mason et al., 1995, 1996, 2022; Benoit et al., 1999; Mason, 2002). Consequently, the passive diffusion of free $MeHg^+$ through biological membranes is not thermodynamically favorable, but frequently mentioned in the literature. The formation of mercury complexes with inorganic analogue of thiol (H_2S) also increase the octanol-water partition coefficient but the coefficient also varies depending on the pH of the media used (Benoit et al., 1999; Mason, 2002). However, the solubility of HgS in aqueous media is very low, thus decreasing the capacity of living cells to absorb the complex.

Figure 4 also indicates that the MeHg-S-Cys complex can be taken up by a transporter (for instance, the L-type amino acid transporter 1 or LAT1) (Aschner and Clarkson, 1988a,b, 1989; Kerper et al., 1992; Simmons-Willis et al., 2002; Yin et al., 2008). The uptake of MeHg-S-Cys complex have been interpreted as molecular mimicry, where the transporter would originally transport methionine amino acid. In fact, methionine can compete with the complex Cys-S-HgMe and decrease the $MeHg^+$ uptake (Aschner, 1989; Kerper et al., 1992; Clarkson, 1993; Zalups and Lash, 1994, 2006; Mokrzan et al., 1995; Zalups, 2000; Bridges et al., 2007; Roos et al., 2011; Zimmermann et al., 2013). Accordingly, the literature has identified different transporter proteins (e.g., LAT1 and LAT2) (Heggland et al., 2009; Mokrzan et al., 1995; Simmons-Willis et al., 2002; Yin et al., 2008; Zimmermann et al., 2013) that can be involved in $MeHg^+$ uptake in different mammalian cell types. However, the contribution of transporters to $MeHg^+$ uptake is only partial, and additional mechanisms must be involved in mercury transport. A representation of three pathways by which $MeHg^+$ can enter enterocytes is summarized in Figure 4.

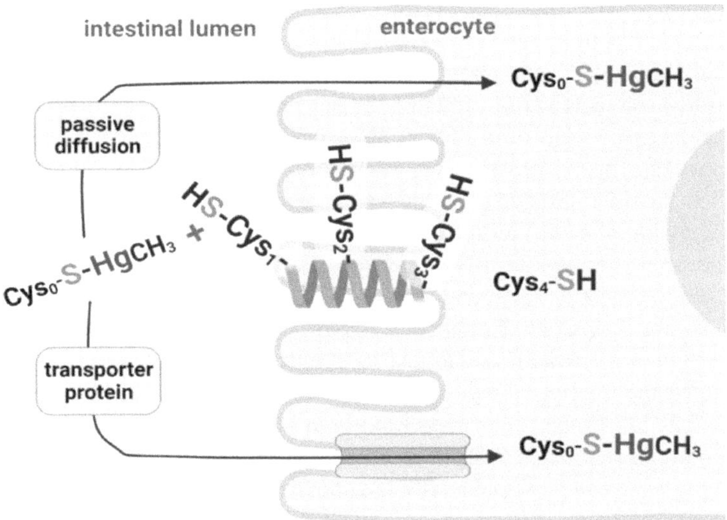

Figure 4. The three main hypothetical mechanisms by which $MeHg^+$ (CH_3Hg^+) can be absorbed in the intestines. One involves passive diffusion through the plasma membrane, the second involves the uptake by amino acid transporters (see text for references) and the third involves a succession of exchange reactions (Rabenstein's Reactions, the third mechanism, is detailed in Figure 5).

In addition to these two processes discussed above, there is a third possibility that will involve approximately 4 or more Rabenstein's Reactions (Figure 5). The critical element in this hypothetical uptake of CH_3Hg^+ is the presence of a transmembrane protein containing at least 3 residues of cysteine (cysteinyl) with free -SH group. The first Rabenstein's reaction will occur at the lumen of the intestine, where the $MeHg^+$ complexed to Cys (from a fish muscle derived source) is moved to the first cysteinyl residue in the membrane protein (first step, Figure 5).

Next, the MeHg complexed with the Cys_1 in the transmembrane protein, exchanges (migrates) from Cys_1 to Cys_2 inside the protein and the plasma membrane (second step or second Rabenstein's Reaction). The last step inside the protein occurs, then by the migration of the MeHg from the Cys_2 to Cys_3, where the cysteinyl residue Cys_3 faces the interior of the cell. Finally, the CH_3Hg^+ complexed in the Cys_3-MeHg can exchange (move) from Cys_3 to a free cysteine residue inside the cytoplasm (Cys_4 in the 4th step or the nth Rabenstein's Reaction). The Cys_4-MeHg can move inside the cell and can make different types of Rabenstein's reactions, for instance, with different HMM-SH (thiol-containing proteins) or with Cys, GSH, etc.

It is important to emphasize that the global process of $MeHg^+$ distribution in the body will depend on a different series of Rabenstein's reactions of the type presented in the Figure 5. However, unfortunately, we have little knowledge on the dominant types of Rabenstein's reactions that occur after the exposure to low or even moderate and heavy intoxication with $MeHg^+$ (and with other electrophilic mercury forms such Hg^{2+} and ethylmercury or $EtHg^+$) (Farina et al., 2011; Dórea et al., 2013).

Specifically, albumin and hemoglobin can form complexes with electrophilic mercurial, but these proteins do not have their main physiological roles disrupted by Hg forms. Thus, albumin and hemoglobin are temporary carriers of electrophilic Hg through the body. The potential toxic targets of E^+Hg will be proteins that will form complexes with either Hg^{2+} or CH_3Hg^+ and will undergo permanent change in structure and will lose their biological function (denaturation). There are 3 possibilities of inactivation of thiol- or selenol-containing proteins (Figure 6): (1) the binding of $MeHg^+$ to a critical -S(e)H will produce a drastic change in the tertiary structure of the protein, loss of its biochemical function and renders the $MeHg^+$ inaccessible for participating in an exchange or Rabenstein's reaction; (2) the formation of a sulfide- or selenide-Hg can facilitate the elimination of the S or Se chalcogen atom and forming a dehydroalanine in the place of the cysteinyl or selenocysteinyl residue (Khan and Wang, 2010; Asaduzzaman and Schreckenbach, 2011; Pickering et al., 2020; Nogara et al., 2021). If this residue is critically involved in the function of the protein, it will lose its activity; (3) the transitory loss of protein function after the binding of $MeHg^+$ to a critical -S(e)H that can be reversed by an exchange reaction (Rabenstein's reaction) with a free -SH. Figure 6 also depicts a group of proteins that bind $MeHg^+$ but are not inactivated by the interaction.

Theoretically, any protein containing either the -SH or the - SeH groups that will have their physiological role compromised after forming a complex with $MeHg^+$ (or Hg^{2+}) may be considered as one of the Molecular Initiating Events, or MIEs, involved in the toxicity elicited by E^+Hg. In fact, Figure 2 depicts a schematic representation of different class of molecules that can be the primary targets of $MeHg^+$ (i.e., the

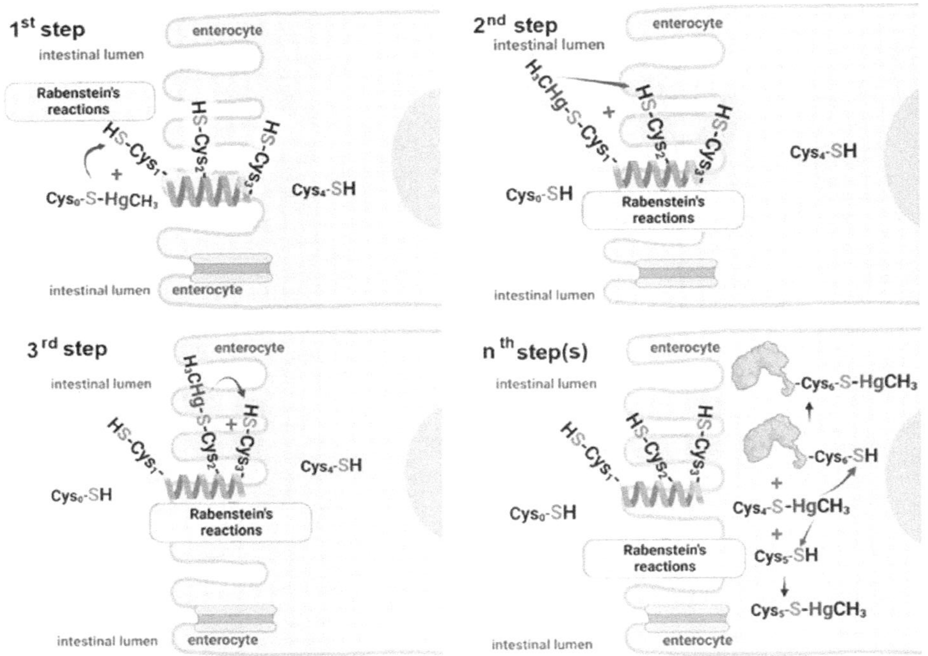

Figure 5. Methylmercury uptake by successive Rabenstein's reactions through a transmembrane alpha-helix containing 3 critical cysteinyl residues with a particular and defined topological distribution along the plasma membrane bilayer. Basically, the MeHg-Cys complex released from a MeHg-contaminated fish protein is exchanged with an external -SH group (Cys_1). Then the MeHg migrates from Cys_1 to Cys_2 in the middle of the alpha-helix. Subsequently, the MeHg will migrate to the last cysteinyl residue in the transmembrane alpha-helix (Cys_3). The MeHg-Cys_3 complex then exchange with a free cysteine (Cys_4 in the figure, which could be a cysteine from GSH or from an HMM-SH). The exchange can also occur with another free intracellular cysteine: for instance, indicated by Cys_5-SH or with an intracellular soluble protein containing a free cysteinyl residue (indicated by the residue Cys_6-SH in the cytoplasm of the enterocyte). Then the LMM-S-HgMe complex can diffuse inside cell and exchange with other LMM-SH molecules or with different HMM-S(e)H, represented by Cys_5 or the cysteinyl Cys_6 in a protein. The 4th step is called the nth step to indicate that the possibilities of exchanges with different types of thiol or selenol groups is very high. The arrows are used to indicate the movement of MeHg- from the intestinal lumen to the cytosol of the enterocyte (and not as arrows are used in conventional organic chemistry to represent the nucleophilic attack).

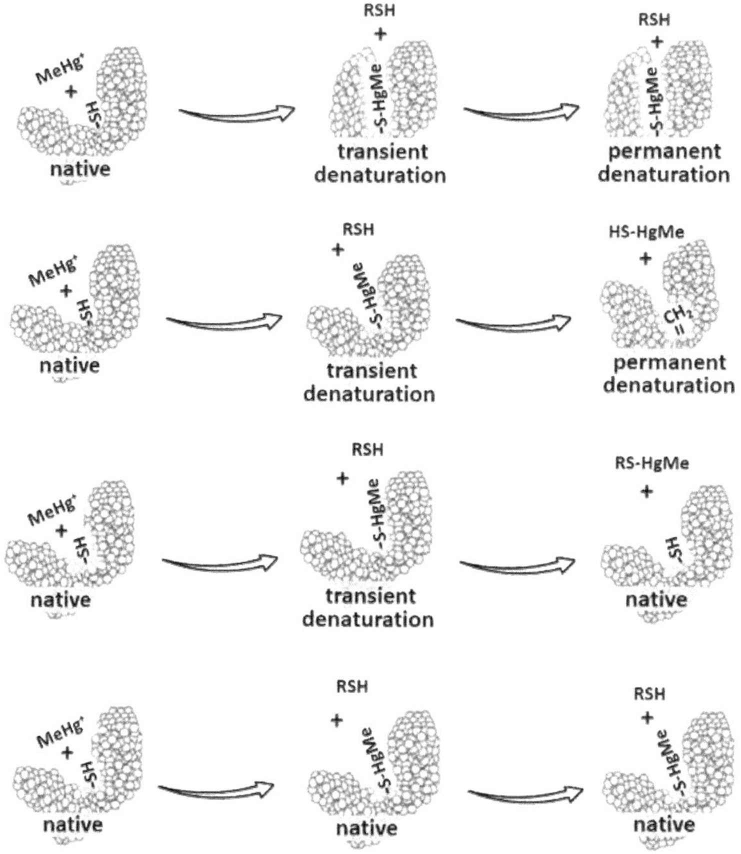

Figure 6. Interaction of $MeHg^+$ with thiol-containing proteins (the interaction can also be with selenoproteins). The interaction can cause a denaturation of the protein that can be permanent (the two first lines, in the first line the protein suffered a drastic change in tertiary structure and it rendered the -S-HgMe cysteinyl residue inaccessible to be reactivated; in the second line, the binding of MeHg facilitated the elimination of the S atom from the cysteinyl residue) or transitory (the third line, where the protein is renatured by a free thiol). The binding of MeHg to a thiol group can also not modify the protein function (4th line). Native indicates a functional, whereas denatured (denaturation) a non-functional protein.

potential MIEs) and how their inactivation by electrophilic mercury (E^+Hg) forms can result in the activation of the Adverse Outcome Pathways associated with Hg toxicity. It is important to highlight that our precise knowledge about the MIEs is still not well defined (FitzGerald et al., 2022).

Hg^0 is the less reactive form of Hg, whereas Hg^{2+}, which can be formed from oxidation of Hg^0, can easily coordinate with 2 free -SH groups (particularly with Cys and GSH). $MeHg^+$ coordinates with one free thiol group and after binding to Cys it can mimicry the amino acid methionine to get access to different tissues. The affinity of Hg^{2+} for thiol groups is higher than that of $MeHg^+$, but the formation constant of the MeHg-SR (where RS can be derived from organic thiol containing molecules, such as Cys, GSH and HMM-SH or proteins) is nearly diffusion controlled (i.e., the velocity of the reaction will depend on the velocity of diffusion of the reactants)

(Rabenstein and Fairhurst, 1975; Rabenstein, 1978a,b; Rabenstein and Evans, 1978; Benoit et al., 1999). Since the overall biological chemistry of E^+Hg will be influenced by both the affinity constants and concentration of reactants (Hg forms, -SH, -SeH, -Hg-S and -Hg-Se-, etc.), more details about the occurrence and abundance of -SH and SeH groups in biomolecules (and their approximate affinities for mercurials) will be discussed in the following sections of this chapter.

3. The relative concentrations of thiol/thiolate and selenol/selenolate in the body: thiol groups as the most abundant functional target groups for methylmercury ($MeHg^+$)

In relation to the $MeHg^+$ targets schematically depicted in Figure 2, the thiol (-SH) groups found both in LMM-SH molecules (e.g., Cys, GSH, lipoic acid, coenzyme A, etc.) and thiol-containing proteins (HMM-SH) are by far the most abundant examples. On average, the concentrations of -SH groups in biological fluids are in the millimolar or mmol/L range. A more detailed discussion about their occurrence in the human body will be presented in the following sections with the intention of providing the reader with greater details on the complexity of the chemistry of $MeHg^+$. Emphasis will be directed to LMM-SH and HMM-S(e)H molecules and their interactions with CH_3Hg^+ in mammalian systems.

Indeed, the complex chemistry of $MeHg^+$ is governed by the Rabenstein's reactions, involving both thiol and selenol groups. Two main points must be highlighted here: (1) in general the -SeH (or H_2Se) has a higher affinity for the mercury atom from MeHg-S(e)-R complexes than the analogues -SH (or H_2S) (Arnold et al., 1986; Nogara et al., 2023); (2) the concentration of free -SH groups in the human body is much higher than that of the -SeH (approximately 10^4 to 10^5 times greater). Consequently, even in view of the lower affinity of -SH (about 10 to 100 lower) to MeHg-S(e)-R complexes than an analogue -SeH, the higher concentration of -SH over -SeH will render the molecules containing -SH as the most probable centers of interaction with $MeHg^+$ *in vivo*. As commented above, one limitation in the study of $MeHg^+$ toxicity is the difficulty to identify the primary targets or the MIEs that will trigger the activation of the adverse outcome pathways involved in the toxicity and neurotoxicity of $MeHg^+$.

In relation to the inorganic analogues of -S(e)H, i.e., H_2S and H_2Se, the differences in the affinity for E^+Hg for H_2S and H_2Se are higher (at least for Hg^{2+}). For instance, Dyrssen and Wedborg (1991) calculated a constant of formation of HgS and HgSe of 10^{39} and 10^{45}, respectively. The difference between selenium and sulfur with inorganic Hg is much higher than that observed by Arnold et al. (1986) with $MeHg^+$ and organic -SH and -SeH.

The concentration of thiol groups in the plasma is lower than that found intracellularly in various tissues (see Table 1). In fact, the most abundant thiol containing molecule in the plasma is albumin (~ 0.4–0.8 mmol of -SH/L), whereas the plasmatic concentrations of GSH and cysteine are around 0.1 mmol/L or less. In erythrocytes, the most abundant containing thiol protein is hemoglobin

Table 1. Estimated total concentration of free thiol or selenol [-S(e)H] groups found in proteins (HMM-S(e)H) and LMM-SH intra- and extracellularly in humans. They represent the potential targets for electrophilic mercury forms, including methylmercury ($MeHg^+$) (N/A-not applicable).

Groups	Molecular Mass	[Intracellular] mmol/L [mM]	[extracellular] mmol/L [mM]	Number codons for Cys/Sec residues	Different types of proteins
-SH	Low (LMM-SH)	1–10	$5–25 \times 10^{-2}$	N/A	N/A
	High (HMM-SH)	5–20	0.2–0.7	$\sim 1.0–1.5 \times 10^5$	$\sim 1–1.5 \times 10^4$
-SeH	Low (LMM-SeH)	0.0	0.0	N/A	N/A
	High (HMM-SeH)	$10^{-5}–10^{-3}$	$10^{-5}–10^{-3}$	35	25

(~ 10–15 mmol -SH/L). The thiol concentration of high molecular mass thiol molecules (HMM-SH, i.e., in proteins) varies depending on the tissue considered, but is always in the millimolar range (mmol/L). The intracellular concentration of GSH (the most abundant LMM-SH in mammalian cells) varies depending on the tissues considered from about 1 to 10 mmol/L (Turell et al., 2013; Nogara et al., 2023). Differing from the LMM-SH, which are stable in physiological media (e.g., plasma, intracellular medium, interstitial fluids, etc.) (Table 1), the LMM-SeH molecules are extremely reactive in the presence of oxygen and water and are easily oxidized to their diselenides (Barbosa et al., 2017; Nogara et al., 2023). Indeed, the fact that selenocysteine (Sec) is synthesized at the level of its t-RNA and is not found as free selenocysteine in the plasma further support the notion of chemical instability of LMM-SeH molecules (Nogara et al., 2023) (Table 1).

In contrast to HMM-SH, which can be found in thousands of different types of proteins (Table 1), the diversity and quantity of -SeH (selenol)-containing molecules (HMM-SeH) are much smaller and found only in 25 types of selenoproteins in humans (Labunskyy et al., 2014; Oliveira et al., 2023). The concentrations of selenoproteins are in the pmol/L to μmol/L range depending on the type of selenoprotein and the tissue considered. The selenoprotein P (SELENOP) is one of the most concentrated selenoprotein in human fluids (about 0.04–0.1 μmol/L) and it is synthesized in the liver and secreted to the plasma. SELENOP is the only selenoprotein with about 10 residues of selenocysteine (selenocysteinyl) and is involved in the transport of selenium to various tissues, including the brain (Labunskyy et al., 2014; Oliveira et al., 2023). The concentration of the other 24 selenoproteins are lower than that of SELENOP, but several of them are essential in the maintenance of the adequate redox state of the cells (for example, glutathione peroxidase 1 and 4, thioredoxin reductase, etc.) (for comprehensive review see Labunskyy et al., 2014). Thus, the inhibition of such important selenoenzymes by $MeHg^+$ can be involved in its toxicity and both direct and indirect interaction of $MeHg^+$ with selenoproteins have been proposed to be involved in the adverse outcome pathways activated by $MeHg^+$ (Farina et al., 2009; Franco et al., 2009; Peterson et al., 2009; Wagner et al., 2010; Oliveira et al., 2017; Ralston and Raymond, 2018; Carvalho et al., 2008; Branco et al., 2014; Branco and Carvalho, 2019).

Of note, the disruption of selenoproteins' function or synthesis by $MeHg^+$ can contribute decisively to the neurotoxicity of $MeHg^+$. To some extent, the interference of $MeHg^+$ with the normal biochemistry of selenoproteins can be comparable with the neurologic dysfunction observed in mutant and developing mice with deletion of SELENOP, a selenoprotein involved in selenium (Se) transport and homeostasis (particularly to the brain) (Caito et al., 2011). In contrast to selenoproteins, which are very rare and usually only have catalytic activity, HMM-SH or thiol-containing proteins have diverse catalytic and structural functions. For instance, the disulfide bonds (-S-S-) are critical in the stabilization of a variety of proteins. Furthermore, the transition from reduced (-SH) to the oxidized forms (SOH or sulfenic acid, SO_2H or sulfinic acid and -S-S- or disulfide) can serve as redox sensitive regulatory sites in proteins (Fomenko et al., 2008; Turell et al., 2009; Trujillo et al., 2016; Oliveira et al., 2018).

The direct interaction of $MeHg^+$ with selenoenzymes has been reported to inhibit glutathione peroxidase (GPX) and thioredoxin reductase (TXNRD) activities (Branco and Carvalho, 2019; Carvalho et al., 2008; Wagner et al., 2010; Franco et al., 2009). As commented in previous sections, another plausible mechanism of $MeHg^+$ toxicity may involve the depletion of the inorganic Se pool (specifically H_2Se) by the formation of a very stable complex (Figure 2), which will deplete the selenium available for selenoprotein synthesis (Usuki et al., 2011). Experimental support to the indirect interaction has been discussed in the literature and the major evidence suggesting that was the demonstration that selenium supplementation can mitigate the experimental toxicity of $MeHg^+$ and other electrophilic forms of Hg. In addition, the presence of insoluble mercury selenide (HgSe) particles in different animal species have also been considered as a clear indication for the stable interaction of electrophilic mercury forms with metabolically relevant selenium-containing molecules (Sumino et al., 1977; Cuvin-Aralar et al., 1991; Rice, 2008; Khan and Wang, 2009; Korbas et al., 2010; Caito et al., 2011; Ralston and Raymond, 2018; Spiller, 2018; Ralston et al., 2019; Manceau et al., 2021a,b; James et al., 2022; Tinggi and Perkins, 2022; Gao et al., 2023).

4. The Rabenstein's Reaction(s) or the fast exchange of $MeHg^+$ from a mercury sulfide bond (CH_3Hg-S-R^1) to a free thiolate ($R^2SH \rightarrow R^2S^-$)

As commented briefly above, though it is common to find reference for the presence of free Hg^{2+} and $MeHg^+$ in living cells, their occurrence as free chemical entities inside biological fluids are unrealistic from the chemical point of view. Indeed, the chemistry and biochemistry of both Hg^{2+} and $MeHg^+$ will depend on their interaction with -SH containing molecules. Here we will discuss only the complex chemistry involved in the absorption, metabolism, and distribution of $MeHg^+$ in terms of the

Rabenstein's Reaction, which is a type of fast exchange reaction first described by the Rabenstein and colleagues. The schematic representation of the Rabenstein's reaction is:

$$\mathbf{R^1S\text{-}HgMe + R^2SH \rightleftarrows R^1SH + R^2S\text{-}HgMe}$$

Scheme 1. Rabenstein's Reaction or Exchange reaction between a thiol/thiolate with MeHg-SR.

Though the affinity of $MeHg^+$ for thiol groups is very high (the formation constant of the complex RS-HgMe from a free $MeHg^+$ and a $R\text{-}S^-$ (thiolate) is very high, approximately 10^{16}), the sulfide-mercury bond (in the complex RS-HgMe) is very labile, and in the presence of another free thiol/thiolate, MeHg easily migrates from the complex to the free thiol/thiolate forming a new complex (in Scheme 1, $MeHg^+$ is moving from R^1S^- to R^2S^- or R^2SH). The symbol "$\rightleftarrows$" indicates that the reaction is reversible and the $MeHg^+$ can move from one to another free thiolate (Rabenstein and Fairhurst, 1975; Rabenstein and Evans, 1978; Rabenstein et al., 1982; Rabenstein and Reid, 1984; Arnold et al., 1986; Nogara et al., 2019). Indeed, the chemistry of $MeHg^+$ is mediated by the formation of highly stable complexes with thiol-containing molecules (both LMM-SH or HMM-SH) by successive exchange reactions (or the Rabenstein's Reaction(s)).

The equilibria of the Rabenstein's Reactions will depend on the chemical nature of the organic moieties R^1- or R^2- (in the Scheme 1). For two similar R- groups the constant is close to 1 but the exchange rate is very high and near to a diffusion-controlled reaction (the constant for the exchange of a complexed glutathione (GS-HgMe) by a free glutathionate ion (GS^-) is near to $10^9\ M^{-1}\cdot sec^{-1}$ (Rabenstein and Fairhurst, 1975; Rabenstein and Evans, 1978). Although Rabenstein's reaction has been a little more explored for thiol/thiolate ligands, the exchange can also occur for ligands containing the selenol/selenolate ligands (Arnold et al., 1986), but the study by Arnold et al., 1986 is the only one found in literature for the exchange of MeHg by selenol/selenolate from MeHg-selenide complexes. Thus, to be more generalizing we can expand the scheme to include the exchange between selenol/selenolate or thiol/thiolate and RSe-HgMe or RS-HgMe:

$$\mathbf{R^1S(e)\text{-}HgMe + R^2S(e)^- \rightleftarrows R^1S(e)^- + R^2S(e)\text{-}HgMe}$$

Scheme 2. Generalizing Rabenstein's or exchange reactions between either thiol/thiolate or selenol/selenolate with MeHg-SR or MeHg-SeR. To simplify, the thiolate or selenolate group is represented by $\text{-}S(e)^-$ and complexes by -S(e)-HgMe.

A very simplified scheme of the potential targets of $MeHg^+$ in the human body was discussed above and depicted in Figure 2. In addition to the LMM-SH, HMM-SH and HMM-SeH molecules, $MeHg^+$ (and Hg^{2+}) can also form complexes with inorganic molecules of H_2S and H_2Se, which are the inorganic analogs of the organic thiol and selenol groups (-SH and -SeH) (Scheme 3). Of note, the chemical reactivity of H_2S and H_2Se is predicted to be higher than their organic thiol and selenol analogues, but their concentrations in biological fluids are very low (Nogara et al., 2023). In fact, the toxicity of E^+Hg can be also mediated by disruption of H_2S cell physiological functions as an extracellular chemical messenger. H_2S is gaseous neurotransmitter or neuromodulator, and a messenger involved in the control of vasodilation in the endothelium (Filipovic et al., 2018; Aschner et al., 2022). In fact, H_2S can also modulate physiological functions via formation of persulfides (Filipovic et al., 2018; Cuevasanta et al., 2022). H_2Se is a critical metabolite of both organic and inorganic selenium and is the only form of selenium that is incorporated into the 21st proteogenic amino acid selenocysteine found in the 25 types of selenoproteins in humans (Nogara et al., 2023; Oliveira et al., 2023). Thus, the formation of complexes between H_2Se and electrophilic mercury forms can deplete selenium availability for proper selenoprotein synthesis (Ralston and Raymond, 2018). Other emerging classes of both inorganic and organic molecules that can react with E^+Hg with high affinity are the persulfides or -SSH (HSSH or H_2S_2) and different types of organic persulfides (R-SSH; for instance, GSSH or cysteine-SSH) (Ihara et al., 2017; Kumagai and Abiko, 2017; Filipovic et al., 2018; Cuevasanta et al., 2022; Horai et al., 2022). The inorganic and organic persulfides are more nucleophilic and reactive than their thiols and they can have protective roles against electrophiles (including $MeHg^+$).

It is important to emphasize that the chemistry and biochemistry of $MeHg^+$ interaction with H_2S or H_2Se in the presence of excess of organic thiol or selenol ligands (LMM-SH and HMM-S(e)H) has yet to be studied. Specifically, the Rabenstein's Reactions between H_2S or H_2Se and LMM-S-HgMe or HMM-S-HgMe have not been explored. The study of their interaction with some abundant LMM-SH (for instance, GSH and Cys) and HMM-SH (for instance, albumin and hemoglobin) are needed all will give important information about the relative affinity of $MeHg^+$ for LMM-SH or HMM-SH in the presence of more nucleophilic inorganic thiol analogs (Nogara et al., 2023).

Note that in view of the higher nucleophilicity of $HS(e)^-$ when compared with the organic thiol or selenol analogues, the equilibria of both equations would be expected to shift leftward (i.e., HS^- and HSe^- are expected to displace MeHg from an organic analogue R-S-HgMe or R-Se-Hg-R). However, these reactions have not been studied experimentally. In addition, the interaction of inorganic $HS(e)^-$ or its complexes HS(e)-HgMe with HMM-S(e)H have not been investigated. In the case of HMM-SeH (or selenoproteins) the problem is further magnified, because the selenoproteins are not readily obtainable in pure forms and they are also extremely sensitive to oxidation in the presence of O_2. In short, the complexity of the Rabenstein's Reactions between reactive $HS(e)^-$ and organic thiolates and selenolates have not been studied in sufficient detail. The detailed study of the reactions described in the Schemes 1–3 will be fundamental for better characterization of kinetic data that will allow the construction of a theoretical framework to predict the distribution of $MeHg^+$ in mammalian body.

$HS(e)\text{-}HgMe + LMM\text{-}S^- \leftrightarrow HS(e)^- + LMM\text{-}S\text{-}HgMe$ (1)
$HS(e)\text{-}HgMe + HMM\text{-}S^- \leftrightarrow HS(e)^- + HMM\text{-}S\text{-}HgMe$ (2)

Scheme 3. Rabenstein's or exchange reactions between either inorganic HS^- (the dissociated H_2S) or HSe^- (the dissociated H_2Se), (which are the inorganic analogues of the organic thiol/thiolate or selenol/selenolate groups, represented by $HS(e)^-$ in the reactions (1) and (2)) with a low-molecular mass thiol (LMM-SH) (equation 1) or a high molecular mass thiol (HMM-SH) (equation 2).

In summary, the targets of $MeHg^+$ can be either inorganic molecules, e.g., H_2S, H_2Se and persulfides; or the targets can be organic molecules containing -SH (either low molecular mass thiol – LMM-SH or thiol-containing proteins or high molecular mass – HMM-SH molecules). The selenoproteins or HMM-SeH molecules can also be preferential targets of $MeHg^+$, because usually the -SeH group has a higher affinity for electrophilic mercury than an analog -SH group (Arnold et al., 1986). In Figure 2, HMM-SH or HMM-SeH are highlighted as the molecular initiating events (MIEs) involved in the toxicity of electrophilic mercury. Though it is intuitive that HMM-SH and HMM-SeH have to be the MIEs of $MeHg^+$ (particularly in view of the strong affinity of -SH and -SeH groups for electrophilic mercury forms), the picture is much more complex (as depicted in Figure 2) and will depend on the distribution of the electrophilic mercury forms to different tissues and on the interaction with H_2S, H_2Se, HSSH, RSSH, LMM-SH and different types of HMM-SH or with a few types of HMM-SeH proteins. The depletion of H_2S, H_2Se, HSSH (inorganic persulfide) or RSSH (organic persulfides) can also interfere in a complex way with the normal cell physiology. However, how electrophilic mercury interferes with these inorganic and organic low-molecular mass compounds has only incipiently been investigated and more studies will be needed to clarify how they can modify the toxicity of mercury. In short, the molecular events involved in toxicity of $MeHg^+$ and Hg^{2+} are rather complex and apparently involve the interaction of electrophilic Hg (E^+Hg) forms with different molecular initiating events (MIEs).

The interaction of Hg^{2+} or RS(e)-Hg^+ or RS(e)-Hg-(e)SR with R-S(e)H or H_2S and H_2Se can follow the same general chemical paths presented in Schemes 1–3. However, the reactions with Hg^{2+} are expected to be more complicated because one cation of Hg^{2+} can participate in two Rabenstein's Reactions (specifically Hg^{2+} coordinates with and binds two thiol ligands) (Scheme 4).

$$\mathbf{R^1S(e)\text{-}Hg\text{-}S(e)R^2 + R^3S(e)^- \rightleftarrows R^1S(e)\text{-}Hg\text{-}S(e)R^3 + R^2S(e)^-}$$
$$\mathbf{R^1S(e)\text{-}Hg\text{-}S(e)R^3 + R^4S(e)^- \rightleftarrows R^1S(e)\text{-}Hg\text{-}S(e)R^4 + R^3S(e)^-}$$
$$\mathbf{R^1S(e)\text{-}Hg\text{-}S(e)R^3 + R^4S(e)^- \rightleftarrows R^4S(e)\text{-}Hg\text{-}S(e)R^3 + R^1S(e)^-}$$
$$\mathbf{R^1S(e)\text{-}Hg\text{-}S(e)R^2 + R^3S(e)^- \rightleftarrows R^3S(e)\text{-}Hg\text{-}S(e)R^2 + R^1S(e)^-}$$

Scheme 4. Generalizing Rabenstein's reactions for Hg^{2+}. Hg^{2+} can make multiples exchanges reactions.

5. Conclusions

The toxicity of electrophilic mercury forms depends predominantly on their interaction with thiol- and selenol-containing molecules. Since the toxic concentrations of Hg in exposed organisms are normally in the nanomolar (i.e., the plasmatic or total blood concentrations of Hg are in the range above 25 nmol of electrophilic forms/L, compare for instance with the Hg in non-exposed subjects, where Hg levels vary from 2–20 nmol Hg/L; Ye et al., 2016). Upon chronic exposures, Hg may be fatal in the micromolar (μmol/L) range (for a report with two fatal cases with Hg ranging from 5–50 μmol/L of blood; Triunfante et al., 2009), and their toxicity will be dictated not by free mercury atoms or molecules. In fact, both $MeHg^+$ and Hg^{2+} have a strong affinity for -SH and SeH groups that upon making contact with biological media, MeHg-S-R complexes will be formed. It is important to re-emphasize that the -SH group concentrations in biological fluids are in molar excess over Hg (usually mmol/L for -SH *vs* nmol to low μmol/L of electrophilic forms of Hg or E^+Hg). Complexation with Se-containing molecules is also expected to participate in the toxicity of electrophilic mercury forms, but since the concentration of the -SeH group is about 10,000 lower than that of thiol groups, the -SeH groups will be somewhat protected from $MeHg^+$ by the excess of -SH. Specifically, the tendency of $MeHg^+$ to form complexes with -SH will be facilitated by the presence of millimolar or mmol/L of free -SH groups, while the concentration of -SeH will be 5 orders lower than this. Thus, in terms of the Rabenstein's Reaction, most of the exchange reactions of the type $R^1S(e)H + R^2S(e)\text{-}HgMe \rightleftarrows R^1S(e)\text{-}HgMe + R^2S(e)H$ will occur between the -S-Hg- with a free thiol than with selenol, because the concentration of the thiol groups is much greater than that of the selenol groups.

Other two aspects that further complicate the biochemical scenario of Rabenstein's reactions in mammalian cells are: (1) the affinity of a selenol group for electrophilic mercury forms are higher than that of an analogue thiol group (for organic selenol there are only limited data in the literature, see for instance Arnold et al., 1986, where they found a maximal of ten times higher affinity of -SeH for $MeHg^+$ when compared with an analogue -SH). For inorganic H_2S and H_2Se, the affinity of H_2Se is expected to be much greater than that of H_2S, but experimental data about the migration of $MeHg^+$ from HS(e)-HgMe complexes are not available. (2) The diversity of thiol groups in proteins are much greater than that of selenol groups in selenoproteins. Indeed, it is estimated that at least 10–20 thousand chemically distinct thiol groups exist in proteins, whereas the maximal number of chemically different selenol groups is 35.

In short, the complex chemistry of $MeHg^+$ migration from a sulfide or selenide to a free thiol or selenol group has been little studied. Consequently, advances in our knowledge about the toxicity of $MeHg^+$ and Hg^{2+} will require much more studies about the exchange or Rabenstein's reaction between relevant thiol- or selenol-containing proteins (HMM-S(e)H) with complexes containing LMM-S-HgMe (for instance, Cys and GSH).

References

Aaseth, J., Wallace, D.R., Vejrup, K. and Alexander, J. 2020. Methylmercury and developmental neurotoxicity: A global concern. Current Opinion in Toxicology 19: 80–87, doi: 10.1016/j.cotox.2020.01.005.

Ahlmark, A. 1948. Poisoning by methyl mercury compounds. British Journal of Industrial Medicine 5(3): 117–119, doi: 10.1136/oem.5.3.117.

Ajsuvakova, O.P., Tinkov, A.A., Aschner, M., Rocha, J.B., Michalke, B., Skalnaya, M. et al. 2020. Sulfhydryl groups as targets of mercury toxicity. Coordination Chemistry Reviews 417: 213343, doi: 10.1016/j.ccr.2020.213343.

AMAP/UN Environment. 2019. Technical Background Report for the Global Mercury Assessment 2018. Arctic Monitoring and Assessment Programme, Oslo, Norway/UN Environment Programme, Chemicals and Health Branch, Geneva, Switzerland. viii, 426 pp including E-Annexes.

Arnold, A.P., Tan, K.S. and Rabenstein, D.L. 1986. Nuclear magnetic resonance studies of the solution chemistry of metal complexes. 23. Complexation of methylmercury by selenohydryl-containing amino acids and related molecules. Inorg. Chem. 25(14): 2433–2437, doi: 10.1021/ic00234a030.

Asaduzzaman, A.M. and Schreckenbach, G. 2011. Degradation mechanism of methyl mercury selenoamino acid complexes: a computational study. Inorg. Chem. 50: 2366–2372.
Aschner, M. and Clarkson, T.W. 1988a. Uptake of methylmercury in the rat brain: effects of amino acids. Brain Research 462(1): 31–39.
Aschner, M. and Clarkson, T.W. 1988b. Distribution of mercury 203 in pregnant rats and their fetuses following systemic infusions with thiol-containing amino acids and glutathione during late gestation. Teratology 38(2): 145–155.
Aschner, M. 1989. Brain, kidney and liver 203Hg-methyl mercury uptake in the rat: relationship to the neutral amino acid carrier. Pharmacology & Toxicology 65(1): 17–20.
Aschner, M. and Clarkson, T.W. 1989. Methyl mercury uptake across bovine brain capillary endothelial cells *in vitro*: the role of amino acids. Pharmacology & Toxicology 64(3): 293–297.
Aschner, M. and Aschner, J.L. 1990. Mercury neurotoxicity: mechanisms of blood-brain barrier transport. Neuroscience & Biobehavioral Reviews 14(2): 169–176, DOI 10.1016/S0149-7634(05)80217-9.
Aschner, M., Skalny, A.V., Ke, T., da Rocha, J.B., Paoliello, M.M., Santamaria, A. et al. 2022. Hydrogen sulfide (H2S) signaling as a protective mechanism against endogenous and exogenous neurotoxicants. Current Neuropharmacology 20(10): 1908–1924.
Atti, S.K., Silver, E.M., Chokshi, Y., Casteel, S., Kiernan, E., Dela Cruz, R. et al. 2020. All that glitters is not gold: Mercury poisoning in a family mimicking an infectious illness. Current Problems in Pediatric and Adolescent Health Care 50(2): 100758, doi: 10.1016/j.cppeds.2020.100758.
Barbosa, N.V., Nogueira, C.W., Nogara, P.A., de Bem, A.F., Aschner, M. and Rocha, J.B. 2017. Organoselenium compounds as mimics of selenoproteins and thiol modifier agents. Metallomics 9(12): 1703–1734, doi: 10.1039/c7mt00083a.
Basu, N., Bastiansz, A., Dórea, J.G., Fujimura, M., Horvat, M., Shroff, E. et al. 2023. Our evolved understanding of the human health risks of mercury. Ambio 52(5): 877–896, doi: 10.1007/s13280-023-01831-6.
Benoit, J.M., Mason, R.P. and Gilmour, C.C. 1999. Estimation of mercury-sulfide speciation in sediment pore waters using octanol-water partitioning and implications for availability to methylating bacteria. Environmental Toxicology and Chemistry 18(10): 2138–2141.
Bernhoft, R.A. 2012. Mercury toxicity and treatment: a review of the literature. J. Environ. Publ. Health 2012: 460508, doi: 10.1155/2012/460508.
Bjørklund, G. 1995. Mercury and acrodynia. Journal of Orthomolecular Medicine 10(3-4): 145–146.
Bjørklund, G., Dadar, M., Mutter, J. and Aaseth, J. 2017. The toxicology of mercury: Current research and emerging trends. Environmental Research 159: 545–554, doi: 10.1016/j.envres.2017.08.051.
Bjørklund, G., Peana, M., Dadar, M., Chirumbolo, S., Aaseth, J. and Martins, N. 2020. Mercury-induced autoimmunity: drifting from micro to macro concerns on autoimmune disorders. Clinical Immunology 213: 108352, doi: 10.1016/j.clim.2020.108352.
Branco, V., Godinho-Santos, A., Gonçalves, J., Lu, J., Holmgren, A. and Carvalho, C. 2014. Mitochondrial thioredoxin reductase inhibition, selenium status, and Nrf-2 activation are determinant factors modulating the toxicity of mercury compounds. Free Radical Biology and Medicine 73: 95–105.
Branco, V. and Carvalho, C. 2019. The thioredoxin system as a target for mercury compounds. Biochimica et Biophysica Acta (BBA)-General Subjects 1863(12): 129255.
Bridges, C.C., Battle, J.R. and Zalups, R.K. 2007. Transport of thiol-conjugates of inorganic mercury in human retinal pigment epithelial cells. Toxicology and Applied Pharmacology 221(2): 251–260, doi: 10.1016/ j.taap.2007.03.004.
Bridges, C.C. and Zalups, R.K. 2017. The aging kidney and the nephrotoxic effects of mercury. J. Toxicol. Environ. Health B Crit. Rev. 20(2): 55–80. doi: 10.1080/10937404.2016.1243501.
Caito, S.W., Milatovic, D., Hill, K.E., Aschner, M., Burk, R.F. and Valentine, W.M. 2011. Progression of neurodegeneration and morphologic changes in the brains of juvenile mice with selenoprotein P deleted. Brain Res. 1398: 1–12, doi: 10.1016/j.brainres.2011.04.046.
Calabrese, E.J., Iavicoli, I., Calabrese, V., Cory-Slechta, D.A. and Giordano, J. 2018. Elemental mercury neurotoxicity and clinical recovery of function: A review of findings, and implications for occupational health. Environmental Research 163: 134–148, doi: 10.1016/j.envres.2018.01.021.
Carvalho, C.M., Chew, E.H., Hashemy, S.I., Lu, J. and Holmgren, A. 2008. Inhibition of the human thioredoxin system: a molecular mechanism of mercury toxicity. Journal of Biological Chemistry 283(18): 11913–11923. doi: 10.1074/jbc.M710133200.

Cheesman, B.V., Arnold, A.P. and Rabenstein, D.L. 1988. Nuclear magnetic resonance studies of the solution chemistry of metal complexes. 25. Hg (thiol) 3 complexes and Hg (II)-thiol ligand exchange kinetics. Journal of the American Chemical Society 110(19): 6359–6364, doi 10.1021/ja00227a014.

Clarkson, T.W. 1972. The pharmacology of mercury compounds. Annu. Rev. Pharmacol. Toxicol. 12: 375–406.

Clarkson, T.W. 1993. Molecular and ionic mimicry of toxic metals. Annu. Rev. Pharmacol. Toxicol. 32: 545–571.

Clarkson, T.W. and Magos, L. 2006. The toxicology of mercury and its chemical compounds. Crit. Rev. Tox. 36(8): 609–662, DOI: 10.1080/10408440600845619.

Clarkson, T.W., Vyas, J.B. and Ballatori, N. 2007. Mechanisms of mercury disposition in the body. Am. J. Ind. Med. 50(10): 757–764, doi: 10.1002/ajim.20476.

Cooke, C.A., Hintelmann, H., Ague, J.J., Burger, R., Biester, H., Sachs, J.P. et al. 2013. Use and legacy of mercury in the Andes. Environmental Science & Technology 47(9): 4181–4188, DOI: 10.1021/es3048027.

Crespo-Lopez, M.E., Arrifano, G.P., Augusto-Oliveira, M., Macchi, B.M., Lima, R.R., do Nascimento, J.L.M. et al. 2023. Mercury in the Amazon: The danger of a single story. Ecotoxicology and Environmental Safety 256: 114895, doi: 10.1016/j.ecoenv.2023.114895.

Cuevasanta, E., Benchoam, D., Möller, M.N., Carballal, S., Banerjee, R. and Alvarez, B. 2022. Hydrogen sulfide and persulfides. pp. 451–486. *In*: Redox Chemistry and Biology of Thiols (Amsterdam, Netherlands: Elsevier), doi: 10.1016/B978-0-323-90219-9.00011-X.

Cuvin-Aralar, M.L.A. and Furness, R.W. 1991. Mercury and selenium interaction: a review. Ecotoxicology and Environmental Safety 21(3): 348–364, doi: 10.1016/0147-6513(91)90074-Y.

Driscoll, C.T., Mason, R.P., Chan, H.M., Jacob, D.J. and Pirrone, N. 2013. Mercury as a global pollutant: sources, pathways, and effects. Environmental Science & Technology 47(10): 4967–4983, doi: 10.1021/es305071v.

Dórea, J.G., Farina, M. and Rocha, J.B. 2013. Toxicity of ethylmercury (and Thimerosal): a comparison with methylmercury. Journal of Applied Toxicology 33(8): 700–711.

Dyrssen, D. and Wedborg, M. 1991. The sulphur-mercury (II) system in natural waters. Water Air & Soil Pollution 56: 507–519.

Evans, S., Smith, J. and Caron, E. 2018. A case of mercury toxicity complicated by acute inflammatory demyelinating polyneuropathy. Journal of Child Neurology 33(13): 817–819, doi: 10.1177/08830738187904.

Farina, M., Campos, F., Vendrell, I., Berenguer, J., Barzi, M., Pons, S. et al. 2009. Probucol increases glutathione peroxidase-1 activity and displays long-lasting protection against methylmercury toxicity in cerebellar granule cells. Toxicological Sciences 112(2): 416–426, doi: 10.1093/toxsci/kfp219.

Farina, M., Aschner, M. and Rocha, J.B. 2011. Oxidative stress in MeHg-induced neurotoxicity. Toxicology and Applied Pharmacology 256(3): 405–417, doi: 10.1016/j.taap.2011.05.001.

Farina, M., Aschner, M. and Rocha, J.B.T. 2024. Mercury. Chapter 36. Patty's Toxicology, Seventh Edition. Edited by Dennis J. Paustenbach, William H. Farland and James E. Klaunig. John Wiley & Sons, Inc.

Ferreira, G., Santander, A., Chavarría, L., Cardozo, R., Savio, F., Sobrevia, L. et al. 2022. Functional consequences of lead and mercury exposomes in the heart. Molecular Aspects of Medicine 87: 101048, doi: 10.1016/j.mam.2021.101048.

Filipovic, M.R., Zivanovic, J., Alvarez, B. and Banerjee, R. 2018. Chemical biology of H2S signaling through persulfidation. Chemical Reviews 118(3): 1253–1337, doi: 10.1021/acs.chemrev.7b00205.

FitzGerald, R., Zurich, M.G., Nunes, C., Sandström, J., Aschner, M., Rocha, J. et al. 2022. Binding of electrophilic chemicals to SH (thiol)-group of proteins and/or to seleno-proteins involved in protection against oxidative stress during brain development leading to impairment of learning and memory. OECD Series on Adverse Outcome Pathways, N°. 20, OECD Publishing, Paris, doi: 10.1787/4df0e9e4-en.

Fomenko, D.E., Marino, S.M. and Gladyshev, V.N. 2008. Functional diversity of cysteine residues in proteins and unique features of catalytic redox-active cysteines in thiol oxidoreductases. Molecules and Cells 26(3): 228–235.

Franco, J.L., Posser, T., Dunkley, P.R., Dickson, P.W., Mattos, J.J., Martins, R. et al. 2009. Methylmercury neurotoxicity is associated with inhibition of the antioxidant enzyme glutathione peroxidase. Free Radical Biology and Medicine 47(4): 449–457, doi: 10.1016/j.freeradbiomed.2009.05.013.

Furieri, L.B., Galán, M., Avendaño, M.S., García-Redondo, A.B., Aguado, A., Martínez, S. et al. 2011. Endothelial dysfunction of rat coronary arteries after exposure to low concentrations of mercury is dependent on reactive oxygen species. British Journal of Pharmacology 162(8): 1819–1831.

Ganguly, J., Kulshreshtha, D. and Jog, M. 2022. Mercury and movement disorders: the toxic legacy continues. Canadian Journal of Neurological Sciences 49(4): 493–501, doi: 10.1017/cjn.2021.146.

Gao, P.C., Wang, A.Q., Chen, X.W., Cui, H., Li, Y. and Fan, R.F. 2023. Selenium alleviates endoplasmic reticulum calcium depletion-induced endoplasmic reticulum stress and apoptosis in chicken myocardium after mercuric chloride exposure. Environmental Science and Pollution Research 30(18): 51531–51541, doi: 10.1007/s11356-023-25970-1.

Halbach, S. 1985. The octanol/water distribution of mercury compounds. Archives of Toxicology 57(2): 139–141.

Heggland, I., Kaur, P. and Syversen, T. 2009. Uptake and efflux of methylmercury in vitro: comparison of transport mechanisms in C6, B35 and RBE4 cells. Toxicology *In Vitro* 23(6): 1020–1027.

Horai, S., Abiko, Y., Unoki, T., Shinkai, Y., Akiyama, M., Nakata, K. et al. 2022. Concentrations of nucleophilic sulfur species in small Indian mongoose (Herpestes auropunctatus) in Okinawa, Japan. Chemosphere 295: 133833.

Houston, M.C. 2011. Role of mercury toxicity in hypertension, cardiovascular disease, and stroke. The Journal of Clinical Hypertension 13(8): 621–627, doi: 10.1111/j.1751-7176.2011.00489.x.

Hughes, W. L. 1957. A physicochemical rationale for the biological activity of mercury and its compounds. Ann. N Y Acad. Sci. 65(5): 454–460, doi: 10.1111/j.1749-6632.1956.tb36650.x.

Ihara, H., Kasamatsu, S., Kitamura, A., Nishimura, A., Tsutsuki, H., Ida, T. et al. 2017. Exposure to electrophiles impairs reactive persulfide-dependent redox signaling in neuronal cells. Chemical Research in Toxicology 30(9): 1673–1684, doi: 10.1021/acs.chemrestox.7b00120.

Jaishankar, M., Tseten, T., Anbalagan, N., Mathew, B.B. and Beeregowda, K.N. 2014. Toxicity, mechanism and health effects of some heavy metals. Interdisciplinary Toxicology 7(2): 60–72, doi: 10.2478/intox-2014-0009.

James, A.K., Dolgova, N.V., Nehzati, S., Korbas, M., Cotelesage, J.J., Sokaras, D. et al. 2022. Molecular fates of organometallic mercury in human brain. ACS Chemical Neuroscience 13(12): 1756–1768, doi: 10.1021/acschemneuro.2c00166.

Johnson-Arbor, K., Tefera, E. and Farrell Jr, J. 2021. Characteristics and treatment of elemental mercury intoxication: A case series. Health Science Reports 4(2): e293, doi: 10.1002/hsr2.293.

Katsuma, A., Hinoshita, F., Masumoto, S., Hagiwara, A. and Kimura, A. 2014. Acute renal failure following exposure to metallic mercury. Clinical Nephrology 82(1): 73–76, doi: 10.5414/cn107669.

Kerper, L.E., Ballatori, N. and Clarkson, T.W. 1992. Methylmercury transport across the blood-brain barrier by an amino acid carrier. American Journal of Physiology-Regulatory, Integrative and Comparative Physiology 262(5): R761–R765.

Khan, M.A.K. and Wang, F. 2009. Mercury-selenium compounds and their toxicological significance: toward a molecular understanding of the mercury-selenium antagonism. Environ. Toxicol. Chem. 28(8): 1567–1577, doi: 10.1897/08-375.1.

Khan, M.A.K. and Wang, F. 2010. Chemical demethylation of methylmercury by selenoamino acids. Chem. Res. Toxicol. 23: 1202–1206, doi: 10.1021/tx100080s.

Korbas, M., O'Donoghue, J.L., Watson, G.E., Pickering, I.J., Singh, S.P., Myers, G.J. et al. 2010. The chemical nature of mercury in human brain following poisoning or environmental exposure. ACS Chem. Neurosci. 1(12): 810–818, doi: 10.1021/cn1000765.

Kumagai, Y. and Abiko, Y. 2017. Environmental electrophiles: protein adducts, modulation of redox signaling, and interaction with persulfides/polysulfides. Chemical Research in Toxicology 30(1): 203–219, doi: 10.1021/acs.chemrestox.6b00326.

Labunskyy, V.M., Hatfield, D.L. and Gladyshev, V.N. 2014. Selenoproteins: molecular pathways and physiological roles. Physiological Reviews 94(3): 739–777.

Lemos, N.B., Angeli, J.K., Faria, T.D.O., Ribeiro Junior, R.F., Vassallo, D.V., Padilha, A.V. et al. 2012. Low mercury concentration produces vasoconstriction, decreases nitric oxide bioavailability and increases oxidative stress in rat conductance artery. PLoS One 7(11): e49005.

Lombardi, G., Lanzirotti, A., Qualls, C., Socola, F., Ali, A.M. and Appenzeller, O. 2012. Five hundred years of mercury exposure and adaptation. J. Biomed. Biotechnol. 2012: 472858, doi: 10.1155/2012/472858.

Manceau, A., Bourdineaud, J.-P., Oliveira, R.B., Sarrazin, S.L.F., Krabbenhoft, D.P., Eagles-Smith, C.A. et al. 2021a. Demethylation of methylmercury in bird, fish, and earthworm. Environ. Sci. Technol. 55: 1527–1534, doi: 10.1021/acs.est.0c04948.

Manceau, A., Gaillot, A.-C., Glatzel, P., Cherel, Y. and Bustamante, P. 2021b. *In vivo* formation of HgSe nanoparticles and Hg-tetraselenolate complex from methylmercury in seabirds - implications for the Hg-Se antagonism. Environ. Sci. Technol. 55: 1515–1526, doi: 10.1021/acs.est.0c06269.

Mason, R.P., Reinfelder, J.R. and Morel, F.M.M. 1995. Bioaccumulation of mercury and methylmercury, Water Air Soil Pollut. 80: 915–921, doi: 10.1007/BF01189744.

Mason, R.P., Reinfelder, J.R. and Morel, F.M.M. 1996. Uptake, toxicity, and trophic transfer of mercury in a coastal diatom. Environ. Sci. Technol. 30: 1835–1845, doi: 10.1021/es950373d.

Mason, R.P. 2002. The bioaccumulation of mercury, methylmercury, and other toxic elements into pelagic and benthic organisms. Chapter 6, pp. 127–149. *In*: In Coastal and Estuarine Risk Assessment, Edited by N.C. Newman, M.H. Roberts, and R.C. Hale; Lewis Publishers, Boca Raton, FL, USA.

Mason, R.P., Buckman, K.L., Seelen, E.A., Taylor, V.F. and Chen, C.Y. 2023. An examination of the factors influencing the bioaccumulation of methylmercury at the base of the estuarine food web. Science of The Total Environment 886: 163996, doi: 10.1016/j.scitotenv.2023.163996.

Mercer, J.J., Bercovitch, L. and Muglia, J.J. 2012. Acrodynia and hypertension in a young girl secondary to elemental mercury toxicity acquired in the home. Pediatric Dermatology 29(2): 199–201.

Mokrzan, E.M., Kerper, L.E., Ballatori, N. and Clarkson, T.W. 1995. Methylmercury-thiol uptake into cultured brain capillary endothelial cells on amino acid system L. Journal of Pharmacology and Experimental Therapeutics 272(3): 1277–1284.

Naija, A. and Yalcin, H.C. 2023. Evaluation of cadmium and mercury on cardiovascular and neurological systems: effects on humans and fish. Toxicology Reports 10: 498–508, doi: 10.1016/j.toxrep.2023.04.009.

Nierenberg, D.W., Nordgren, R.E., Chang, M.B., Siegler, R.W., Blayney, M.B., Hochberg, F. et al. 1998. Delayed cerebellar disease and death after accidental exposure to dimethylmercury. N. Eng. J. Med. 338(23): 1672–1686, doi: 10.1056/ NEJM199806043382305.

Nogara, P.A., Oliveira, C.S., Schmitz, G.L., Piquini, P.C., Farina, M., Aschner, M. et al. 2019. Methylmercury's chemistry: From the environment to the mammalian brain. Biochimica et Biophysica Acta (BBA)-General Subjects 1863(12): 129284, doi: 10.1016/j.bbagen.2019.01.006.

Nogara, P.A., Madabeni, A., Bortoli, M., Teixeira Rocha, J.B. and Orian, L. 2021. Methylmercury can facilitate the formation of dehydroalanine in selenoenzymes: insight from DFT molecular modeling. Chemical Research in Toxicology 34(6): 1655–1663, doi: 10.1021/acs.chemrestox.1c00073.

Nogara, P.A., Pereira, M.E., de Oliveira, C.S., Orian, L. and Rocha, J.B.T. 2023. Organic Selenocompounds: are they the panacea to human illnesses? New Journal of Chemistry 47: 9959–9988, doi: 10.1039/D2NJ05694A.

O'Connor, D., Hou, D., Ok, Y.S., Mulder, J., Duan, L., Wu, Q. et al. 2019. Mercury speciation, transformation, and transportation in soils, atmospheric flux, and implications for risk management: A critical review. Environment International 126: 747–761, doi: 10.1016/j.envint.2019.03.019.

Oliveira, C.S., Piccoli, B.C., Aschner, M. and Rocha, J.B.T. 2017. Chemical speciation of selenium and mercury as determinant of their neurotoxicity. Neurotoxicity of Metals 18: 53–83, doi: 10.1007/978-3-319-60189-2_4.

Oliveira, C.S., Nogara, P.A., Ardisson-Araújo, D.M., Aschner, M., Rocha, J.B. and Dórea, J.G. 2018. Neurodevelopmental effects of mercury. pp. 27–86. In Advances in Neurotoxicology, Academic Press, v. 2, doi: 10.1016/bs.ant.2018.03.005.

Oliveira, C.S., Piccoli, B.C., Nogara, P.A., Pereira, M.E., de Carvalho, K.A.T., Skalny, A.V. et al. 2023. Selenium neuroprotection in neurodegenerative disorders. pp. 2489–2523. *In*: Handbook of Neurotoxicity, Springer, doi: 10.1007/978-3-031-15080-7_238.

Park, J.D. and Zheng, W. 2012. Human exposure and health effects of inorganic and elemental mercury. Journal of Preventive Medicine and Public Health 45(6): 344–352, doi: 10.3961/jpmph.2012.45.6.344.

Peterson, S.A., Ralston, N.V., Whanger, P.D., Oldfield, J.E. and Mosher, W.D. 2009. Selenium and mercury interactions with emphasis on fish tissue. Environmental Bioindicators 4(4): 318–334.

Pickering, I.J., Cheng, Q., Rengifo, E.M., Nehzati, S., Dolgova, N.V., Kroll, T. et al. 2020. Direct observation of methylmercury and auranofin binding to selenocysteine in thioredoxin reductase. Inorganic Chemistry 59(5): 2711–2718, doi: 10.1021/acs.inorgchem.9b03072.

Rabenstein, D.L. and Fairhurst, M.T. 1975. Nuclear magnetic resonance studies of the solution chemistry of metal complexes. XI. Binding of methylmercury by sulfhydryl-containing amino acids and by glutathione. Journal of the American Chemical Society 97(8): 2086–2092, doi: 10.1021/ja00841a015.

Rabenstein, D.L. 1978a. The aqueous solution chemistry of methylmercury and its complexes. Accounts of Chemical Research 11(3): 100–107, doi: 10.1021/ar50123a004.

Rabenstein, D.L. 1978b. Chemistry of methylmercury toxicology. Journal of Chemical Education 55(5): 292, doi: 10.1021/ed055p292.

Rabenstein, D.L. and Evans, C.A. 1978. The mobility of methylmercury in biological systems. Bioinorganic Chemistry 8(2): 107–114, doi: 10.1016/S0006-3061(00)80237-9.

Rabenstein, D.L., Isab, A.A. and Reid, R.S. 1982. A proton nuclear magnetic resonance study of the binding of methylmercury in human erythrocytes. Biochimica et Biophysica Acta (BBA)-Molecular Cell Research 720(1): 53–64, doi: 10.1016/0167-4889(82)90038-6.

Rabenstein, D.L., Reid, R.S. and Isab, A.A. 1983. 1H NMR study of the effectiveness of various thiols for removal of methylmercury from hemolyzed erythrocytes. Journal of Inorganic Biochemistry 18(3): 241–251, doi: 10.1016/0162-0134(83)85006-5.

Rabenstein, D.L. and Reid, R.S. 1984. Nuclear magnetic resonance studies of the solution chemistry of metal complexes. 20. Ligand-exchange kinetics of methylmercury (II)-thiol complexes. Inorganic Chemistry 23(9): 1246–1250, doi: 10.1021/ic00177a016.

Rabenstein, D.L., Arnold, A.P. and Guy, R.D. 1986. 1H-NMR study of the removal of methylmercury from intact erythrocytes by sulfhydryl compounds. Journal of Inorganic Biochemistry 28(2-3): 279–287, doi: 10.1016/0162-0134(86)80092-7.

Ralston, N.V. and Raymond, L.J. 2018. Mercury's neurotoxicity is characterized by its disruption of selenium biochemistry. Biochimica et Biophysica Acta (BBA)-General Subjects 1862(11): 2405–2416, doi: 10.1016/j.bbagen.2018.05.009.

Ralston, N.V., Kaneko, J.J. and Raymond, L.J. 2019. Selenium health benefit values provide a reliable index of seafood benefits vs. risks. Journal of Trace Elements in Medicine and Biology 55: 50–57, doi: 10.1016/j.neuro.2008.07.007.

Rice, D.C. 2008. Overview of modifiers of methylmercury neurotoxicity: chemicals, nutrients, and the social environment. Neurotoxicology 29(5): 761–766, doi: 10.1016/j.neuro.2008.07.004.

Rice, K.M., Walker Jr, E.M., Wu, M., Gillette, C. and Blough, E.R. 2014. Environmental mercury and its toxic effects. Journal of Preventive Medicine and Public Health 47(2): 74, doi: 10.3961/jpmph.2014.47.2.74.

Rizzetti, D.A., Martín, Á., Corrales, P., Fernandez, F., Simões, M.R., Peçanha, F.M. et al. 2017. Egg white-derived peptides prevent cardiovascular disorders induced by mercury in rats: role of angiotensin-converting enzyme (ACE) and NADPH oxidase. Toxicology Letters 281: 158–174.

Roos, D.H., Puntel, R.L., Farina, M., Aschner, M., Bohrer, D., Rocha, J.B.T. et al. 2011. Modulation of methylmercury uptake by methionine: prevention of mitochondrial dysfunction in rat liver slices by a mimicry mechanism. Toxicology and Applied Pharmacology 252(1): 28–35.

Schartup, A.T. 2022. Methylmercury as a molecular imposter. Nat. Chem. 14: 240, doi: 10.1038/s41557-021-00885-x.

Scheuhammer, A.M., Meyer, M.W., Sandheinrich, M.B. and Murray, M.W. 2007. Effects of environmental methylmercury on the health of wild birds, mammals, and fish. AMBIO: A Journal of the Human Environment 36(1): 12–19, DOI: 10.1579/0044-7447(2007)36.

Simmons-Willis, T.A., Koh, A.S., Clarkson, T.W. and Ballatori, N. 2002. Transport of a neurotoxicant by molecular mimicry: the methylmercury–L-cysteine complex is a substrate for human L-type large neutral amino acid transporter (LAT) 1 and LAT2. Biochemical Journal 367(1): 239–246.

Sonke, J.E., Angot, H., Zhang, Y., Poulain, A., Björn, E. and Schartup, A. 2023. Global change effects on biogeochemical mercury cycling. Ambio 52(5): 853–876, doi: 10.1007/s13280-023-01855-y.

Spiller, H.A. 2018. Rethinking mercury: the role of selenium in the pathophysiology of mercury toxicity. Clinical Toxicology 56(5): 313–326, doi: 10.1080/15563650.2017.1400555.

Sumino, K., Yamamoto, R. and Kitamura, S. 1977. A role of selenium against methylmercury toxicity. Nature 268(5615): 73–74, doi: 10.1038/268073a0.

Takahata, N., Hayashi, H., Watanabe, S. and Anso, T. 1970. Accumulation of mercury in the brains of two autopsy cases with chronic inorganic mercury poisoning. Psychiatry and Clinical Neurosciences 24(1): 59–69, doi: 10.1111/j.1440-1819.1970.tb01457.x.

Tinggi, U. and Perkins, A.V. 2022. Selenium status: its interactions with dietary mercury exposure and implications in human health. Nutrients 14(24): 5308, doi: 10.3390/nu14245308.

Triunfante, P., Soares, M.E., Santos, A., Tavares, S., Carmo, H. and de Lourdes Bastos, M. 2009. Mercury fatal intoxication: two case reports. Forensic Science International 184(1-3): e1–6, doi: 10.1016/j.forsciint.2008.10.023.

Trujillo, M., Alvarez, B. and Radi, R. 2016. One- and two-electron oxidation of thiols: mechanisms, kinetics and biological fates. Free Radical Research 50(2): 150–171, doi: 10.3109/10715762.2015.1089988.

Turell, L., Carballal, S., Botti, H., Radi, R. and Alvarez, B. 2009. Oxidation of the albumin thiol to sulfenic acid and its implications in the intravascular compartment. Brazilian Journal of Medical and Biological Research 42: 305–311, doi: 10.1590/S0100-879X2009000400001.

Turell, L., Radi, R. and Alvarez, B. 2013. The thiol pool in human plasma: the central contribution of albumin to redox processes. Free Radical Biology and Medicine 65: 244–253, doi: 10.1016/j.freeradbiomed.2013.05.050.

UNEP - United Nations Environment Programme. 2019. Global Mercury Assessment 2018: Key Findings. https://wedocs.unep.org/20.500.11822/29830.

Unoki, T., Akiyama, M., Shinkai, Y., Kumagai, Y. and Fujimura, M. 2022. Spatio-temporal distribution of reactive sulfur species during methylmercury exposure in the rat brain. The Journal of Toxicological Sciences 47(1): 31–37, doi: 10.2131/JTS.47.31.

Usuki, F., Yamashita, A. and Fujimura, M. 2011. Post-transcriptional defects of antioxidant selenoenzymes cause oxidative stress under methylmercury exposure. Journal of Biological Chemistry 286(8): 6641–6649, doi: 10.1074/jbc.M110.168872.

Wagner, C., Sudati, J.H., Nogueira, C.W. and Rocha, J.B. 2010. *In vivo* and *in vitro* inhibition of mice thioredoxin reductase by methylmercury. Biometals 23: 1171–1177, doi: 10.1007/s10534-010-9367-4.

Woolf, A.D. 2022. Three methylmercury poisoning disasters. pp. 15–33. In History of Modern Clinical Toxicology, Academic Press, doi: 10.1016/B978-0-12-822218-8.00037-5.

Wunder, M. and Bhati, P. 2021. A case of methylmercury poisoning. University of Western Ontario Medical Journal 90(1): 1–11, doi: 10.5206/uwomj.v90i1.13516.

Ye, B.J., Kim, B.G., Jeon, M.J., Kim, S.Y., Kim, H.C., Jang, T.W. et al. 2016. Evaluation of mercury exposure level, clinical diagnosis and treatment for mercury intoxication. Annals of Occupational and Environmental Medicine 28(1): 1–8, doi: 10.1186/s40557-015-0086-8.

Yin, Z., Jiang, H., Syversen, T., Rocha, J.B., Farina, M. and Aschner, M. 2008. The methylmercury-L-cysteine conjugate is a substrate for the L-type large neutral amino acid transporter. Journal of Neurochemistry 107(4): 1083–1090.

Young, A.C., Wax, P.M., Feng, S.Y., Kleinschmidt, K.C. and Ordonez, J.E. 2020. Acute elemental mercury poisoning masquerading as fever and rash. Journal of Medical Toxicology 16: 470–476, doi: 10.1007/s13181-020-00792-6.

Zalups, R.K. and Lash, L.H. 1994. Advances in understanding the renal transport and toxicity of mercury. Journal of Toxicology and Environmental Health, Part A Current Issues 42(1): 1–44, doi: 10.1080/15287399409531861.

Zalups, R.K. 2000. Molecular interactions with mercury in the kidney. Pharmacological Reviews 52(1): 113–144.

Zalups, R.K. and Lash, L.H. 2006. Cystine alters the renal and hepatic disposition of inorganic mercury and plasma thiol status. Toxicology and Applied Pharmacology 214(1): 88–97, doi: 10.1016/j.taap.2005.12.007.

Zimmermann, L.T., Santos, D.B., Naime, A.A., Leal, R.B., Dórea, J.G., Barbosa Jr, F. et al. 2013. Comparative study on methyl-and ethylmercury-induced toxicity in C6 glioma cells and the potential role of LAT-1 in mediating mercurial-thiol complexes uptake. Neurotoxicology 38: 1–8.

Chapter 11

Gold

Structure, Reactivity, Biological Activities, and Toxicity

Pablo A. Nogara,[1,*] *Folorunsho B. Omage,*[2,3] *Caroline S. Schiavon,*[1] *Karise F. Nogara,*[4] *Wilian C. da Rosa,*[5] *Pamela C. da Rosa*[6] and *Lisandro von Mühlen*[7,*]

1. Introduction

Gold (Au, from the Latin *aurum*, "shining dawn"), a chemical element with unique properties, has been captivated humans for millennia. It can be shaped into various forms, including spherical, cage-like, and rod-like. When Au is reduced to the nanometer scale (1–100 nm), it exhibits entirely different properties from bulk Au due to a combination of various effects that are not yet fully understood (Frank Shaw, 1999; Jin and Higaki, 2021; Perez-Potti et al., 2023; Rabiee et al., 2022a).

The gold abundance in the earth's crust is around 0.004 ppm and lower than that of copper (Cu) and silver (Ag), which are 68 and 0.08 ppm respectively. Naturally, gold occurs mainly as grains of metal disseminated in quartz veins, which is also

[1] Federal Institute of Education, Science and Technology Sul-rio-grandense (IFSul), Av. Leonel de Moura Brizola, 2501, Bagé, 96418-400, RS, Brazil.
[2] Biological Chemistry Laboratory, Department of Organic Chemistry, Institute of Chemistry, University of Campinas (UNICAMP), Campinas, SP, Brazil.
[3] Computational Biology Research Group, Embrapa Agricultural Informatics, Campinas, SP, Brazil.
[4] Graduate Program on Animal Science, Department of Animal Science, Federal University of Paraná (UFPR), Curitiba, 80035-050, PR, Brazil.
[5] Colégio Marista Aparecida, Rua Ramiro Barcelos, 307, Bento Gonçalves, 95700-074, RS, Brazil.
[6] Federal University of Paraná, Department of Physical Education, Campus Centro Politécnico Jardim das Américas, Curitiba, 82590-300, PR, Brazil.
[7] Federal University of Rio Grande do Sul (UFRGS), Faculty of Veterinary Medicine, Av. Bento Gonçalves, 9090, Porto Alegre, 91540-000, RS, Brazil.
* Corresponding authors: pbnogara@gmail.com; lisandrovm@gmail.com

found as powdered rock sediment in riverbeds. The metal gold is extracted with mercury (Hg) or cyanide (CN^-). Mercury is added to water containing powdered gold rock, dissolving only the gold, and forming an amalgam with Hg^0 (ranging from $AuHg_2$ to Au_8Hg). The amalgam is distilled with heating, and the metals Au and Hg partially recovered for reuse (Catherine E. Housecroft and Alan G. Sharpe, 2012; John David Lee, 2008; Peter W. Atkins et al., 2010). However, in this process Hg is also released in the atmosphere and environment, poisoning rivers, plants, and animals. For instance, as more recently observed in the Amazon River in Brazil (Crespo-Lopez et al., 2023; Gerson et al., 2022). In the cyanide process, 0.1–0.2% of KCN solution is used to treat the rocks. The dissolution of gold in cyanide solutions in the presence of air are used during the metal extraction from crude ores ($4\ Au + 8\ KCN + 2\ H_2O + O_2 \rightarrow 4\ K[Au(CN)_2] + 4\ KOH$). Habitually, the salt is not isolated, but solutions of the dicyanoaurate ion ($[Au(CN)_2]^-$) are generated on a large scale in the extraction of gold (Catherine E. Housecroft and Alan G. Sharpe, 2012; John David Lee, 2008; Peter W. Atkins et al., 2010). From the environmental point of view, the recycling of gold from discarded jewelry and electronic waste can have an important impact on the conservation of specific ecosystems and human and animal health. An option that needs to be studied more and explored is gold biomineralization by *Delftia acidovorans*, which could be used for industrial waste treatment and gold recovery. The bacteria produce secondary metabolites to generate solid gold forms from soluble and toxic gold (Funari et al., 2019; Gwynne, 2013; Johnston et al., 2013).

Since ancient times, gold has been used in medicine and as an indicator of social and economic power. Nowadays, the main use of gold is as gold bullion (used as international currency); it is also used for jewelry, coinage, art, and a small amount of Au is used in electrical devices, such as computers and smartphones (Christopher Corti and Richard Holliday, 2009; Englinger et al., 2019; Schoenberger, 2011). There are important gold drugs for rheumatoid arthritis treatment: auranofin, sodium aurothiosulfate (sanocrysin), aurothiomalate (myochrysine), and aurothioglucose (solganal) (Figure 1). Auranofin is monomeric and is orally administered, whereas aurothiomalate and aurothioglucose are polymeric and are administered by injection (Abdou et al., 2009; Englinger et al., 2019; Ott, 2009). Another interesting class of organogold compounds is the Au-NHCs complexes, which have N-heterocyclic carbene ligands (NHC) in the place of phosphines. These molecules have been reported to be potential anticancer agents (Porchia et al., 2018). However, gold compounds could be toxic, causing DNA damage, reduction in cell viability, oxidative stress, and present adverse effects, such as skin rash, oral ulcers, proteinuria, thrombocytopenia, bone marrow suppression, diarrhea, allergic reaction, pulmonary injury, eye problems, and hepatoxicity (Barnard and Berners-Price, 2007; Kean et al., 1992; Sani et al., 2021).

There are many experimental reports of gold as organocompounds and nanoparticles (AuNPs) presenting biological activities, such as immunotherapeutic, antimicrobial, anti-inflammatory, analgesic, anti-arthritic, anti-parasitic, antitumor and anticancer, by interacting with various biomolecules such as proteins, DNA and RNA (Englinger et al., 2019; Kean et al., 1992; Kot et al., 2023; Perez-Potti et al.,

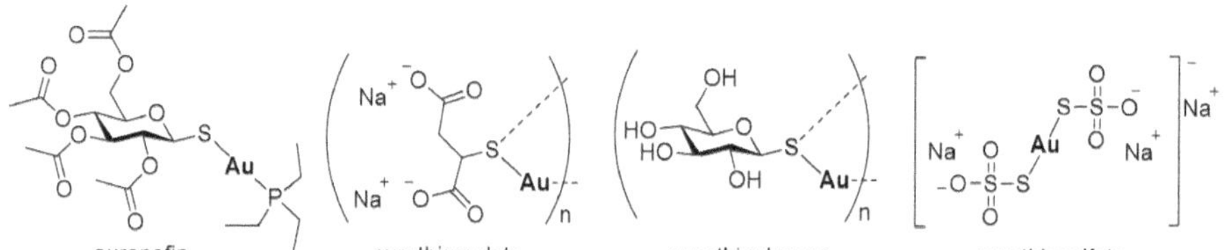

auranofin aurothiomalate aurothioglucose aurothiosulfate

Figure 1. Chemical structure of auranofin, aurothiomalate, aurothioglucose and aurothiosulfate.

2023; Rabiee et al., 2022b; Suárez-Moreno et al., 2022; Tolbatov et al., 2021; Veselov et al., 2022; Yeo et al., 2018a). In addition, gold is also used as food additive and/or decoration (E 175, edible gold) (EFSA Panel on Food Additives and Nutrient Sources added to Food, 2016). Edible gold has no nutritional value, and it is reported that the ingestion of gold leaf is nontoxic because the gold is excreted without reaction with the digestive system. However, there is the possibility that when small pieces of gold leaf remain in the human digestive tract for a long-time ionic gold can be formed and released, which could lead to chronic inflammation and carcinogenesis (EFSA Panel on Food Additives and Nutrient Sources added to Food, 2016; Freyberg et al., 1941; Imai, 2018). People with a history of short-term edible gold consumption did not develop chrysiasis (a dermatological condition induced by gold) (AlHargan et al., 2019), however, studies have shown that gold accumulation can cause cytotoxicity in the intestinal epithelial cells (Yao et al., 2015). Thus, the side effects of gold leaf consumption cannot be ruled out.

Gold compounds can also be used as catalysts for bio-orthogonal transformations, which are selectively chemical reactions that can occur in living systems without interfering with the native cell function (Liang et al., 2022; Thomas and Casini, 2020a). Some of the factors that affect the reactivity of gold compounds with biological molecules are the oxidation state, the ligand type, the solvent, the pH, and the redox potential (Yeo et al., 2018b). For example, Au(III) compounds are more prone to reduction and ligand exchange than Au(I) compounds, which makes them less stable and more reactive in biological environments (Đurović et al., 2017). The ligand type can also influence the reactivity and selectivity of gold compounds towards different biomolecules. For example, some ligands can enhance the affinity of gold compounds for specific receptors or enzymes, while others can modulate the catalytic activity or fluorescence properties of gold complexes (Thomas and Casini, 2020b). The solvent, pH, and redox potential can affect the solubility, stability, and speciation of gold compounds, which in turn can alter their reactivity and bioavailability (Yeo et al., 2018b).

The photoelectric properties of gold molecules have been applied in fields such as sensing probes and drug delivery systems in biological and pharmaceutical applications. In relation to AuNPs, they can be functionalized with specific ligands for binding selectively to receptors on the surface of target cells. When illuminated with light at the appropriate wavelength, the AuNPs can generate heat through a process known as photothermal therapy. This can be used to release therapeutic agents at the target site (Nejati et al., 2021). The functionalization of gold with bioactive molecules, such as drugs, genes, and cell-specific targeting ligands, is facilitated by the presence of localized negative charge on AuNPs.

Studies have shown that several AuNPs are biocompatible and non-toxic, with a potent ability to interact with and enter cells. This has encouraged researchers to attach various compounds and biological macromolecules to gold in an effort to combine functionality with transport. The potential of AuNPs to stabilize and protect DNA, RNA, and other conjugates in solution offers additional benefits over multistep methods that require separate approaches to delivery and stabilization (Kong et al., 2017).

Due to the biological importance of organogold molecules and AuNPs, we proposed to report a general panorama of gold compounds, properties, biological applications, and toxicity. Particular emphasis will be given to those organogold compounds and AuNPs that are used clinically to treat human and animal diseases.

2. The inorganic and organic chemistry of gold

For a long time, gold was considered biologically inert, however, in recent decades, several points of evidence have indicated the pharmacological usefulness of gold and its broad applications in various research areas, ranging from nanotechnology to medicinal applications. The discovery of new chemical transformations catalyzed by gold continues to drive research in the field.

Gold (electron configuration: ${}_{79}Au$: [Xe] $4f^{14}$ $5d^{10}$ $6s^{1}$) is generally inert and is not attacked by oxygen (O_2) or non-oxidizing acids. The dominant stable oxidation states of gold are Au(III) and Au(I). Numerous Au(III) compounds have been synthetized presenting square planar coordination, such as AuF_3, presenting a d^8 configuration. For Au(I) molecules, linear coordination is common, however Au···Au interactions in the solid state are found, and trigonal planar and tetrahedral complexes are reported. The hydrated oxide $Au_2O{\cdot}H_2O$ is the only established oxide of Au. According to the Hard Acid Soft Base (HSAB) Theory, proposed by Dr Ralph G. Pearson in 1963 (Pearson, 1963, 1990), which qualitatively explains the mechanism and stability of metal complexes, Au(I) prefers soft donor atoms, thus, is common to find complexes with Au–P and Au–S bonds. Molecules of R_3PAuCl and R_2SAuCl (for which many different R groups are known) contain linear Au(I), but in the solid state by virtue of aggregation, the Au···Au interactions are often observed. Au(I) is used to treat rheumatoid arthritis, where the containing-Au(I) drugs are linear complexes of the type RS-Au-SR, R_3P-Au-PR_3 or R_3P-Au-SR (R_3P and RS^- are organophosphines and organothiolates, in which the phosphorus (P) and sulfur (S) atoms, respectively, bind to the Au metal). The R group has a significant role because it can change the complex solubility and affects how readily the drug is spread in the body. Au(V) species are rare, and an example is in AuF_5, a highly reactive compound which presents dimeric structure in the solid state (Catherine E. Housecroft and Alan G. Sharpe, 2012; John David Lee, 2008; Peter W. Atkins et al., 2010; Rocchigiani and Bochmann, 2021).

The application of gold in catalysis is more recent. In fact, Au was initially considered inert, but now it has emerged as an active catalytic agent and the exploitation of gold as a catalytic agent is a promising field. With its unique properties and the ability to selectively activate π-bonds, gold chemistry offers numerous opportunities for scientific advancements and practical applications in various areas of modern chemistry. The gold-based complexes are interesting due to their different possible oxidation states (e.g., Au(I) and Au(III)), stability, and ligand exchange reactions, which confer different mechanisms of activity. These structures are formed by ligand coordination with Au(I) or Au(III) ions, which contain nitrogen (N), phosphorus (P), sulfur (S), carbon (C), and other atoms. The geometric configuration of gold complexes is changeable, mostly in compositions of two, three, and four ligands (Frank Shaw, 1999; Rocchigiani and Bochmann, 2021; Zhang et al., 2022).

The studies on gold complexes were expanded with the discovery of auranofin in the 1970s. Auranofin is an Au(I) complex containing thiolate as ligand, which is obtained by the reaction of thiols (compounds containing the -SH group) with gold molecules. The compound is used to treat rheumatoid arthritis and was synthesized with the aim of combining the beneficial pharmacological properties of gold with greater therapeutic efficacy and lower toxicity compared to inorganic gold compounds (Massai et al., 2022; Roder and Thomson, 2015).

Another class of interesting gold complexes is the N-Heterocyclic Carbene ligands (NHC), which serve as optimal ligands for Au(I). NHC ligands are excellent sigma donors and possess a planar coordinative environment due to the substituents at the 1st and 3rd positions. The chemical, steric, and electronic properties of NHC ligands resemble those of triaryl or trialkyl phosphines (R_3P). According to the nature of the ligands bound to Au(I), the Au-NHC compounds can be divided into two classes: the neutral and the cationic, which determines the linear coordination of Au(I). According to the general structure NHC–Au–L, when L represents a negatively charged ligand, neutral and lipophilic structures are obtained. On the other hand, when L is a neutral ligand, such as the carbene ligand itself or a phosphine, the cationic and more hydrophilic is obtained (Estrada-Ortiz et al., 2017; Galassi et al., 2022; Porchia et al., 2018; Tialiou et al., 2022). Thus, due to the large number of different ligands that can be combined with gold ions, many organogolds can be designed and synthesized aiming to generate potential drugs.

3. Gold nanoparticles (AuNPs): properties and applications

Nanoscience is currently undergoing rapid development. The production of nano-scale structures has become the focus of many studies and research projects, mainly due to the various properties and applications of nanoparticles. The physical properties such as the structure and morphology of nanoparticles allow different applications, ranging from drug encapsulation (Thambiraj et al., 2021) to changes in surface wettability, providing superhydrophobic characteristics (Han et al., 2022), to the production of catalysts for energy generation (Attia et al., 2014). Gold compounds have been studied and applied in medicine, pharmaceutical, and veterinary fields, for instance, as biosensors, disease diagnosis, and transport of molecules to certain cells. Thus, the interest and development of new gold molecules for the potential use in the treatment of chronic inflammation, infections, oncology, degenerative and autoimmune diseases is growing (Cancino et al., 2014; Ko et al., 2022a; Porchia et al., 2018; Saha et al., 2012).

The electronic and optical properties of AuNPs depends heavily on their shape and size. They exhibit a distinct surface effect, where the interaction of light with electrons on the surface of AuNPs results in the collective oscillation of electrons. This produces surface plasmon resonance bands and a distinct macroscopic quantum tunneling effect, leading to substantial attenuation of light through absorption and scattering (Zhang et al., 2020). The specific wavelength or frequency of this effect is highly dependent on the size, shape, surface, and aggregation condition of the gold nanoparticles (Kong et al., 2017; Zhang et al., 2020).

As well known, gold nanomaterials (AuNM) have been playing a crucial role on medicine (Ashraf et al., 2016; Dreaden et al., 2012). When properly functionalized, those materials may enter living cells, working as efficient deliverers for various forms of pharmaceuticals, targeted/controlled drug and genes release, and many other biological purposes (Giljohann et al., 2010). AuNM have always been considered as non-toxic for biological systems, but recent studies have shown that the presence of AuNM may cause several toxic effects.

The acquisition of the necessary characteristics for each application is achieved by different nanoparticle synthesis techniques. Depending on the technique used, it is possible to form cubic, spherical, nanorod, nanoplate, among other structures. These structures may be obtained by applying different temperature conditions, exposure to radiation and pressure, as well as doping with other substances. All these factors can directly influence the biological properties of the nanoparticles.

In the case of AuNPs themselves, gold is already a material of great interest as it is a noble metal that normally prevents oxidation of materials and molecules. When obtained in the form of nanoparticles, the behavior of gold becomes even more intriguing due to the appearance of new physical, chemical, and optical properties, as well as its high stability. These properties can be explored by incorporating AuNPs into various materials and organisms, which can have distinct biological actions in different animal models. In medicine, AuNPs have been used to produce drug capsules, enabling better targeting of medication to specific cellular groups of interest and reducing toxicity to the rest of the body. Additionally, they are used in diagnostic imaging in medicine and in cancer treatment as a coadjutant in radiotherapy, among other applications (Ko et al., 2022; Perez-Potti et al., 2023; Rabiee et al., 2022b; Saha et al., 2012; Sani et al., 2021).

Nosrati et al. conducted research using mice with breast cancer and demonstrated that AuNPs acting as drug carriers and radiosensitizers exhibited great efficacy in cancer treatment (Nosrati et al., 2022). Ding et al. utilized a hybrid nanomedicine with AuNPs plus camptothecin (a topoisomerase inhibitor and anticancer molecule tested in clinical trials) as a chemotherapeutic agent. The combination of chemotherapy with the new AuNP medication has a synergistic anticancer effect, preventing tumor growth (Ding et al., 2020).

Another important application of AuNPs is in the development of new materials, for instance, the construction of superhydrophobic surfaces. Superhydrophobicity is the characteristic of repelling water. When a water droplet falls on a superhydrophobic surface, it immediately rolls or slides off. Therefore, superhydrophobicity prevents water adhesion to the surface, providing important features such as self-cleaning, corrosion resistance, and contamination prevention. Corrosion resistance is fundamental to the naval and automotive industries. For the photovoltaic energy generation industry, self-cleaning can increase the efficiency of solar panels, as water droplets that hit the surface of the panels slide off, carrying away dust particles and reducing light reflection. In medicine, the production of superhydrophobic prostheses and surgical instruments can inhibit the adhesion of fungi and bacteria to surfaces, preventing contamination. Superhydrophobicity is achieved by the combination of low surface energy with a specific nano- and/or microstructure or hierarchical structure (Demann et al., 2005; Falde et al., 2016; Liu et al., 2021). AuNPs have shown effectiveness in the production and assembling of materials that can be used for building superhydrophobic surfaces. This applies to both surfaces constructed with nanoscale structures and surfaces structured hierarchically (Lee et al., 2013; Von White et al., 2012; Yeh et al., 2012). In the latter case, AuNPs serve as the second level of structures that prevent water penetration, enhancing surface repellency through capillary force.

Many studies have been conducted using AuNPs in the production of superhydrophobic surfaces. Liu et al. developed a graphene foam for oil absorption and water-oil separation, aiming to address potential oil leakage issues. AuNPs were deposited via cathodic spraying to structure the graphene sponge surface. The material exhibited high superhydrophobicity, with contact angles exceeding 150°, and excellent oil absorption capacity (Liu et al., 2021). Wang et al. developed a new flexible, superhydrophobic, and highly sensitive substrate for surface-enhanced Raman spectroscopy (SERS) detection, utilizing films with various thicknesses of gold nanoislands. These vertically deposited nanoislands generated a tip-enhancement effect, producing a very high electromagnetic field, consequently amplifying the Raman signal. This technique allows for biodetection, especially for hydrophobic biomolecules (Wang et al., 2022). Therefore, AuNPs can be used in various applications, such as coatings through thin films for surface structuring and functionalization.

Nanoparticles, in general, can be obtained through the extraction of matter from larger structures (a technique known as top-down), or by organizing the components of the material at the atomic scale (a technique known as bottom-up) (Zhang et al., 2004). For AuNPs, the bottom-up approach is commonly used, employing various chemical, physical, and biological processes (Ali et al., 2022), such as electrodeposition and successive reduction-oxidation cycles (Bhattarai et al., 2018). Additionally, biosynthesis using natural materials such as bacteria, fungi, algae, and plant extracts have been widely employed in the production of AuNPs, aiming for environmentally friendly techniques (Iravani, 2011). The different synthesis routes provide specific characteristics for the nanoparticles (Attia et al., 2014; Patil et al., 2023).

The field of nanotoxicology, which studies the potential interactions of AuNPs with biological systems and their possible toxic effects in living organisms, has been growing considerably in the last decades. The main objective of nanotoxicology is to better understand the safety and the potential long-term toxicity of gold nanomaterials (Girgis et al., 2012; Huang et al., 2015; Umair et al., 2016).

In effect, AuNPs are examples of AuNMs, and the absorption and distribution of AuNPs in biological systems is directly influenced by their particle size (Liu and Peng, 2017a). Frequently, the small particles are better absorbed because they can overcome cellular barriers more efficiently than large nanoparticles. As a corollary, they normally are distributed more easily in the body of living organisms. For instance, small nanoparticles can overcome the blood-brain barrier (BBB), leading to accumulation of AuNPs inside the brain cells. In contrast, after the intravenous administration of AuNPs, the increase in particle size leads to a higher accumulation in spleen and liver tissues, and a reduced accumulation in lung and brain tissues (Sonavane et al., 2008).

Of toxicological significance, different types of AuNPs can have reduced applications due to interactions and adsorption of proteins on the nanoparticle surface, which results in the formation of supramolecular structure called protein corona (PC). AuNPs that can form PC are expected to have unpredictable pharmacological or toxicological action after *in vivo* administration (Liu and Peng, 2017b). The AuNP-PC formation can induce conformational changes in the adsorbed proteins, which can cause either a gain or loss in the protein biochemical function (Saptarshi et al., 2013). The AuNP-PC complexes can also produce abnormal unfolding-re-folding of the adsorbed protein, exposing hydrophobic areas and novel epitopes. Consequently, the exposure of immunoreactive epitopes can result in autoimmunotoxicity.

In addition to the potential toxicity of AuNPs to humans and animals, they can also disrupt the environmental health. For instance, the presence of AuNPs in aquatic environments have shown to be toxic and accumulated in fish tissues (Bai and Tang, 2020). Experiments conducted with zebra fish (*Danio rerio*) in presence of aquatic AuNPs resulted in genome modifications (repression of *sod2*, *cox1*, *rad51* and *gaad* genes) and alteration in neurotransmissions increasing brain acetylcholinesterase activity (Dedeh et al., 2015). In female zebrafish, the exposure to AuNPs caused an increase in the incidence of strand breaks in ovarian cells (Dayal et al., 2016). AuNPs exposure in adult zebrafish has been shown to cause brain and muscle mitochondrial dysfunction, which varies depending on AuNPs size, concentration, and exposure time (Geffroy et al., 2012). Of potential neurodevelopmental significance, exposure of zebrafish embryos to AuNPs has been associated with accumulation of AuNPs, redox cellular imbalance and oxidative stress damage, determined by dichlorofluorescein diacetate (DCFDA) assay (Verma et al., 2018).

There are also reports about the toxicity of AuNPs in cockroaches. The ingestion of AuNPs by German cockroaches (*Blattella germanica*) lead to a 40% reduction on the number of nymphs from the first and third ootheca and reduced the viability of ootheca from cockroaches treated with AuNPs. The nymph's survival rate was disrupted until the third instar (Small et al., 2016). Though the deleterious effects of AuNPs can be seen as an indication of the general toxicity of AuNPs in non-target organisms, the potential selective toxicity of AuNPs to some groups of insects may also be exploited to develop new classes of pesticides.

In accordance with the data discussed above, *in vitro* studies have demonstrated that cells exposed to AuNPs absorbed the nanoparticles, which was associated with oxidative-related cytotoxicity as determined by DNA damage, cell death, and cell cycle arrest (Jia et al., 2017). The *in vivo* treatment of mice with AuNPs caused pathological changes in the spleen (white pulp aberrations). In addition, the exposure of mice to AuNPs with particle sizes lower than 37 nm was lethal, indicating that the injected AuNPs could damage major organs (Chen et al., 2009). Chronic exposure to AuNPs caused progressive bioaccumulation of AuNPs in the brain, in a dose-dependent manner, with indications of non-saturable uptake of AuNPs in cerebral tissues. The accumulation of AuNPs in the liver, kidneys, lungs, and spleen also tended to increase with the dose, but the percentage of AuNPs accumulated in relation to the dose administered decreased as the dose of AuNPs injected increased (Lasagna-Reeves et al., 2010). The results from hepatic, renal pulmonary, and splenic tissues indicated a saturation of the uptake as a function of the dose of AuNPs.

Different studies have evaluated the toxicity of AuNPs in human cells. For instance, sperm mobility and viability were decreased by AuNPs, possibly by interacting with membrane receptors or with cell signaling involved in the tail mobility machinery maintenance (Moretti et al., 2013). In small airway epithelial cells (SAECs), which are the first cells of contact for inhaled NPs, and human lung fibroblast cells (MRC5) the exposure to AuNPs resulted in the internalization of gold particles by endocytosis, which was associated with increased oxidative stress, cytotoxicity, and genotoxicity (Ng et al., 2013). In another study with MRC5 lung fibroblast, 72 h of exposure to AuNPs caused DNA strand breaks, and chromosomal breaks, which was associated with differential expression of genes (mRNA) of 16 proteins with functions that range from the oxidative stress response, regulation of cell cycle, cytoskeleton, and DNA repair (Li et al., 2011).

In the medical diagnosis field, the detection of the inflammatory process, antigens, cancer cells, and tumors is extremely important, consequently, AuNPs conjugated with immunofluorescence staining and/or primary antibodies have been explored. Its use allows the real-time detection of gold absorption in living cells and/or has the ability to increase the intensity of fluorescence of molecules (Dykman and Khlebtsov, 2011; Garcia et al., 2019; Guirgis et al., 2012; Page Faulk and Malcolm Taylor, 1971; Perez-Potti et al., 2023; Versiani et al., 2016). AuNPs-based contrast agents are used in imaging modalities for several applications; for example, targeting the nanoparticles into the biological system aiming to make it interact with specific cells and molecules, creating conjugates, and enhancing or enabling the use of imaging techniques. Historically the use of AuNPs in electron microscopy is predominant, especially with transmission electron microscopy (TEM), for the detection of biospecific interactions using colloidal gold particles. Modern applications involve the use of high-resolution transmission electron microscope (HRTEM) and systems of digital recording and processing (Dykman and Khlebtsov, 2011). For cancer studies, targeting the AuNPs to cancer cells results in an enhanced fluorescence of conjugated molecules, enabling detection by indirect immunofluorescence and early detection of early neoplasia based on molecular signatures specific for cancer (Garcia et al., 2019). Also, fluorescence immunoassay for the detection of malaria antigen in infected mice was successfully applied using two AuNP-conjugates (the adsorption of antibody cells on the surface of AuNPs and the covalent bonding using cross-linking agents) (Guirgis et al., 2012).

The use of AuNPs has been speculated as a potential for treating allergic inflammatory processes such as asthma, based on a rodent model where bacterial lipopolysaccharide (LPS) was used to induce lung inflammation. Treatment with AuNPs presented anti-inflammatory and antioxidant activity without causing any overt sign of toxicity (Haupenthal et al., 2020). Other reported uses of AuNPs were the microbial detection of porcine pleuropneumonia, caused by *Actinobacillus pleuropneumoniae*, and *Cryptococcus gattii* isolated from humans and animals. The most used tests for the detection of microbials disease are monovalent and polyvalent antigens by enzyme-linked immunosorbent assay (ELISA) and PCR (polymerase chain reaction). In relation to the PCR methodology, the use of AuNPs in the detection of infected animals was considered faster since there is no need of prior amplification of DNA for diagnosis. Consequently, AuNPs allowed a quick, easy, and low-cost analysis (Brandão et al., 2014; Maruyama et al., 2017). AuNPs were also used to detect canine visceral leishmaniasis, which is a zoonosis caused by the protozoan *Leishmania* (L.) *infantum chagasi*. The results showed that the method obtained greater sensitivity and specificity when compared to PCR (Rosa et al., 2013).

Osteoarthritis is a chronic disease, which leads to functional disability of joints due to progressive degeneration of joint cartilage and can afflict humans and animals. The use of AuNPs managed to inhibit inflammation and clinical signs of the disease in mice (Campos et al., 2017; Huang et al., 2012). Studies in the treatment of skin tumors in dogs using brachytherapy with gold-198 (^{198}Au) leaves proved to be effective. Of note, the association of gold with the conventional standard treatment with brachytherapy presented superior radiobiological results when compared with the standard method (Fernandes et al., 2003).

The discussed examples indicated the promising action of AuNPs in the treatment and diagnosis of diseases. In part, the effectiveness of gold is related to its relative biologically inertness in humans and animals. However, even though AuNPs demonstrate lower toxicity to organisms compared to other materials, long-term AuNPs could potentially generate toxicity in humans and animals. For this reason, research on the potentially harmful effects of AuNPs should be continued both *in vitro* and *in vivo* animal models.

4. Reactivity of gold molecules

To study the reactivity of gold compounds with biological molecules, various experimental and computational methods can be used. Some of the experimental methods include spectroscopic techniques (such as UV-vis, IR, NMR, X-ray), electrochemical techniques (such as cyclic voltammetry), mass spectrometry (such as ESI-MS), chromatography (such as HPLC), and biological assays (such as cell viability, enzyme inhibition, DNA cleavage) (Đurović et al., 2017; Thomas and Casini, 2020b, 2020a; Yeo et al., 2018b). The use of *in silico* methodologies can also be helpful. For instance, some of the computational methods include quantum chemical calculations (such as DFT), molecular dynamics simulations (MD), docking studies (such as AutoDock), and pharmacokinetic modeling (such as ADME) (Ali et al., 2023; Chen et al., 2022; Uzonwanne et al., 2022). These methods can help to elucidate the structure, stability, kinetics, thermodynamics, and mechanism of reaction of gold compounds with biological molecules. In relation to the reactivity of gold, the focus is mainly on the DFT (Density Function Theory) studies.

DFT is a powerful computational tool that can be used to study the electronic structure, reactivity, and spectroscopic properties of Au compounds with biological molecules. DFT calculations provide insight into the bonding, stability, and mechanism of reaction of gold compounds with various biomolecules such as proteins, RNA and DNA. We have studies such as the one by Rodríguez et al. reporting the photophysical properties of several Au(I) complexes with different phosphine ligands. Though the study of Rodríguez et al. was not the earliest study involving DFT, it was possibly the first that implemented the time-dependent DFT (TD-DFT). The geometry of the complexes was optimized using DFT (B3LYP), and TD-DFT calculations were used to assign the lowest energy absorption bands of the ligand-to-metal charge transfer (LMCT) transitions. The article also discusses the potential applications of these Au(I) complexes as luminescent materials and sensors (Rodríguez et al., 2008). Another notable study discussed Au(I/III)-phosphine complexes as potent antiproliferative agents (Kim et al., 2019). The study reported the anticancer activity of several Au(I/III) complexes with chiral and non-chiral phosphine ligands. The electrochemical properties of the complexes were studied by cyclic voltammetry, and theoretical insight into the complexes was acquired by DFT and TD-DFT calculations. These gold-phosphine complexes effectively kill K562, H460, and OVCAR8 cancer cell lines with IC_{50} in the range of 0.1–2.5 μM, triggering early – late stage apoptosis through potential disruption of redox homeostasis.

Gold complexes, both coordination and organometallic, have emerged as promising tools for bio-orthogonal transformations. These transformations are chemical reactions that can occur selectively in living systems without interfering with the native biological processes. Gold complexes are endowed with excellent reactivity and selectivity, compatibility within aqueous reaction medium, fast kinetics of ligand exchange reactions, under mild reaction conditions. Recent advances in the use of gold complexes for catalysis and metal-mediated transformations in living systems have been reported in the scientific literature. For example, Thomas and Casini reviewed the design, synthesis, and characterization of various gold complexes, including coordination complexes, organometallic complexes, and supramolecular assemblies (Thomas and Casini, 2020a). They also highlighted the potential applications of these gold compounds in bio-orthogonal transformations, such as bioconjugation, bioimaging, biosensing, and drug delivery. Vidal et al. (Vidal et al., 2018) reported the use of designed water-activatable gold chloride complexes to achieve a gold-mediated C–C bond formation inside living mammalian cells. They also demonstrated the viability of achieving this gold-promoted process in parallel with a ruthenium-mediated reaction inside living cells in a bioorthogonal and mutually orthogonal manner. The studies by Thomas and Vidal suggested that gold complexes have great potential as tools for bio-orthogonal transformations in living systems. Further research is needed to fully understand the mechanisms of action of these complexes and to optimize their design for specific applications.

DFT quantum chemistry calculations have been instrumental in the elucidation of chemical and biological properties of different Au complexes (Rocchigiani and Bochmann, 2021). The DFT investigation of the interaction between Au(I), as $[Au(PMe_3)]^+$, with the thioredoxin reductase (TrxR) active site models (the tetrapeptides H_2N-GlyCysCysGly-CO_2H and H_2N-GlyCysSecGly-CO_2H) indicated a stronger binding of gold to selenium than to sulfur (Howell, 2009). Docking and DFT analyses showed that the adamantane–azole Au(I) complexes bind in the TrxR active site and the Au(I) complexes interacted with high affinity with the Se atom of Sec. The study also demonstrated the cytotoxicity against different cell lines (kidney normal cell (BHK-21), colon cancer (CT26WT), mammary adenocarcinoma (4T1), and metastatic skin melanoma (B16F10)) and TrxR inhibition by Au(I) complexes. The authors highlight that the type of ligand (R) bound to Au has an interference in biological activity, where the one with triethylphosphine and thiazolidine ring was the most cytotoxic and active against TrxR (Garcia et al., 2016). In relation to the interaction of AuNPs and DNA, it was reported that an Au13 nanocluster model had high affinity for nucleic bases following the trend: Adenine > Cytosine > Guanine > Thymine. The nature of the interaction between AuNPs with the purine and pyrimidine nitrogenous bases was partially covalent and highly polar (Hashemkhani Shahnazari and Darvish Ganji, 2021).

The use of experimental and *in silico* methods are essential to unravel and understand the biological properties of gold molecules, and many studies still are necessary to elucidate their reactivity, targets, metabolites, safety, and applications.

5. Gold's putative molecular mechanism of action

Clinical studies, cellular and enzymatic assays demonstrated that gold compounds have potential therapeutic action by inhibiting enzymes and killing cells and microorganisms (Tables 1 and 2), however, to better understand the physiological and macromolecular effects of gold it is important to analyze the behavior of this metal at micromolecular and atomic level. In this sense, crystallographic, molecular simulations, and quantum-chemistry calculations are of paramount importance. Crystal structures of Au bound to proteins and DNA demonstrated that Au makes covalent bonds mainly with the sulfur atom from Cys residues, nitrogen of His, and oxygen from Asp and Glu residues (Figure 2). In many cases, the Au geometry is linear, however, the Au(I) in the active site of MCR-1 is coordinated with Glu246, Asp465, His466 and TPO285 (phosphorylated Thr285) presenting a distorted tetrahedral geometry (Figure 2G) (Zhang et al., 2023). Interestingly, the Au(III) can interact with two thymidine (dT) residues of DNA besides 6 water molecules. Thus, the Au atom is octacoordinated and exhibited a distorted square antiprismatic geometry (Figure 2H) (Kondo and Iwase, 2022). In fact, the biological action of Au appears to be related to the nature of its interactions and coordination environment (Nobili et al., 2010). The complexed Au atoms can bind with selectivity to certain cavities, active and allosteric sites of enzymes, proteins, and DNA, blocking their functions, which could lead to cell death (Adhireksan et al., 2017; Englinger et al., 2019; Gurba et al., 2022; Messori et al., 2013; Nobili et al., 2010).

In addition, the redox properties of Au also have an important role in their biological action. Au(I) compounds are thermodynamically more stable than Au(III), thus proteins could reduce Au(III) to Au(I) leading to a redox system that can scavenge

Table 1. Biological activity of some gold compounds.

Compound	Biological activity/assay	Values	Reference
Auranofin	mTrxR inhibition	IC_{50} = 20 nM	(Barnard and Berners-Price, 2007)
Aurothiomalate	mTrxR inhibition	IC_{50} = 280 nM	(Barnard and Berners-Price, 2007)
Auranofin	GR inhibition	IC_{50} = 15 μM	(Porchia et al., 2018)
Auranofin	Hexokinase inhibition	IC_{50} ~ 5 μM	(Hou et al., 2018)
Auranofin	MCF-7 (breast cancer cell line)	EC_{50} = 1.1 μM	(Porchia et al., 2018)
Auranofin	A2780 (ovarian cancer cell line)	EC_{50} = 1.25 μM	(Porchia et al., 2018)
$Au(PEt_3)I$	Cytotoxicity (Calu-3 cell line)	CC_{50} = 12 μM	(Cirri et al., 2021)
AuNPs	Antiviral (Vero cell line infected with Measles virus)	EC_{50} = 8.8 μg/mL	(Meléndez-Villanueva et al., 2019)
$HAuCl_4$	Antiviral (Vero cell line infected with Measles virus)	EC_{50} = 31.4 μg/mL	(Meléndez-Villanueva et al., 2019)
AuNPs	Antimicrobial (*Pseudomonas aeruginosa*)	EC_{50} = 68.6 ppm	(Yu et al., 2016)

IC_{50} (concentration that inhibits 50% of enzyme activity); EC_{50} (effective concentration that produce 50% inhibition of cell growth); CC_{50} (half-maximal cytotoxic concentration); mTrxR (mitochondrial thioredoxin reductase); GR (glutathione reductase).

Table 2. Examples of clinical studies of some gold compounds.

Compound	Condition or disease	Clinical trials ID
Auranofin	Chronic Lymphocytic Leukemia (CLL)	NCT01419691
Auranofin	Epithelial ovarian, primary peritoneal, or fallopian tube cancer	NCT01747798
AuNPs	Salivary gland tumours	NCT04907422
AuNPs	Antimicrobial (*Streptococcus mutans* and *Candida albicans*)	NCT05816512
AuNPs	Vaccine against SARS-CoV-2	NCT05113862

reactive oxygen species. As Au(III) is more reactive than Au(I), it is suggested that Au(III) is responsible for toxicity and adverse effects of gold (Englinger et al., 2019; Gurba et al., 2022; Nobili et al., 2010). In this sense, the search for ligands able to stabilize the Au(III) under physiological conditions is essential (Gurba et al., 2022).

From the toxicological point of view, it is important to highlight some similarities between Au and Hg besides of the thiol affinity, once Hg is highly toxic (Oliveira et al., 2017, 2019). Au(I) and Hg(II) present similar electron configuration ([Xe] $4f^{14}$ $5d^{10}$), linear and tetrahedral geometries (Ajsuvakova et al., 2020; Jalilehvand et al., 2013), inhibit selenium metabolism and selenoprotein activities and synthesis (Englinger et al., 2019; Gromer et al., 1998; Nobili et al., 2010; Nogara et al., 2019; Pia Rigobello et al., 2004), and have thiol-ligand exchange reactions (Englinger et al., 2019; Garcia et al., 2016; Kean et al., 1992; Madabeni et al., 2020; Messori et al., 2013; Nogara et al., 2019). In fact, the ligand exchange reactions of Hg (R_1–S(e)–HgMe + R_2–S(e)–H ⇋ R_1–S(e)–H + R_2–S(e)–HgMe), known as

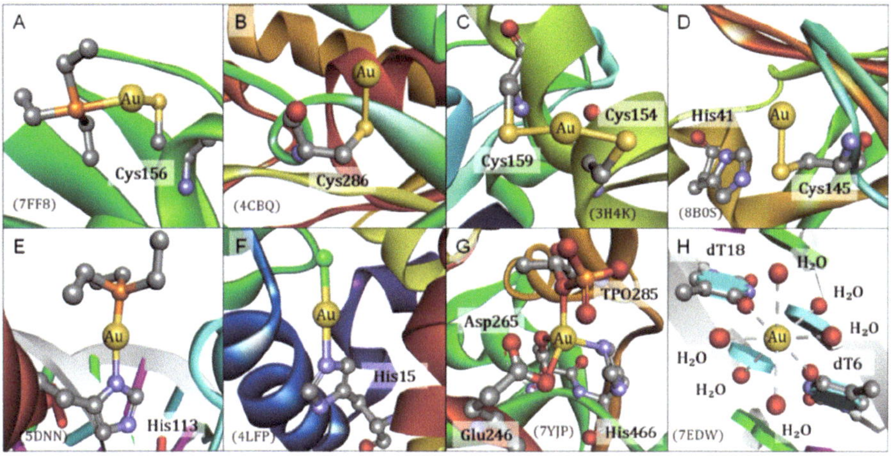

Figure 2. Examples of gold bound to proteins and DNA. The Au atom is bound to the sulfur atom from Cys residues, the oxygen atom from Asp, Glu, Thr, and water, and the nitrogen atom from His and thymidine (dT) residues. (A) *Pseudomonas aeruginosa* virulence factor regulator (PDB ID 7FF8), (B) *Entamoeba histolytica* thioredoxin reductase (PDB ID 4CBQ), (C) *Schistosoma mansoni* thioredoxin glutathione reductase (PDB ID 3H4K), (D) SARS-COV-2 Main Protease (M^{pro}) (PDB ID 8B0S), (E) histone proteins (PDB ID 5DNN), (F) egg white lysozyme (PDB ID 4LFP), (G) *Escherichia coli* MCR-1 (mcr-1-encoding transmembrane phosphoethanolamine transferase) (PDB ID 7YJP), and (H) DNA duplex containing T-T base pairs (PDB ID 7EDW). The main molecules/atoms are represented by the ball-and-stick model with the Au atom highlighted.

Rabenstein's reaction, could explain the Hg toxicity and mobility in biological systems (Nogara et al., 2019; Rabenstein, 1978a, 1978b; Rabenstein and Evans, 1978), and in some way is similar to the equilibrium observed in some organogold molecules as prodrugs: $R_1S–Au–L + R_2S–H \leftrightharpoons R_1S–H + R_2S–Au–L$ (Barnard and Berners-Price, 2007; Dos Santos, 2014; Ecker et al., 1986; Englinger et al., 2019; Frank Shaw, 1999; Tolbatov et al., 2021).

For example, auranofin can react with the Cys residue from bovine serum albumin (BSA) replacing the thiosugar ligand and not the phosphine group. In addition, this adduct underwent thiol exchange with Cys and glutathione (GSH) molecules. In the presence of metallothionein, small proteins rich in Cys residues, auranofin have the thiosugar and phosphine ligands dissociated (Ecker et al., 1986; Frank Shaw, 1999). This interaction with metallothioneins could lead to resistance of cells towards organogold (Ott, 2009). Thus, these thiol exchange reactions could be involved in the gold uptake by cells and mobility (Figure 3) (Frank Shaw, 1999).

These observations support the hypothesis that the Au atom/ion is chiefly responsible for the biological effect of organogold complexes and the ligands (L), or organic moiety; is necessary for the selectivity (i.e., to determine with which macromolecule inside of the cell the organogold will interact); Au(I)/(III) stability, and ADMET (absorption, distribution, metabolism, excretion, and toxicity) properties of the molecule.

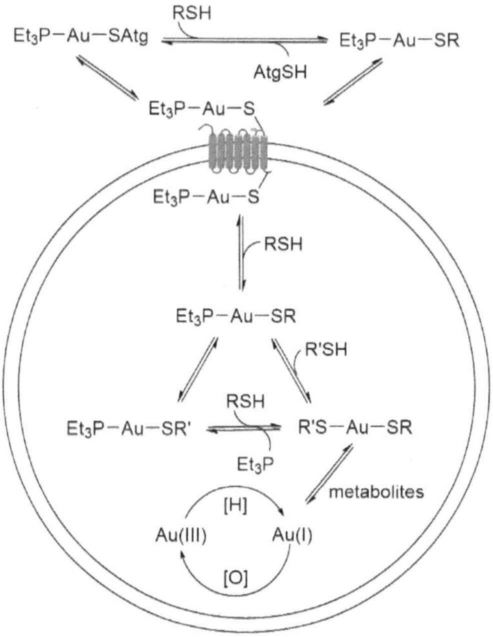

Figure 3. Proposed mechanism of gold molecules uptake by cells and biological redox cycling of Au(I) and Au(III). By ligand exchange reactions, gold binds to many low- and/or high-molecular-mass thiol (or selenol) molecules which allows its uptake, mobility, and biological activity in cells. As general example, auranofin ($Et_3PAuSAtg$) is used. SAtg = acetylthioglucose; RSH and R'SH = low- and/or high-molecular-mass thiol (or selenol).

6. Conclusion

Gold has been reported to be a safe and inert metal, however, it is not totally true. All the structure of the molecule or particle of gold is important for its pharmacological or toxic effects. AuNPs are promising therapeutic agents, with many different applications, however, more studies are needed to determine their selectivity, stability, and safety. A challenge is to determine and understand the potential non-specific interactions and adsorption of proteins (and other molecules) on the AuNPs' surface, which could lead to unpredictable biological effects. In general, gold molecules appear to be prodrugs that require activation by thiol/selenol ligand exchange reactions ($R_1S(e)–Au–L + R_2S(e)–H \leftrightharpoons R_1S(e)–H + R_2S(e)–Au–L$), thus, the main biological targets of gold are thiol- and seleno-containing proteins. Theoretical studies have provided valuable insights into the reactivity of gold compounds with biological macromolecules. These studies have used computational methods to investigate the electronic structure, bonding, and mechanism of the reaction of gold compounds with various biomolecules. Further research is needed to fully understand the mechanisms of action of these compounds and to optimize their design for specific applications. A deep investigation of the molecular initiating events (MIEs) involved in gold pharmacology and toxicity is still needed, requiring coordinated efforts from biochemists, toxicologists, and theoretical and analytical chemists.

References

Abdou, H.E., Mohamed, A.A., Fackler, J.P., Burini, A., Galassi, R., López-de-Luzuriaga, J.M. et al. 2009. Structures and properties of gold(I) complexes of interest in biochemical applications. In Coordination Chemistry Reviews 253(11–12): 1661–1669. https://doi.org/10.1016/j.ccr.2009.02.010.

Adhireksan, Z., Palermo, G., Riedel, T., Ma, Z., Muhammad, R., Rothlisberger, U. et al. 2017. Allosteric cross-talk in chromatin can mediate drug-drug synergy. Nature Communications 8. https://doi.org/10.1038/ncomms14860.

Ajsuvakova, O.P., Tinkov, A.A., Aschner, M., Rocha, J.B.T., Michalke, B., Skalnaya, M.G. et al. 2020. Sulfhydryl groups as targets of mercury toxicity. Coordination Chemistry Reviews 417: 213343. https://doi.org/10.1016/j.ccr.2020.213343.

AlHargan, A., AlTalhab, S., Ghobara, Y., Alissa, A. and AlJasser, M.I. 2019. Effect of quality-switched laser exposure in patients with history of edible gold consumption. Lasers in Medical Science 34(9): 1813–1817. https://doi.org/10.1007/s10103-019-02780-x.

Ali, A.M., Makki, A.A., Ibraheem, W., Abdelrahman, M., Osman, W., Sherif, A.E. et al. 2023. Design of novel phosphatidylinositol 3-kinase inhibitors for non-Hodgkin's Lymphoma: Molecular docking, molecular dynamics, and density functional theory studies on gold nanoparticles. Molecules 28(5). https://doi.org/10.3390/MOLECULES28052289.

Ali, F., Hamza, M., Iqbal, M., Basha, B., Alwadai, N. and Nazir, A. 2022. State-of-art of silver and gold nanoparticles synthesis routes, characterization and applications: a review. Zeitschrift Für Physikalische Chemie 236(3): 291–326. https://doi.org/10.1515/zpch-2021-3084.

Ashraf, S., Pelaz, B., del Pino, P., Carril, M., Escudero, A., Parak, W.J. et al. 2016. Gold-Based Nanomaterials for Applications in Nanomedicine (pp. 169–202). https://doi.org/10.1007/978-3-319-22942-3_6.

Attia, Y.A., Buceta, D., Blanco-Varela, C., Mohamed, M.B., Barone, G. and López-Quintela, M.A. 2014. Structure-directing and high-efficiency photocatalytic hydrogen production by Ag clusters. Journal of the American Chemical Society 136(4): 1182–1185. https://doi.org/10.1021/ja410451m.

Bai, C. and Tang, M. 2020. Toxicological study of metal and metal oxide nanoparticles in zebrafish. Journal of Applied Toxicology 40(1): 37–63. https://doi.org/10.1002/jat.3910.

Barnard, P.J. and Berners-Price, S.J. 2007. Targeting the mitochondrial cell death pathway with gold compounds. In Coordination Chemistry Reviews 251(13-14) SPEC. ISS.: 1889–1902. https://doi.org/10.1016/j.ccr.2007.04.006.

Bhattarai, J., Neupane, D., Nepal, B., Mikhaylov, V., Demchenko, A. and Stine, K. 2018. Preparation, modification, characterization, and biosensing application of nanoporous gold using electrochemical techniques. Nanomaterials 8(3): 171. https://doi.org/10.3390/nano8030171.

Brandão, L.N.S., Pitchenin, L.C., Maruyama, F.H., Chitarra, C.S., Silva, G.F.R. da, Klein, C. et al. 2014. Standardization of unmodified gold nanoparticle (AuNPs) for detection of Actinobacillus pleuropneumoniae in swine lungs. Pesquisa Veterinária Brasileira 34(7): 621–625. https://doi.org/10.1590/S0100-736X2014000700002.

Campos, W.N. da S., Leite, A.E.T., Sonego, D.A., Andrade, M.A. de, Pizzinatto, F.D., Marangoni, V.S. et al. 2017. Synthesis and characterization of gold nanoparticles combined with curcumin and its effects on experimentally induced osteoarthritis. Ciência Rural 47(7). https://doi.org/10.1590/0103-8478cr20161001.

Cancino, J., Marangoni, V.S. and Zucolotto, V. 2014. Nanotechnology in medicine: concepts and concerns. Química Nova 37(3). https://doi.org/10.5935/0100-4042.20140086.

Catherine, E. Housecroft and Alan G. Sharpe. 2012. Inorganic Chemistry (4th ed.). Pearson.

Chen, X., Yang, L., Tang, J., Wen, X., Zheng, X., Chen, L. et al. 2022. An AuNPs-based fluorescent sensor with truncated aptamer for detection of sulfaquinoxaline in water. Biosensors 12(7). https://doi.org/10.3390/bios12070513.

Chen, Y.-S., Hung, Y.-C., Liau, I. and Huang, G.S. 2009. Assessment of the *in vivo* toxicity of gold nanoparticles. Nanoscale Research Letters 4(8): 858. https://doi.org/10.1007/s11671-009-9334-6.

Christopher Corti and Richard Holliday. 2009. Gold: Science and Applications (1st ed.). CRC Press.

Cirri, D., Marzo, T., Tolbatov, I., Marrone, A., Saladini, F., Vicenti, I. et al. 2021. *In vitro* anti-sars-cov-2 activity of selected metal compounds and potential molecular basis for their actions based on computational study. Biomolecules 11(12). https://doi.org/10.3390/biom11121858.

Crespo-Lopez, M.E., Arrifano, G.P., Augusto-Oliveira, M., Macchi, B.M., Lima, R.R., do Nascimento, J.L.M. et al. 2023. Mercury in the Amazon: The danger of a single story. In Ecotoxicology and Environmental Safety (Vol. 256). Academic Press. https://doi.org/10.1016/j.ecoenv.2023.114895.

Dayal, N., Thakur, M., Patil, P., Singh, D., Vanage, G. and Joshi, D.S. 2016. Histological and genotoxic evaluation of gold nanoparticles in ovarian cells of zebrafish (*Danio rerio*). Journal of Nanoparticle Research 18(10): 291. https://doi.org/10.1007/s11051-016-3549-0.

Dedeh, A., Ciutat, A., Treguer-Delapierre, M. and Bourdineaud, J.-P. 2015. Impact of gold nanoparticles on zebrafish exposed to a spiked sediment. Nanotoxicology 9(1): 71–80. https://doi.org/10.3109/17435390.2014.889238.

Demann, E.T.K., Stein, P.S. and Haubenreich, J.E. 2005. Gold as an implant in medicine and dentistry. Journal of Long-Term Effects of Medical Implants 15(6): 687–698. https://doi.org/10.1615/JLongTermEffMedImplants.v15.i6.100.

Ding, Y., Xu, H., Xu, C., Tong, Z., Zhang, S., Bai, Y. et al. 2020. A nanomedicine fabricated from gold nanoparticles-decorated metal–organic framework for cascade chemo/chemodynamic cancer therapy. Advanced Science 7(17): 2001060. https://doi.org/10.1002/advs.202001060.

Dos Santos, H.F. 2014. Reactivity of auranofin with S-, Se- and N-containing amino acids. Computational and Theoretical Chemistry 1048: 95–101. https://doi.org/10.1016/j.comptc.2014.09.005.

Dreaden, E.C., Alkilany, A.M., Huang, X., Murphy, C.J. and El-Sayed, M.A. 2012. The golden age: gold nanoparticles for biomedicine. Chem. Soc. Rev. 41(7): 2740–2779. https://doi.org/10.1039/C1CS15237H.

Đurović, M.D., Bugarčić, Ž.D. and van Eldik, R. 2017. Stability and reactivity of gold compounds – From fundamental aspects to applications. Coordination Chemistry Reviews 338: 186–206. https://doi.org/10.1016/J.CCR.2017.02.015.

Dykman, L.A. and Khlebtsov, N.G. 2011. Gold nanoparticles in biology and medicine: recent advances and prospects. Acta Naturae 3(2): 34–55.

Ecker, D.J., Hempel, J.C., Sutton, B.M., Kirsch, R. and Crooke, S.T. 1986. Reactions of the metallodrug auranofin [(1-thio-.beta.-D-glucopyranose-2,3,4,6-tetraacetato-S)(triethylphosphine)gold] with biological ligands studied by radioisotope methodology. Inorganic Chemistry 25(18): 3139–3143. https://doi.org/10.1021/ic00238a009.

EFSA Panel on Food Additives and Nutrient Sources added to Food. 2016. Scientific Opinion on the re-evaluation of gold (E 175) as a food additive. EFSA Journal 14(1): 4362. https://doi.org/10.2903/j.efsa.2016.4362.

Englinger, B., Pirker, C., Heffeter, P., Terenzi, A., Kowol, C.R., Keppler, B.K. et al. 2019. Metal drugs and the anticancer immune response. In Chemical Reviews 119(2): 1519–1624. American Chemical Society. https://doi.org/10.1021/acs.chemrev.8b00396.

Estrada-Ortiz, N., Guarra, F., de Graaf, I.A.M., Marchetti, L., de Jager, M.H., Groothuis, G.M.M. et al. 2017. Anticancer gold-heterocyclic carbene complexes: a comparative *in vitro* and *ex vivo* study. ChemMedChem 12(17): 1429–1435. https://doi.org/10.1002/cmdc.201700316.

Falde, E.J., Yohe, S.T., Colson, Y.L. and Grinstaff, M.W. 2016. Superhydrophobic materials for biomedical applications. Biomaterials 104: 87–103. https://doi.org/10.1016/j.biomaterials.2016.06.050.

Fernandes, M.A.R., Andrade, A.L. de, Biazzono, L., Luvizotto, M.C.R., Santos, A. dos and Correa, C. 2003. Gold (198Au) foils brachytherapy use on canine skin tumor. Brazilian Journal of Veterinary Research and Animal Science 40(5): 321–327. https://doi.org/10.1590/S1413-95962003000500002.

Frank Shaw, C. 1999. Gold-based therapeutic agents. Chemical Reviews 99(9): 2589–2600. https://doi.org/10.1021/cr980431o.

Freyberg, R.H., Block, W.D. and Levey, S. 1941. Metabolism, toxicity and manner of action of gold compounds used in the treatment of arthritis. I. Human plasma and synovial fluid concentration and urinary excretion of gold during and following treatment with gold sodium thiomalate, gold sodium thiosulfate, and colloidal gold sulfide. Journal of Clinical Investigation 20(4): 401–412. https://doi.org/10.1172/jci101235.

Funari, R., Ripa, R., Söderström, B., Skoglund, U. and Shen, A.Q. 2019. Detecting gold biomineralization by delftia acidovorans biofilms on a quartz crystal microbalance. ACS Sensors 4(11): 3023–3033. https://doi.org/10.1021/acssensors.9b01580.

Galassi, R., Luciani, L., Wang, J., Vincenzetti, S., Cui, L., Amici, A. et al. 2022. Breast cancer treatment: the case of gold(I)-based compounds as a promising class of bioactive molecules. In Biomolecules 12(1): 80. MDPI. https://doi.org/10.3390/biom12010080.

Garcia, A., Machado, R.C., Grazul, R.M., Lopes, M.T.P., Corrêa, C.C., Dos Santos, H.F. et al. 2016. Novel antitumor adamantane-azole gold(I) complexes as potential inhibitors of thioredoxin reductase. Journal of Biological Inorganic Chemistry 21(2): 275–292. https://doi.org/10.1007/s00775-016-1338-y.

Garcia, V.B., de Carvalho, T.G., da Silva Gasparotto, L.H., da Silva, H.F.O., de Araújo, A.A., Guerra, G.C.B. et al. 2019. Environmentally compatible bioconjugated gold nanoparticles as efficient contrast agents for inflammation-induced cancer imaging. Nanoscale Research Letters 14(1): 166. https://doi.org/10.1186/s11671-019-2986-y.

Geffroy, B., Ladhar, C., Cambier, S., Treguer-Delapierre, M., Brèthes, D. and Bourdineaud, J.-P. 2012. Impact of dietary gold nanoparticles in zebrafish at very low contamination pressure: The role of size, concentration and exposure time. Nanotoxicology 6(2): 144–160. https://doi.org/10.3109/17435390.2011.562328.

Gerson, J.R., Szponar, N., Zambrano, A.A., Bergquist, B., Broadbent, E., Driscoll, C.T. et al. 2022. Amazon forests capture high levels of atmospheric mercury pollution from artisanal gold mining. Nature Communications 13(1). https://doi.org/10.1038/s41467-022-27997-3.

Giljohann, D.A., Seferos, D.S., Daniel, W.L., Massich, M.D., Patel, P.C. and Mirkin, C.A. 2010. Gold nanoparticles for biology and medicine. Angewandte Chemie International Edition 49(19): 3280–3294. https://doi.org/10.1002/anie.200904359.

Girgis, E., Khalil, W.K.B., Emam, A.N., Mohamed, M.B. and Rao, K.V. 2012. Nanotoxicity of gold and gold–cobalt nanoalloy. Chemical Research in Toxicology 25(5): 1086–1098. https://doi.org/10.1021/tx300053h.

Gromer, S., Arscott, L.D., Williams, C.H., Schirmeri, R.H. and Becker, K. 1998. Human placenta thioredoxin reductase. Isolation of the selenoenzyme, steady state kinetics, and inhibition by therapeutic gold compounds. Journal of Biological Chemistry 273(32): 20096–20101. https://doi.org/10.1074/jbc.273.32.20096.

Guirgis, B.S.S., Sá e Cunha, C., Gomes, I., Cavadas, M., Silva, I., Doria, G. et al. 2012. Gold nanoparticle-based fluorescence immunoassay for malaria antigen detection. Analytical and Bioanalytical Chemistry 402(3): 1019–1027. https://doi.org/10.1007/s00216-011-5489-y.

Gurba, A., Taciak, P., Sacharczuk, M., Młynarczuk-biały, I., Bujalska-zadrożny, M. and Fichna, J. 2022. Gold (III) derivatives in colon cancer treatment. In International Journal of Molecular Sciences 23(2). MDPI. https://doi.org/10.3390/ijms23020724.

Gwynne, P. 2013. Microbiology: There's gold in them there bugs. Nature 495(7440): S12–S13. https://doi.org/10.1038/495S12a.

Han, Y., Han, Y., Sun, J., Liu, H., Luo, X., Zhang, Y. et al. 2022. Controllable nanoparticle aggregation through a superhydrophobic laser-induced graphene dynamic system for surface-enhanced raman scattering detection. ACS Applied Materials & Interfaces 14(2): 3504–3514. https://doi.org/10.1021/acsami.1c21159.

Hashemkhani Shahnazari, G. and Darvish Ganji, M. 2021. Understanding structural and molecular properties of complexes of nucleobases and Au13 golden nanocluster by DFT calculations and DFT-MD simulation. Scientific Reports 11(1). https://doi.org/10.1038/s41598-020-80161-z.

Haupenthal, D.P.S., Mendes, C., Bem Silveira, G., Zaccaron, R.P., Corrêa, M.E.A.B., Nesi, R.T. et al. 2020. Effects of treatment with gold nanoparticles in a model of acute pulmonary inflammation induced by lipopolysaccharide. Journal of Biomedical Materials Research Part A 108(1): 103–115. https://doi.org/10.1002/jbm.a.36796.

Hou, G.X., Liu, P.P., Zhang, S., Yang, M., Liao, J., Yang, J. et al. 2018. Elimination of stem-like cancer cell side-population by auranofin through modulation of ROS and glycolysis article. Cell Death and Disease 9: 89. https://doi.org/10.1038/s41419-017-0159-4.

Howell, J.A.S. 2009. DFT investigation of the interaction between gold (I) complexes and the active site of thioredoxin reductase. Journal of Organometallic Chemistry 694(6): 868–873. https://doi.org/10.1016/j.jorganchem.2008.10.029.

Huang, D., Zhou, H., Liu, H. and Gao, J. 2015. The cytotoxicity of gold nanoparticles is dispersity-dependent. Dalton Transactions 44(41): 17911–17915. https://doi.org/10.1039/C5DT02118A.

Huang, Y.-J., Shiau, A.-L., Chen, S.-Y., Chen, Y.-L., Wang, C.-R., Tsai, C.-Y. et al. 2012. Multivalent structure of galectin-1-nanogold complex serves as potential therapeutics for rheumatoid arthritis by enhancing receptor clustering. European Cells and Materials 23: 170–181. https://doi.org/10.22203/eCM.v023a13.

Imai, K. 2018. Concern of carcinogenic risk of eating gold leaf (gold foil) - In relation to asbestos carcinogenesis mechanism. Nano Biomedicine 10(1): 26–30. https://doi.org/10.11344/nano.10.26.

Iravani, S. 2011. Green synthesis of metal nanoparticles using plants. Green Chemistry 13(10): 2638. https://doi.org/10.1039/c1gc15386b.

Jalilehvand, F., Parmar, K. and Zielke, S. 2013. Mercury(ii) complex formation with N-acetylcysteine. Metallomics 5: 1368–1376. https://doi.org/10.1039/c3mt00173c.

Jia, Y.-P., Ma, B.-Y., Wei, X.-W. and Qian, Z.-Y. 2017. The *in vitro* and *in vivo* toxicity of gold nanoparticles. Chinese Chemical Letters 28(4): 691–702. https://doi.org/10.1016/j.cclet.2017.01.021.

Jin, R. and Higaki, T. 2021. Open questions on the transition between nanoscale and bulk properties of metals. Communications Chemistry 4(1): 1–4. https://doi.org/10.1038/s42004-021-00466-6.

John David Lee. 2008. Concise Inorganic Chemistry (5th ed.). Wiley.

Johnston, C.W., Wyatt, M.A., Li, X., Ibrahim, A., Shuster, J., Southam, G. et al. 2013. Gold biomineralization by a metallophore from a gold-associated microbe. Nature Chemical Biology 9(4): 241–243. https://doi.org/10.1038/nchembio.1179.

Kean, W.F., Lock, C.J.L., Watson Buchanan, W., Howard-Lock, H. and Hogan, M.G. 1992. Gold toxicity: chemical, structural, biological and clinical experimental issues. pp. 321–343. *In*: Rainsford, K.D. and Velo, G.P. (eds.). Side-Effects of Anti-Inflammatory Drugs 3.

Kim, J.H., Reeder, E., Parkin, S. and Awuah, S.G. 2019. Gold(I/III)-phosphine complexes as potent antiproliferative agents. Scientific Reports 9(1): 1–18. https://doi.org/10.1038/s41598-019-48584-5.

Ko, W.C., Wang, S.J., Hsiao, C.Y., Hung, C.T., Hsu, Y.J., Chang, D.C. et al. 2022a. Pharmacological role of functionalized gold nanoparticles in disease applications. In Molecules 27(5). MDPI. https://doi.org/10.3390/molecules27051551.

Ko, W.C., Wang, S.J., Hsiao, C.Y., Hung, C.T., Hsu, Y.J., Chang, D.C. et al. 2022b. Pharmacological role of functionalized gold nanoparticles in disease applications. Molecules 27: 1551. https://doi.org/10.3390/molecules27051551.

Kondo, J. and Iwase, E. 2022. DNA duplex containing T-T base pairs in complex with Au(III). Protein Data Bank.

Kong, F.Y., Zhang, J.W., Li, R.F., Wang, Z.X., Wang, W.J. and Wang, W. 2017. Unique roles of gold nanoparticles in drug delivery, targeting and imaging applications. Molecules: A Journal of Synthetic Chemistry and Natural Product Chemistry 22(9). https://doi.org/10.3390/MOLECULES22091445.

Kot, M., Kalińska, A., Jaworski, S., Wierzbicki, M., Smulski, S. and Gołębiewski, M. 2023. *In vitro* studies of nanoparticles as a potentially new antimicrobial agent for the prevention and treatment of lameness and digital dermatitis in cattle. International Journal of Molecular Sciences 24(7). https://doi.org/10.3390/ijms24076146.

Lasagna-Reeves, C., Gonzalez-Romero, D., Barria, M.A., Olmedo, I., Clos, A., Sadagopa Ramanujam, V.M. et al. 2010. Bioaccumulation and toxicity of gold nanoparticles after repeated administration in mice. Biochemical and Biophysical Research Communications 393(4): 649–655. https://doi.org/10.1016/j.bbrc.2010.02.046.

Lee, H.-Y., Shin, S.H.R., Abezgauz, L.L., Lewis, S.A., Chirsan, A.M., Danino, D.D. et al. 2013. Integration of gold nanoparticles into bilayer structures via adaptive surface chemistry. Journal of the American Chemical Society 135(16): 5950–5953. https://doi.org/10.1021/ja400225n.

Li, J.J., Lo, S.-L., Ng, C.-T., Gurung, R.L., Hartono, D., Hande, M.P. et al. 2011. Genomic instability of gold nanoparticle treated human lung fibroblast cells. Biomaterials 32(23): 5515–5523. https://doi.org/10.1016/j.biomaterials.2011.04.023.

Liang, T., Chen, Z., Li, H. and Gu, Z. 2022. Bioorthogonal catalysis for biomedical applications. Trends in Chemistry 4(2): 157–168. https://doi.org/10.1016/j.trechm.2021.11.008.

Liu, J. and Peng, Q. 2017a. Protein-gold nanoparticle interactions and their possible impact on biomedical applications. Acta Biomaterialia 55: 13–27. https://doi.org/10.1016/j.actbio.2017.03.055.

Liu, J. and Peng, Q. 2017b. Protein-gold nanoparticle interactions and their possible impact on biomedical applications. Acta Biomaterialia 55: 13–27. https://doi.org/10.1016/j.actbio.2017.03.055.

Liu, S., Wang, S., Wang, H., Lv, C., Miao, Y., Chen, L. et al. 2021. Gold nanoparticles modified graphene foam with superhydrophobicity and superoleophilicity for oil-water separation. Science of The Total Environment 758: 143660. https://doi.org/10.1016/j.scitotenv.2020.143660.

Madabeni, A., Dalla Tiezza, M., Omage, F.B., Nogara, P.A., Bortoli, M., Rocha, J.B.T. et al. 2020. Chalcogen–mercury bond formation and disruption in model Rabenstein's reactions: A computational analysis. Journal of Computational Chemistry 41: 2045–2054. https://doi.org/10.1002/jcc.26371.

Maruyama, F.H., de Paula, D.A.J., Favalessa, O.C., Hahn, R.C., Cezarino, P.G., Rosa, J.M.A. et al. 2017. Rapid detection of *Cryptococcus gattii sensu lato* using gold nanoparticles. Revista Iberoamericana de Micología 34(2): 122–123. https://doi.org/10.1016/j.riam.2016.07.001.

Massai, L., Grifagni, D., De Santis, A., Geri, A., Cantini, F., Calderone, V. et al. 2022. Gold-based metal drugs as inhibitors of coronavirus proteins: the inhibition of SARS-CoV-2 main protease by auranofin and its analogs. Biomolecules 12(11). https://doi.org/10.3390/biom12111675.

Meléndez-Villanueva, M.A., Morán-Santibañez, K., Martínez-Sanmiguel, J.J., Rangel-López, R., Garza-Navarro, M.A., Rodríguez-Padilla, C. et al. 2019. Virucidal activity of gold nanoparticles synthesized by green chemistry using garlic extract. Viruses 11(12). https://doi.org/10.3390/v11121111.

Messori, L., Scaletti, F., Massai, L., Cinellu, M.A., Gabbiani, C., Vergara, A. et al. 2013. The mode of action of anticancer gold-based drugs: A structural perspective. Chemical Communications 49(86): 10100–10102. https://doi.org/10.1039/c3cc46400h.

Moretti, E., Terzuoli, G., Renieri, T., Iacoponi, F., Castellini, C., Giordano, C. et al. 2013. *In vitro* effect of gold and silver nanoparticles on human spermatozoa. Andrologia 45(6): 392–396. https://doi.org/10.1111/and.12028.

Nejati, K., Dadashpour, M., Gharibi, T., Mellatyar, H. and Akbarzadeh, A. 2021. Biomedical applications of functionalized gold nanoparticles: a review. Journal of Cluster Science 33(1): 1–16. https://doi.org/10.1007/S10876-020-01955-9.

Ng, C.-T., Li, J.J., Gurung, R.L., Hande, M.P., Ong, C.-N., Bay, B.-H. et al. 2013. Toxicological profile of small airway epithelial cells exposed to gold nanoparticles. Experimental Biology and Medicine 238(12): 1355–1361. https://doi.org/10.1177/1535370213505964.

Nobili, S., Mini, E., Landini, I., Gabbiani, C., Casini, A. and Messori, L. 2010. Gold compounds as anticancer agents: chemistry, cellular pharmacology, and preclinical studies. In Medicinal Research Reviews 30(3): 550–580. https://doi.org/10.1002/med.20168.

Nogara, P.A., Oliveira, C.S., Schmitz, G.L., Piquini, P.C., Farina, M., Aschner, M. et al. 2019. Methylmercury's chemistry: From the environment to the mammalian brain. Biochimica et Biophysica Acta - General Subjects 1863(12): 129284. https://doi.org/10.1016/j.bbagen.2019.01.006.

Nosrati, H., Seidi, F., Hosseinmirzaei, A., Mousazadeh, N., Mohammadi, A., Ghaffarlou, M. et al. 2022. Prodrug polymeric nanoconjugates encapsulating gold nanoparticles for enhanced X-ray radiation therapy in breast cancer. Advanced Healthcare Materials 11(3): 2102321. https://doi.org/10.1002/adhm.202102321.

Oliveira, C.S., Piccoli, B.C., Aschner, M. and Rocha, J.B.T. 2017. Chemical speciation of selenium and mercury as determinant of their neurotoxicity. pp. 53–83. *In*: Aschner, M. and Costa, L.G. (eds.). Neurotoxicity of Metals, Advances in Neurobiology (Vol. 18). Springer International Publishing. https://doi.org/10.1007/978-3-319-60189-2_4.

Oliveira, C.S., Nogara, P.A., Garlet, Q.I., Rieder, G.S. and Rocha, J.B.T. 2019. Biological thiols and their interaction with mercury. pp. 1–60. *In*: McAlpine, C.C. (ed.). Thiols: Structure, Properties and Reactions. Nova Science Publishers.

Ott, I. 2009. On the medicinal chemistry of gold complexes as anticancer drugs. Coordination Chemistry Reviews 253(11–12): 1670–1681. https://doi.org/10.1016/j.ccr.2009.02.019.

Page Faulk, W. and Malcolm Taylor, G. 1971. An immunocolloid method for the electron microscope. Immunochemistry 8(11): 1081–1083. https://doi.org/10.1016/0019-2791(71)90496-4.

Patil, T., Gambhir, R., Vibhute, A. and Tiwari, A.P. 2023. Gold nanoparticles: synthesis methods, functionalization and biological applications. Journal of Cluster Science 34(2): 705–725. https://doi.org/10.1007/s10876-022-02287-6.

Pearson, R.G. 1963. Hard and soft acids and bases. Journal of the American Chemical Society 85(22): 3533–3539. https://doi.org/10.1021/ja00905a001.

Pearson, R.G. 1990. Hard and soft acids and bases-the evolution of a chemical concept. Coordination Chemistry Reviews 100(C): 403–425. https://doi.org/10.1016/0010-8545(90)85016-L.

Perez-Potti, A., Rodríguez-Pérez, M., Polo, E., Pelaz, B. and del Pino, P. 2023. Nanoparticle-based immunotherapeutics: From the properties of nanocores to the differential effects of administration routes. In Advanced Drug Delivery Reviews (Vol. 197). Elsevier B.V. https://doi.org/10.1016/j.addr.2023.114829.

Peter W. Atkins, Tina L. Overton, Jonathan P. Rourke, Mark T. Weller and Fraser A. Armstrong. 2010. Inorganic Chemistry (5th ed.). Oxford University Press.

Pia Rigobello, M., Messori, L., Marcon, G., Agostina Cinellu, M., Bragadin, M., Folda, A. et al. 2004. Gold complexes inhibit mitochondrial thioredoxin reductase: Consequences on mitochondrial functions. Journal of Inorganic Biochemistry 98(10 SPEC. ISS.): 1634–1641. https://doi.org/10.1016/j.jinorgbio.2004.04.020.

Porchia, M., Pellei, M., Marinelli, M., Tisato, F., Del Bello, F. and Santini, C. 2018. New insights in Au-NHCs complexes as anticancer agents. European Journal of Medicinal Chemistry 146: 709–746. https://doi.org/10.1016/j.ejmech.2018.01.065.

Rabenstein, D.L. 1978a. The aqueous solution chemistry of methylmercury and its complexes. Accounts of Chemical Research 11(3): 100–107. https://doi.org/10.1021/ar50123a004.

Rabenstein, D.L. 1978b. The chemistry of methylmercury toxicity. The Journal of Chemical Education 37(6): 292–296. https://doi.org/10.1021/ed055p292.

Rabenstein, D.L. and Evans, C.A. 1978. The mobility of methylmercury in biological systems. Bioinorganic Chemistry 8(2): 107–114. https://doi.org/10.1016/S0006-3061(00)80237-9.

Rabiee, N., Ahmadi, S., Iravani, S. and Varma, R.S. 2022a. Functionalized silver and gold nanomaterials with diagnostic and therapeutic applications. In Pharmaceutics 14(10). MDPI. https://doi.org/10.3390/pharmaceutics14102182.

Rabiee, N., Ahmadi, S., Iravani, S. and Varma, R.S. 2022b. Functionalized silver and gold nanomaterials with diagnostic and therapeutic applications. In Pharmaceutics 14(10). MDPI. https://doi.org/10.3390/pharmaceutics14102182.

Rocchigiani, L. and Bochmann, M. 2021. Recent advances in Gold(III) chemistry: Structure, bonding, reactivity, and role in homogeneous catalysis. In Chemical Reviews 121(14): 8364–8451. American Chemical Society. https://doi.org/10.1021/acs.chemrev.0c00552.

Roder, C. and Thomson, M.J. 2015. Auranofin: Repurposing an old drug for a golden new age. Drugs in R and D 15(1): 13–20. https://doi.org/10.1007/s40268-015-0083-y.

Rodríguez, L., Lodeiro, C., Lima, J.C. and Crehuet, R. 2008. Neutral gold(I) metallosupramolecular compounds: Synthesis and characterization, photophysical properties, and density functional theory studies. Inorganic Chemistry 47(11): 4952–4962. https://doi.org/10.1021/IC800266M/SUPPL_FILE/IC800266M-FILE002.PDF.

Rosa, J.M.A., Almeida, A.D.B.P.F. de, Carvalho, R.C.T., Brandão, L.N.S., Pitchenin, L.C., Rocha, I.S.M. et al. 2013. Diagnóstico molecular da leishmaniose visceral canina através da técnica de sonda de nanopartículas de ouro (AuNPprobes). Semina: Ciências Agrárias 34(6Supl2): 3777. https://doi.org/10.5433/1679-0359.2013v34n6Supl2p3777.

Saha, K., Agasti, S.S., Kim, C., Li, X. and Rotello, V.M. 2012. Gold nanoparticles in chemical and biological sensing. In Chemical Reviews 112(5): 2739–2779. https://doi.org/10.1021/cr2001178.

Sani, A., Cao, C. and Cui, D. 2021. Toxicity of gold nanoparticles (AuNPs): A review. In Biochemistry and Biophysics Reports (Vol. 26). Elsevier B.V. https://doi.org/10.1016/j.bbrep.2021.100991.

Saptarshi, S.R., Duschl, A. and Lopata, A.L. 2013. Interaction of nanoparticles with proteins: Relation to bio-reactivity of the nanoparticle. Journal of Nanobiotechnology 11(1). https://doi.org/10.1186/1477-3155-11-26.

Schoenberger, E. 2011. Why is gold valuable? Nature, social power and the value of things. Cultural Geographies 18(1): 3–24. https://doi.org/10.1177/1474474010377549.

Small, T., Ochoa-Zapater, M.A., Gallello, G., Ribera, A., Romero, F.M., Torreblanca, A. et al. 2016. Gold-nanoparticles ingestion disrupts reproduction and development in the German cockroach. Science of The Total Environment 565: 882–888. https://doi.org/10.1016/j.scitotenv.2016.02.032.

Sonavane, G., Tomoda, K. and Makino, K. 2008. Biodistribution of colloidal gold nanoparticles after intravenous administration: Effect of particle size. Colloids and Surfaces B: Biointerfaces 66(2): 274–280. https://doi.org/10.1016/j.colsurfb.2008.07.004.

Suárez-Moreno, G.V., Hernández-Romero, D., García-Barradas, Ó., Vázquez-Vera, Ó., Rosete-Luna, S., Cruz-Cruz, C.A. et al. 2022. Second and third-row transition metal compounds containing

benzimidazole ligands: An overview of their anticancer and antitumour activity. In Coordination Chemistry Reviews (Vol. 472). Elsevier B.V. https://doi.org/10.1016/j.ccr.2022.214790.
Thambiraj, S., Vijayalakshmi, R. and Ravi Shankaran, D. 2021. An effective strategy for development of docetaxel encapsulated gold nanoformulations for treatment of prostate cancer. Scientific Reports 11(1): 2808. https://doi.org/10.1038/s41598-020-80529-1.
Thomas, S.R. and Casini, A. 2020a. Gold compounds for catalysis and metal-mediated transformations in biological systems. Current Opinion in Chemical Biology 55: 103–110. https://doi.org/10.1016/J.CBPA.2019.12.007.
Thomas, S.R. and Casini, A. 2020b. Gold compounds for catalysis and metal-mediated transformations in biological systems. Current Opinion in Chemical Biology 55: 103–110. https://doi.org/10.1016/J.CBPA.2019.12.007.
Tialiou, A., Chin, J., Keppler, B.K. and Reithofer, M.R. 2022. Current developments of N-Heterocyclic Carbene Au(I)/Au(III) complexes toward cancer treatment. In Biomedicines 10(6). MDPI. https://doi.org/10.3390/biomedicines10061417.
Tolbatov, I., Marrone, A., Coletti, C. and Re, N. 2021. Computational studies of Au(I) and Au(III) anticancer metallodrugs: a survey. Molecules 26: 7600. https://doi.org/10.3390/molecules26247600.
Umair, M., Javed, I., Rehman, M., Madni, A., Javeed, A., Ghafoor, A. et al. 2016. Nanotoxicity of inert materials: the case of gold, silver and iron. Journal of Pharmacy & Pharmaceutical Sciences 19(2): 161. https://doi.org/10.18433/J31021.
Uzonwanne, V.O., Navabi, A., Obayemi, J.D., Hu, J., Salifu, A.A., Ghahremani, S. et al. 2022. Triptorelin-functionalized PEG-coated biosynthesized gold nanoparticles: Effects of receptor-ligand interactions on adhesion to triple negative breast cancer cells. Biomaterials Advances 136. https://doi.org/10.1016/J.BIOADV.2022.212801.
Verma, S.K., Jha, E., Panda, P.K., Kumari, P., Pramanik, N., Kumari, S. et al. 2018. Molecular investigation to RNA and protein based interaction induced *in vivo* biocompatibility of phytofabricated AuNP with embryonic zebrafish. Artificial Cells, Nanomedicine, and Biotechnology 46(sup3): 671–684. https://doi.org/10.1080/21691401.2018.1505746.
Versiani, A.F., Andrade, L.M., Martins, E.M., Scalzo, S., Geraldo, J.M., Chaves, C.R. et al. 2016. Gold nanoparticles and their applications in biomedicine. Future Virology 11(4): 293–309. https://doi.org/10.2217/fvl-2015-0010.
Veselov, V.V., Nosyrev, A.E., Jicsinszky, L., Alyautdin, R.N. and Cravotto, G. 2022. Targeted delivery methods for anticancer drugs. In Cancers 14(3). MDPI. https://doi.org/10.3390/cancers14030622.
Vidal, C., Tomás-Gamasa, M., Destito, P., López, F. and Mascareñas, J.L. 2018. Concurrent and orthogonal gold(I) and ruthenium(II) catalysis inside living cells. Nature Communications 9(1): 1–9. https://doi.org/10.1038/s41467-018-04314-5.
Von White, G., Chen, Y., Roder-Hanna, J., Bothun, G.D. and Kitchens, C.L. 2012. Structural and thermal analysis of lipid vesicles encapsulating hydrophobic gold nanoparticles. ACS Nano 6(6): 4678–4685. https://doi.org/10.1021/nn2042016.
Wang, K.-S., Tseng, Z.-L., Liu, C.-Y., Kuan, T.-Y., Jeng, R.-J., Yang, M.-C. et al. 2022. Novel strategy for flexible and super-hydrophobic SERS substrate fabricated by deposited gold nanoislands on organic semiconductor nanostructures for bio-detection. Surface and Coatings Technology 435: 128251. https://doi.org/10.1016/j.surfcoat.2022.128251.
Yao, M., He, L., McClements, D.J. and Xiao, H. 2015. Uptake of gold nanoparticles by intestinal epithelial cells: impact of particle size on their absorption, accumulation, and toxicity. Journal of Agricultural and Food Chemistry 63(36): 8044–8049. https://doi.org/10.1021/acs.jafc.5b03242.
Yeh, Y.-C., Creran, B. and Rotello, V.M. 2012. Gold nanoparticles: preparation, properties, and applications in bionanotechnology. Nanoscale 4(6): 1871–1880. https://doi.org/10.1039/C1NR11188D.
Yeo, C.I., Ooi, K.K. and Tiekink, E.R.T. 2018a. Gold-based medicine: A paradigm shift in anti-cancer therapy? Molecules 23: 1410. https://doi.org/10.3390/molecules23061410.
Yeo, C.I., Ooi, K.K. and Tiekink, E.R.T. 2018b. Gold-based medicine: a paradigm shift in anti-cancer therapy? Molecules 23(6): 1410. https://doi.org/10.3390/MOLECULES23061410.
Yu, Q., Li, J., Zhang, Y., Wang, Y., Liu, L. and Li, M. 2016. Inhibition of gold nanoparticles (AuNPs) on pathogenic biofilm formation and invasion to host cells. Scientific Reports 6: 26667. https://doi.org/10.1038/srep26667.

Zhang, J., Mou, L. and Jiang, X. 2020. Surface chemistry of gold nanoparticles for health-related applications. Chemical Science 11(4): 923–936. https://doi.org/10.1039/C9SC06497D.
Zhang, J., Li, Y., Fang, R., Wei, W., Wang, Y., Jin, J. et al. 2022. Organometallic gold(I) and gold(III) complexes for lung cancer treatment. In Frontiers in Pharmacology (Vol. 13). Frontiers Media S.A. https://doi.org/10.3389/fphar.2022.979951.
Zhang, Q., Wang, M., Hu, X., Yan, A., Ho, P.L., Li, H. et al. 2023. Gold drugs as colistin adjuvants in the fight against MCR-1 producing bacteria. Journal of Biological Inorganic Chemistry 28(2): 225–234. https://doi.org/10.1007/s00775-022-01983-y.
Zhang, X., Sun, C. and Fang, N. 2004. Manufacturing at nanoscale: top-down, bottom-up and system engineering. Journal of Nanoparticle Research 6(1): 125–130. https://doi.org/10.1023/B:NANO.0000023232.03654.40.

Index

M

N

O

P

R

S

T

V

Y

Z